# Education

Core Programming Topic at the 2011 AIChE Annual Meeting

## Minneapolis, Minnesota, USA
## 16-21 October 2011

ISBN: 978-1-61839-725-6

**Printed from e-media with permission by:**

Curran Associates, Inc.
57 Morehouse Lane
Red Hook, NY 12571



Printed by Curran Associates, Inc. (2012)

For permission requests, please contact AIChE
at the address below.

AIChE
3 Park Avenue
New York, NY 10016-5991

Phone: (203) 702-7660
Fax: (203) 775-5177

www.aiche.org

**Additional copies of this publication are available from:**

Curran Associates, Inc.
57 Morehouse Lane
Red Hook, NY 12571 USA
Phone: 845-758-0400
Fax: 845-758-2634
Email: curran@proceedings.com
Web: www.proceedings.com

# TABLE OF CONTENTS

# (3a) Multiscale Modeling for the Fundamental Understanding of Heterogeneous Catalysis for Energy Applications

**Michail Stamatakis**, Chemical Engineering, University of Delaware, Newark, DE

In view of recent economic and environmental pressures, the necessity of a paradigm shift in the production and utilization of energy has emerged. Key challenges associated with this shift include the development of efficient hydrogen production, storage and distribution technologies; the utilization of alternative and/or renewable energy sources, such as biomass; and the capture, storage and utilization of $CO_2$. Computational catalysis has and will continue to play an important role in overcoming these challenges, as the demand for a fundamental understanding of realistic catalyst structures necessitates the development of novel simulation methodologies and multiscale theoretical approaches.

My research will address these challenges through the development of a framework capable of simulating complex catalytic structures, such as heterogeneous populations of supported nanoparticles. This multiscale modeling approach will combine quantum chemistry calculations at the electronic scale, with kinetic Monte Carlo simulation at the molecular and catalyst scales. The framework will be used to simulate chemistries of interest in energy applications such as hydrogen production through the water-gas shift reaction, which is relevant to biomass reforming, and $CO_2$ hydrogenation to methanol, a challenging problem in $CO_2$ utilization for fuel production. This work will introduce a novel component on the rational catalyst design principle, namely that of the structural optimization of the material, and will also provide directions for future research on catalyst deactivation through nanoparticle sintering.

See more of this Session: Meet the Faculty Candidate Poster Session
See more of this Group/Topical: Education

# (3aa) Photo/Electrochemical Charge Transfer Processes: Fundamentals to Devices for Energy Scavenging and Storage

**Vidhya Chakrapani**, School of Chemical & Biomolecular Engineering, Georgia Institute of Technolgy, Atlanta, GA

Today, semiconductor electrochemistry plays vital roles in several technological processes such as energy scavenging by photoelectrochemical solar cells to energy storage by Li ion batteries. Improvements in device performance require understanding the physics of the semiconductor excitonic transitions to the chemistry of the interfacial phenomena. The proposed research project is highly interdisciplinary in nature, which spans fundamental studies of interfacial phenomenon, spectroscopic techniques in ultrafast time scale, which probes exited state transitions, and synthesis of nanostructure to fabrication of photoelectrochemical solar cells and Li ion batteries. Specific projects are discussed below.

"Electrochemical doping" is recently discovered phenomenon in which the oxygen redox couple in adsorbed water films tends to pin the Fermi level in semiconductor at the electrochemical potential of the redox couple.[1,2] The effect, which was first noted with undoped diamond, makes it conductive. Electrochemically mediated charge transfer process is a very general phenomenon that has often been unrecognized. Since its discovery in diamond, the effect has been observed in other technologically important materials such as in single-walled semiconducting carbon nanotubes, graphene[3]; gallium nitride and zinc oxide.[4] The effect can occur whenever semiconductors or other solids are exposed to humid air. We propose to explore this effect in metal oxide and nitride nanowires. Novel surface conducting devices and sensors that exploit this effect would be fabricated by microfabrication technique.

Photocatalytic and photoelectrochemical activity especially in nanostructure is strongly influenced by the surface electronic properties of the semiconductor. These properties in turn depend on the nature of defect states and band edge positions of the semiconductor with respect to the redox couple. Thus by controlling these properties one can tailor the performance of photoelectrodes.

The proposed studies will couple the synthesis and characterization of inorganic nanowires with fundamental studies of energy absorption and transport, for high performance solar cells and Li ion batteries. Using "band edge engineering," novel nanowire based hybrid electrode would be designed that are stable against degradation.


1. Chakrapani, *et al. Science* **2007**, *318*, 1424-1430.

2. Maier, *et al. Phys. Rev. Lett.* **2000**, *85*, 3472-3475.

3. Sque, *et al. Phys. Stat. Sol. (A)* **2007**, *204,* 3078-3084.

4. Chakrapani, *et al. J. Am. Chem. Soc.* **2008, *130***, 12944-12952.



See more of this Session: Meet the Faculty Candidate Poster Session
See more of this Group/Topical: Education

# (3ab) Catalytic Synthesis Gas Conversion to Produce Chemical Products From Non-Petroleum Resources

**Fuat E. Celik**, Chemical and Biological Engineering, University of Wisconsin-Madison, Madison, WI

The production of many fuel and chemical products begins with petroleum. As concerns over the price of and access to petroleum increase, interest in natural gas, coal, and biomass resources as an alternative feed stocks also increases. All three can all be reformed or gasified to produce synthesis gas (syngas) a reactive mixture of CO and H-$_2$. While the production of syngas has been commercialized, the development of simple and cost effective processes to utilize syngas to make a wide range of products is still under investigation.

I will present on three strategies to utilize syngas to synthesize chemical products from syngas by 1) reactions with $H_2$, 2) reactions with CO, 3) reactions with both $H_2$ and CO. In each of these cases, emphasis was placed on novel processes and catalysts that lead to higher selectivities to the desired products while also simplifying processing conditions relative to more traditional methods.

1) Direct synthesis of $H_2O_2$ from $H_2$ and $O_2$ over near-surface alloys

Commercial production of $H_2O_2$ separates the O--$_2$ reduction and $H_2$ oxidation steps by oxidizing and reducing an anthraquinone instead, which produces the desired $H_2O_2$ but requires separation and recycling of the athraquinone catalyst. The two steps are separated to avoid the thermodynamically favored formation of $H_2O$. The direct synthesis of $H_2O_2$ over a heterogeneous catalyst holds promise by simplifying the process and eliminating many costly separation steps, if high selectivity to $H_2O_2$ over $H_2O_2$ can be achieved. I am investigating Au/Ni near-surface alloys (NSAs) using periodic, self-consistent Density Functional Theory (DFT) calculations. While Au may have higher O-H bond formation activity, it is a poor $O_2$-dissociation catalyst, and likewise Ni is very effective at $O_2$-

dissociation but not $O_2$ hydrogenation. NSAs can combine the high activity of one metal with the higher selectivity of another, less-reactive metal catalyst, and Au atoms substituted into the surface of Ni (111) have already been shown to possess higher steam reforming selectivity than pure Ni by impeding the formation of carbon.

2) Carbon-carbon bond formation in syngas derivatives

Syngas contains no carbon-carbon bonds, whereas many desirable products contain one or more C-C bonds. This represents a challenge to finding alternatives to petroleum, which already contains C-C bonds. Many $C_1$ molecules containing no C-C bonds such as methanol, formaldehyde, and dimethoxymethane (DMM) can be made readily from syngas today. Coupling $C_1$ compounds by solid acid-catalyzed carbonylation reactions can produce desirable fuel and chemical products, such as ethylene glycol (monoethylene glycol – MEG), a polyester monomer currently produced from petroleum products. A high pressure liquid-phase reaction to produce MEG from syngas derivatives was known which gave very poor selectivity at low pressure. I discovered a novel low-pressure gas-phase process to couple DMM and CO to produce methyl methoxyacetate (MMAc), a MEG precursor, with high selectivity using acidic zeolites. Among common zeolites, H-FAU was found to possess the highest selectivity because its large cage structure inhibited the rate of the undesired disproportionation reaction. The Si/Al ratio of the zeolite, which determines the density of acid sites within the catalyst pore structure, has a strong effect on the rate of MMAc formation. By reducing the density of acid sites, the rate of MMAc formation increased for both FAU and acidic ZSM-5 (MFI) up to Si/Al ratios of ~30, and then did not increase further. At a Si/Al ratio of 30, the active sites in the zeolites are placed far enough from each other so as to avoid interactions between them. An extensive FTIR spectroscopic investigation provided evidence for the proposed mechanism, which was also supported by DFT calculations performed by a colleague.

3) Supported liquid-phase Rh hydroformylation catalysts

Hydroformylation reactions are used to upgrade olefins to their next higher homolog aldehydes. Propene hydroformylation produces *n*- and isobutanal, and homogeneous Rh catalysts are preferred for their tunable selectivity between isomers. The process depends on efficient capture and recycling of expensive Rh-phosphine catalysts, and so a heterogeneous catalyst and gas-phase process is preferred. I investigated Rh particles supported on $SiO_2$ for gas-phase propene hydroformylation activity. It became apparent that while butanals where produced with 100% selectivity the addition of triphenylphosphine to the catalyst was essential. Kinetic rate data together with FTIR and NMR evidence led to the discovery that the phosphine was melting and producing a liquid phase inside the $SiO_2$ pores that was facilitating the dissolution of the Rh to produce essentially the same Rh-phosphine catalysts as the commercial homogenous process. In this way, a homogeneous catalyst was supported inside a solid support, combining the high selectivity of the liquid-phase process with the simplicity and high throughput of the gas-phase process.

See more of this Session: Meet the Faculty Candidate Poster Session
See more of this Group/Topical: Education

# (3ac) Engineering InorganicOrganic Porous Nanohybrids for Energy and Environment-Related Gases Storage and Separation

**Watcharop Chaikittisilp**, School of Chemical & Biomolecular Engineering, Georgia Institute of Technology, Atlanta, GA

The main focus of my research is to develop novel nanoporous materials, that is, zeolites, siloxane-based materials, metal–organic frameworks, and porous inorganic–organic polymers, for solving energy and environment-related problems and for improving chemical processes as well. In particular, I have been integrating synthetic chemistry (both inorganic and organic methods) and structural chemistry with the aids of characterization techniques and computer modeling to engineer such nanoporous materials. Gas adsorption/storage, membrane separation, chiral adsorptive separation, and heterogeneous catalysis are the targeted uses of the developed materials.

In my PhD work at The University of Tokyo, I developed the novel synthetic route leading to designed inorganic–organic framework materials from the strategic molecular building units [1,2]. Considering secondary building units existing in zeolite frameworks, the cubic double four-ring (D4R) siloxane cage was selected as a starting precursor. Utilization of intermolecular interaction ranging from weak interactions, i.e., hydrogen bond, $\pi$–$\pi$ stacking, and van der Waals force, through coordination bond, to strong carbon–carbon covalent bond, siloxane-based inorganic–organic hybrid materials were synthesized with partially targeted manners. By combination of structural, synthetic organic, and materials chemistry, a series of porous poly(organosiloxane) networks with controlled over molecular homogeneity in a certain degree was achieved; they possessed narrow pore size distribution and high surface area. This synthetic strategy can expand to synthesize other porous hybrid materials.

My postdoctoral research has involved the development of porous inorganic–organic hybrids containing active amine moieties as a robust and effective adsorbent for $CO_2$ capture from moderately dilute gas streams such as flue gas and ultra-dilute concentrations such as ambient air. This includes engineering of amine-based adsorbents that are stable upon steam exposure, development of novel synthetic route to introduce active amines into structural objects such as supported membranes and monoliths, and preparation of a novel class of active adsorbents.

**Postdoctoral Advisor:** Professor Christopher W. Jones, School of Chemical & Biomolecular Engineering, Georgia Institute of Technology, Atlanta, GA.

**Doctoral Thesis Advisor**: Professor Tatsuya Okubo, Department of Chemical System Engineering, The University of Tokyo, Tokyo, Japan.

**References**

[1] W. Chaikittisilp, A. Sugawara, A. Shimojima, T. Okubo, *Chem. Eur. J.* **2010**, *16*, 6006-6014.

[2] W. Chaikittisilp, A. Sugawara, A. Shimojima, T. Okubo, *Chem. Mater.* **2010**, *22*, 4841-4843.

See more of this Session: Meet the Faculty Candidate Poster Session
See more of this Group/Topical: Education

# (3ad) Understanding the Antibiotic Resistance Game: New Secrets of Gene Regulation Revealed

**Anushree Chatterjee**, Chemical Engineering and Materials Science, University of Minnesota, Minneapolis, MN

The recent rise in microbial drug resistance is a growing challenge for future therapy of bacterial infections. Increased prevalence of antibiotic resistance in bacteria is an outcome of evolution via natural selection. However, the built-in design feature of bacteria to transfer DNA containing antibiotic resistance both within the same species and across species is the main culprit for the spread of drug resistance. In order to combat microbial drug resistance, novel strategies need to be developed to block such transmission of antibiotic resistance. My research interest lies in understanding the molecular mechanisms responsible for antibiotic resistance transfer among bacteria and genetic factors that control propagation of infectious diseases using both mathematical modeling and experimental techniques, with the goal of discovering novel drug targets. In this poster, I will be describing the biological function of antisense transcription with focus on the regulatory role of RNA polymerase traffic, the main enzyme responsible for producing RNA (pre-cursors for proteins), and antisense RNA, especially their role in conferring biological systems with robust switch-like behavior. The regulatory role of antisense transcription in controlling drug resistance transfer in the nosocomial pathogen *Enterococcus faecalis* and antibiotic production in *Streptomyces coelicolor* would be discussed. As an application, synthetic gene-networks with engineered-antisense RNA regulation would be discussed.

See more of this Session: Meet the Faculty Candidate Poster Session
See more of this Group/Topical: Education

# (3ae) A Property Based Approach for Simultaneous Process and Molecular Design

**Nishanth G. Chemmangattuvalappil** and Mario Richard Eden, Department of Chemical Engineering, Auburn University, Auburn, AL

The selection of product/product mixtures that give the optimum performance of a process is a critical issue for a design engineer. The process performance is usually understood in terms of physical properties and on many occasions, the physical properties of the product rather than their chemical structure determine the suitability of a specific product as the input to the process. Molecular design algorithms generally require target properties to design the molecules. Therefore, for obtaining the optimal solution for this type of problem, it is necessary to have a methodology to represent the product performance in terms of measurable physical properties and identify the molecule/mixture that gives the property targets corresponding to the optimum process performance.

In my dissertation work, systematic algorithms have been developed for the design of molecules corresponding to the optimum performance of a process. The concept of property clustering has been extended to the formation of molecular clusters based on second and third order group contribution methods. An algebraic approach has been developed utilizing higher order groups built from first order groups subject to the constraints of overlapping. The advantage of using an algebraic approach is that it can handle any number of molecular groups or properties and can generate all possible compounds within the required range of properties. The most significant aspect of the aforementioned method is that both the application range and reliability of the molecular property clustering technique are considerably increased by incorporating second and third order estimation. A methodology has been developed for incorporating the property contribution predicted using combined group contribution and connectivity indices into the design framework in case the property contributions of any of the molecular groups of interest are not available in literature. For the design of simple monofunctional molecules, a modified visual approach has been used where as for the design of more complicated structures and/or for treating more than three properties at a time, an algebraic method has been used.

The developed algorithm can be used as a tool for the solution of integrated process and product design problems. Since an algebraic approach for the solution of a process design problem is already available, it is now possible to solve the integrated problem irrespective of the number of properties involved. The process design problem can be solved in terms of constitutive variables without committing to any component beforehand. The property targets estimated in the process design step along with the molecular groups will form the input to the molecular design algorithm. The algebraic equations formed using the molecular operators are solved simultaneously to generate the possible molecular structures that match the optimum process performance defined by the property targets.

However, apart from the targets set by the process design, the molecule has to satisfy a number of other environmental and safety constraints in order to be used in an actual process. In addition, there are many properties which can not be estimated through group contribution methods because it is not always possible to find a correlation between the molecular groups and properties. Similarly, not all possible atomic arrangements are represented in group contribution methods. So, there is a need of an efficient methodology for the design of molecules with a more diverse property targets. In my thesis, I have developed an algorithm that uses the concept of molecular signature descriptors for molecular design. The signature is a systematic coding system of atom types and the signature of a molecule can be obtained as a linear combination of its atomic signatures. It has been proved that any of the topological indices of molecules can be represented in terms of molecular signatures and it is possible to correlate the topological indices to the actual properties and biological activities. Here, the new algorithm utilizes molecular property operators based on signatures for solving the reverse problem of obtaining the molecular structures that satisfy the property targets estimated in the process design step. A new set of equations will be employed to ensure that the molecule meets the safety and environmental constraints as well. The principles in graph theory are incorporated to avoid the generation of infeasible structures. The property models required to describe the target properties may be based on QSPR/QSAR (quantitative structure-property/activity relationship) models or in the form of group contribution methods. The accuracy of this method depends only on two factors- how well the actual property-topological index relationships are estimated and the height of atomic signatures used to describe the topological indices. Since, many topological indices can be used to describe each property, this algorithm generally provide reliable results. Finally, a framework has been developed to incorporate the design of process flow sheets into the molecular design problems. The recently developed computer aided flowsheet design techniques are utilized along with the molecular design techniques to obtain the required process performance.

As a future faculty candidate, my primary interest is in the area of process and product design. However, different tools used and developed in the research I have conducted so far have the ability to be applied in many related areas as well. Similarly, many of the methods I have developed in my thesis can be applied in different area of chemical engineering and a number of multi-disciplinary fields.

The initial projects I will be working on are:

- Extend the applicability of the developed algorithms into more rigorous industrial problems by including the effects of varying process conditions and incorporating more flowsheet properties. This approach will be helpful for the design of materials with good accuracy.
- The iterative nature of the molecular design problems will be reduced into a systematic generation process by introducing three dimensional descriptors in the molecular design step to expand the applicability range of reverse problem formulations and by including constraints in the form of signatures that ensure the stability of the designed structures

My long term projects are:

- The extension of molecular signature concept into the exploration of biochemical reaction pathways
- A property based approach to designing optimum molecules by designing the reaction networks using the concept of molecular signatures in process design

See more of this Session: Meet the Faculty Candidate Poster Session
See more of this Group/Topical: Education

# (3af) Multi-Scale and Holistic Integration Approaches for Systems Including Nanoscale Phenomena

**Pil Seung Chung**, Chemical Engineering, Carnegie Mellon University, Pittsburgh, PA

During the past decades, along with remarkable scientific and engineering achievements, the world has faced immense challenges in efficiency and system scaling, which motivates the investigation of novel materials and nanoscale system integration. Mathematical models have been widely studied to satisfy the demand by tuning physical properties, chemical structures, and functionalities of the new materials as a part of the eye-opening progress. The remarkable progress in nanotechnology provides us the understanding of nanoscale physics to atomistically control the system properties. However, advances in nano materials and simulation methodologies direct us to the critical issue of the interaction to meso/macroscopic scales since most systems comprise phenomena at different time and length scales, and often described via a hierarchy of scale-specific models. Consequently, the multi-scale approach based on a clear understanding of nanoscale phenomena becomes an essential multidisciplinary analysis paradigm in science and engineering fields in order to hybridize the new nano materials and systems. Here, the mechanism of information communication can be obtained from one scale and passed to another scale allowing one to predict the system scale behavior from first principles in a bottom-up approach. The device scale design criteria, nonetheless, can be established via nanoscale resolution using the top-down approach.

My research interest covers the broad area of both theoretical and experimental multi-scale approaches in novel materials and nanoscale phenomena originating from tribology, energy, and nano electronics applications by using novel atomistic/ molecular and mesoscale/continuum theory and developing methodologies for hierarchical integration of the systems at the different scales, which will be expanded to multi-phenomena/physics area. The device scale analysis will be coupled with optimization for the process system integration. I performed experiments and vigorously collaborate with experimentalists to complement our ongoing full-scale simulation. In this poster session, I will introduce and discuss the details of my current and future research and teaching plan.

# (3ag) Wrinkling and Cracking Instabilities As Tools for Material Characterization of Polymer Thin Films and Membranes

**Jun Young Chung**, Polymers Division, National Institute of Standards and Technology, Gaithersburg, MD

Nanotechnology promises to revolutionize a growing set of materials applications ranging from electronics to energy production to water purification. However, the quest to engineer materials on the nanoscale is met with the daunting task of measuring the material properties of these systems at these same length scales. For polymers, the challenge is even greater since conventional materials testing platforms lack the resolution for such soft systems. In this session, I will discuss our current progress and advances in wrinkling and cracking-based methodologies for characterizing the physical and mechanical properties, as well as the relaxation dynamics in nanoscale polymeric materials.

See more of this Session: Meet the Faculty Candidate Poster Session
See more of this Group/Topical: Education

# (3ah) Integrative Microengraving Process and Immuno-HCR for Ex-Vivo, Highly Sensitive and Multifunctional Characterization of Rare Immune Cells In Periphery

**Jonghoon Choi**, Department of Chemical Engineering, The David H. Koch Institute for Integrative Cancer Research, Massachusetts Institute of Technology, Cambridge, MA

The human immune system comprises a diverse set of cells that contribute to a productive immunological response. Characterizing these dynamic behaviors for individual cells is critical in gaining insights how immune system operates. As yet, current bioanalytical technologies suffer from challenges in resolving the diverse functional activities and identities of individual subsets of cells as well as limits of signal detection and number of output parameters concurrently detectable in the assay. Recent success in nano and micro technologies has brought new insights to improve bioanalytical assays to utilize them in studying single cells. Miniaturization of biological assays using nano and micro technologies has enabled single cell resolution assays overcoming many challenges that traditional biological assays carry on. However, there is still a need to improve the integration of present single-cell analytical tools based on nano and micro technologies that would offer new insights into their operational variability and phenotypic diversity that are difficult to define with stand-alone analytical tools and bulk correlations. This talk will describe a new technology integrating microengraving-a soft lithographic method for printing arrays of secreted proteins from thousands of single cells-with an *in situ* signal amplification strategy called immuno-HCR, which antibodies are covalently modified with specific oligonucleotides that initiate hybridization chain reaction (HCR) with hairpins labeled with fluorescent molecules that the intensity of the resulting signal doubles with each extension. This new method makes it possible to detect up to 16 different cytokines and chemokines secreted from individual, viable human immune cells with subpicogram sensitivity. In application of this new technology, we will demonstrate measurements of the antigen-dependent, HLA-restricted responses by viable $CD4^+$ T cells co-loaded with antigen-pulsed dendritic cells. Specifically, these multiplexed measures allow quantitative measures of the rates of secretion and number of cells responding for eight different secreted proteins concurrently (MIP-1β, IL-2, IL-4, IL-9, IL-10, IL-17, TNF-α, and IFN-γ). This integrated approach for multidimensional single-cell analysis significantly expands the breadth of multifunctional states enumerated for single cells and suggests a new strategy for ex vivo, high-throughput screening of antigen-specific human T cells and may provide new insights in understanding and characterizing the contributions from rare cells in various immune diseases.

See more of this Session: Meet the Faculty Candidate Poster Session
See more of this Group/Topical: Education

# (3ai) Surface Functionalization of Nanomaterials to Elicit "Smart" Properties

**Allan E. David**, Department of Pharmaceutical Sciences, The University of Michigan, Ann Arbor, MI

Rapid advances in the field of nanotechnology have brought enormous potential for improving the quality of life. The effects of this growth are now being felt in such varied industries as drug delivery, catalysis, tissue engineering, renewable energy, and optical coatings, to name just a few. While the control of length scales at the nanometer level alone yields some exciting properties, further functionalization of the surface with "smart" ligands that provide environment-responsive performance is generating tremendous interest. My research goals are towards the development of novel, response nanomaterials for a broad field of applications. During this poster session, I will present some of the work done in developing functional nanoparticles and nanocomposites for applications in biocatalysis, drug delivery, optical coatings and renewable energy, and discuss the future potential of this technology.

My graduate research was focused on the development of a nanoporous organic-inorganic composite matrix, using sol-gel techniques, for enzyme immobilization. With proper management of surface energies, we developed an organic-inorganic hybrid material with the following desirable properties: 1) high surface area (~650 m2/g) and uniform surface functionalization which together yielded an extremely high loading capacity (>700 mg protein/g matrix); 2) complete retention of enzyme's catalytic efficiency; 3) good pH and thermal stability; and 4) open pore morphology to facilitate substrate/production transport. The extremely high enzymatic activity of this material allowed for reactor miniaturization, which was validated with large-scale, GLP production of a therapeutic peptide (currently undergoing preclinical evaluation for Investigational New Drug approval under a $2.4 million NIH SBIR Phase II Continuation grant).

This material technology was further enhanced during my postdoctoral research with the inclusion of microenvironment-responsive biopolymers within the inorganic matrix for drug delivery applications. Particles formed with this nanocomposite displayed pH-dependent mucoadhesive and swelling properties – allowing for targeted and controlled delivery of drugs to the oral cavity, stomach, or intestinal tract. We have specifically pursued the utilization of this material for oral delivery of insulin and antibiotics. Additionally it was found that this nanocomposite could be utilized to modulate water activity close to the surface, which in turn would provide control of immobilized enzyme activity – yielding control of both reaction extent and side-product generation. There is also potential for this material in wound healing and tissue engineering, as well as in separation processes. I also served as group leader for the MION sub-group, of Dr. Victor Yang's lab (Department of Pharmaceutical Sciences, University of Michigan, Ann Arbor MI), which focused on the utilization of magnetic nanoparticles for simultaneous imaging and therapy of brain tumors. In this capacity, I mentored and trained an undergraduate student, five PhD students, and four postdoctoral fellows, a group which has already produced more than 15 peer-reviewed papers.

Overall, I have an extensive background, and had success, in the development of novel nanomaterials – both functional nanoparticles and nanocomposites (yielding 4 patent applications). I have displayed leadership abilities by maximize the productivity of a diverse group of students/researchers. Additionally, I am very familiar with the grant funding mechanism - having served as ad hoc member for eight different NIH grant review committees, and successfully obtained more than $12 million in research funding from both industrial and governmental sources. The common thread I have going through all my research interests is the surface modification of nanomaterials to elicit "smart" properties. I fully expect to continue my success in acquiring funds to further advance our fundamental understanding of the science of nanotechnology, and to continue its development into life-changing products.

See more of this Session: Meet the Faculty Candidate Poster Session
See more of this Group/Topical: Education

# (3aj) Control of Electronic Properties of Supported Catalysts and Design of Novel Multilayer Catalytic and Photocatalytic Materials

**Prashant Deshlahra**, Chemical and Biochemical Engineering Department, University of Notre Dame, Notre dame, IN, William Schneider, Chemical and Biomolecular Enigineering Department, University of Notre Dame, Notre dame, IN and Eduardo E. Wolf, Chemical and Biomolecular Engineering Department, University of Notre Dame, Notre dame, IN

Design of new catalysts with a precise control over structure and function is extremely important as it can help meet future energy needs and reduce environmental impact of several industrial processes. We have carried out research to design novel catalysts in which the electronic properties can be tuned by applying external voltage to the metal-support catalytic junctions and thus control catalytic activity. This new approach to gas-phase heterogeneous catalysis involves use of nanofabrication techniques to prepare supported catalysts with controlled structure, development of new spectroscopy technique and multiscale simulations.

Density Functional Theory (DFT) based molecular simulations were used to study the effect of electrostatic interactions resulting from support induced electron transfer and from co-adsorbed dipoles on a range of adsorbed molecules on catalysts surface. These studies revealed that such interactions can significantly affect the metal-adsobate bond as indicated by shifts in vibration frequencies of adsorbate and corresponding changes in adsorption energies.

For experimental demonstration of such effect on supported catalysts, a new multilayer enhanced infrared spectroscopy (MEIRAS) technique, capable of sensing small vibrational signals from adsorbed molecules on low area model catalysts was developed. Theoretical electromagnetic simulations show that the metal-dielectric-metal multilayer substrate enhances the surface intensity of incident infrared by a constructive interference, which is optimized by matching the dielectric thickness to infrared wavelength. This enhanced surface intensity and diminished reflectance of the substrate, were shown to cause the observed large signal enhancement. We have prepared thin film and nanofabricated nanowire platinum catalysts and conducted polarization dependent MEIRAS studies of CO adsorption. These experiments combined with theoretical simulations have demonstrated several potential advantages of this simple new technique in catalysis including the capability to selectively probe adsorbates near metal-support interfaces of supported catalysts.

Furthermore, we designed a Pt-$TiO_2$-Au multilayer catalytic diode in which an electron transfer to CO molecules adsorbed on Pt-$TiO_2$ junction can be induced by applying external bias voltage, and its effect can be studied using MEIRAS technique. These experiments have led to a first direct demonstration of the effect of tunable external voltage induced electron transfer in supported gas-phase catalysis on the adsorbed reactants and opened up several new areas of investigation in this field.

The techniques and approaches developed in this work, combined with novel material synthesis methods and insights from computational studies can lead to novel advances in materials for energy needs. Current multilayer structures used for infrared wavelength can be redesigned to match visible wavelengths and select desirable frequency windows and concentrate its intensity on the top surface by interference effects. External voltage effects combined with optical interference effects and plasmonic effects by the top metal layer of well defined size and structure can have extremely important applications to enhance activity and sensitivity in catalysis, photocatalysis and sensors applications. A research plan including application based materials with fundamental experimental investigation and theoretical studies will be presented.

See more of this Session: Meet the Faculty Candidate Poster Session
See more of this Group/Topical: Education

# (3ak) Deoxygenation and Fundamental Catalysis for Sustainable Energy Applications

**Prashant Reuben Daggolu**, The Gene and Linda Voiland School of Chemical Engineering, Washington State University, Pullman, WA

**Introduction:**

The General Theme of my research is *Catalysis for Sustainable Energy*. I received my PhD in chemical Engineering from Mississippi State University in December 2010. I am presently working as a Postdoctoral Research Associate at the Washington State University. My PhD project dealt with converting oxygenates derived from syngas to hydrocarbon. The particular oxygenates under consideration were acetic acid, acetone, methyl acetate, ethanol and acetaldehyde. Deoxygenation was performed using titania based catalytic systems. Most of my PhD research was applied catalysis in nature. However, in my postdoc work I am concentrating on fundamental catalysis to give myself a well-rounded catalysis experience. I am studying and analyzing Pt-Re bimetallic catalysts used for aqueous phase reforming of biomass. The influence of Rhenium is being investigated to understand the catalyst functioning. Several characterization techniques are being employed such as DRIFTS, TPRS, PXRD, XPS and TGA. Since this is a 'work in progress', I shall not be able to give more details at present. However, if you need any specific details, you can contact me using the information at the end of the abstract. Also, I have a Master of Science degree in Chemical Engineering from Mississippi State University. My MS research was concerned with *Dielectrophoresis of Erythrocytes*. Since this is not relevant to my future research, I am not providing details. However, I have a publication from that research.

**Doctor of Philosophy (Ph.D.) Research:**

*Background*

My PhD Project concerns with converting short chain organic oxygenates to fuel range hydrocarbon. This project was part of a campus wide effort to develop sustainable energy at Mississippi State University. Under this program a collaborative research group was established called Sustainable Energy Research Center (SERC), www.serc.msstate.edu, comprising of researchers in the departments of Chemical Engineering, Forest Products, Chemistry, Agricultural and Biological Engineering to name a few. As part of this endeavor, our group under my advisor, Prof. Mark White was developing non-Fisher Tropsch technology for biomass derived syngas to be converted to gasoline. One approach to achieve this goal is to first convert syngas to alcohols and then convert alcohols to gasoline. This approach while effective, yields oxygenate by-products. My Project was to study the use of Titanium dioxide, $TiO_2$ (titania), as a catalyst to convert these byproducts to fuel range hydrocarbon to increase the viability of the biomass-to-gasoline technology.

*Accomplishments*

The main oxygenate by-products formed during conversion of syngas to alcohols are acetic acid, acetaldehyde, methyl acetate and acetone. Hence initially, I performed batch reactor studies using both acid catalysts like H+/ZSM-5 and silica alumina, and base catalysts like titania and magnesium oxide. It was concluded that base catalysts yield almost no coke when reacted with these oxygenates. Then I studied interactions between these oxygenates by reacting pairs of these compounds on titania in a batch reactor. The results were documented and showed no prohibitive interactions. The data from these experiments will be submitted to the journal, *Biomass and Bioenergy*.

Acetic acid has been shown in literature as one of the most coke-causing compounds in both syngas derived products as well as in bio-mass pyrolysis products when tried to convert to hydrocarbon. Hence titania as a catalyst has all the more significance as it can convert acetic acid to acetone and subsequently acetone to mesitylene. I decided to study the acetone condensation to mesitylene much more closely as it forms an important link in converting acetic acid to hydrocarbon. Also, Acetone itself can be converted to liquid hydrocarbon. I studied this reaction in high pressure conditions similar to those employed in syngas conversion to alcohols. This was done keeping in mind the possibility of using a single bed to convert syngas to gasoline including all by-products. I had excellent success in this endeavor and reported high yield and selectivity of mesitylene from acetone. I have also made some mechanistic studies which underline the importance of the dual acid-base nature of titania to catalyze this reaction. The data was presented in national meetings and I have submitted a paper to the journal, *Applied Catalysis A: General*.

As alcohols can be easily converted to gasoline range hydrocarbon on zeolite catalyst, I wanted to show that alcohols were essentially inert on titania. This would ensure that only by-products are converted to hydrocarbon on titania and alcohols are retained as such. In course of making this study, when ethanol was reacted on titania, I found that tiny amounts of 1-butanol was found. I immediately saw an opportunity to upgrade alcohols in carbon number. This would make them better fuel additives than ethanol. Also, previous research in our lab showed that higher alcohols can give better yields of gasoline when reacted on zeolites than methanol or ethanol. Hence I decided to pursue making 1-butanol from ethanol as well. Pacific Northwest National Laboratories (PNNL) has been working on direct conversion of syngas to higher alcohols but was successful only to make ethanol and not substantial yields of any higher ones. Hence our group collaborated with PNNL and I travelled to Richland, WA to work as an intern at PNNL for 10 weeks to show that titania can be used to convert ethanol to 1-butanol. I was successful in doing so. Also, my interaction with scientists at PNNL was a rich experience for me. After returning to Mississippi State University, I continued to work on converting ethanol to 1-butanol not only on titania but also on catalysts with similar known surface chemistry. I found that Zirconium dioxide (Zirconia) could convert ethanol to 1-butanol with much better yields than titania. Also, while titania made 1-butanol as the highest alcohol from ethanol, zirconia could convert ethanol to not only 1-butanol but also various pentanols, hexanols and even octanol. Mechanistic studies revealed that alcohols higher then 1-butanol do not self condense on zirconia. Instead, ethanol reaction with intermediate alcohols such as 1-propanol or 1-butanol or hydration of oligomerized olefins could yield higher alcohols. Also, it was known that only primary alcohols are reactive on this metal oxide. The results of this work are submitted as a paper to the journal, *Journal of Molecular Catalysis A:Chemical.*

From the work that I performed for the last 36 months, I gave 5 presentations, received 1 provisional patent and was selected for 1 award. I produced 2 research papers as corresponding author which are currently in review. One more paper will be submitted shortly. I am also presently drafting a review paper on the general theme of deoxygenation catalysis for sustainable energy as a co author with my advisor, Prof. Mark White. This is an invited paper from the editors of the book series, *Advances in* Catalysis. I gave a brief overview of my accomplishments during my Ph.D. I shall be excited to present these in great detail if invited to your campus.

**Contact Information:**

Prashant Daggolu, PhD

Postdoctoral research Associate

Washington State University

305 NE Spokane Street, Dana 118

Pullman, WA – 99164

Phone: 662 518 1100

Fax: 509 335 4806

Email: prashant.daggolu@wsu.edu

See more of this Session: Meet the Faculty Candidate Poster Session
See more of this Group/Topical: Education

# (3al) Multi-Scale Modelling and Control of Fluidized Beds for Solar Grade Silicon Production

**Juan Du**, Department of Chemical Engineering, Carnegie Mellon University, Pittsburgh, PA

In this work, we present a multi-scale modeling approach for solar grade silicon production in fluidized bed reactor (FBR) by thermal decomposition of silane. The FBR process has been commercialized by several companies and promises to deliver poly-silicon at reasonable and stable price. Silane decomposes to form hydrogen and silicon and seed particles grow by heterogeneous chemical vapor deposition and by scavenging silicon powder produced in homogeneous gas phase reaction. The subsequent particle growth process is represented by a discretized version of population balance. The model uses ordinary differential and algebraic equations to track particle movement through discrete size intervals to simulate changes in the size distribution. The dynamics of gas phase is modeled Computational Fluid Dynamics (CFD) model. The CFD module shows the hydrodynamics via momentum balance, mass and heat transfer between different phases. The volume fraction of phase, concentration of each components and temperature of the bed is generated and are imported into the population balance. The particle size distribution calculated by population balance is used as the input of the CFD module and thus the complex interplay between particulate phase and continuous phase is captured by integrating those two modules.

Based on the multi-scale model, a novel method is proposed to derive the global stability condition to guarantee the stable operation of FBR. The resulting analytical stability condition will be verified by numerical simulations. The particulate process in FBR is complex and typically has very few measurements, inventory control is a simple method for control of complex systems and thus has potential for industrial application. We apply inventory control strategy to control particle size distribution to improve the performance of the silicon production process. Due to the lack of knowledge of the reaction coefficient, which is used in population balance to describe the mass transfer rate from gas phase to dispersed phase, an on-line estimator based on adaptive mechanism will be developed to track experimental behavior more accurately and in addition to enhance robustness of the inventory control system.

# (3am) Engineering *Saccharomyces Cerevisiae* for Biofuel Production From Lignocellulosic Biomass

**Jing Du**, Department of Chemical and Biomolecular Engineering, Institute for Genomic Biology, University of Illinois Urbana Champaign, Urbana, IL and Huimin Zhao, Department of Chemical and Biomolecular Engineering, Department of Chemistry and Biochemistry, Institute for Genomic Biology, University of Illinois at Urbana-Champaign, Urbana, IL

Bioconversion of plant derived lignocellulosic materials into biofuels has drawn increasing attention as an attractive strategy for replacement of fossil fuels. *Saccharomyces cerevisiae*, also known as baker's yeast, is considered one of the most promising organisms for ethanol production from lignocellulosic feedstock. Unfortunately, pentose sugars, which make up to 30% of biomass hydrolysate, cannot be utilized by *S. cerevisiae*. Heterologous pentose utilization pathways from bacterial and fungal resources have been introduced into *S. cerevisiae* to enable the assimilation of pentose sugars. However, pentose utilization of recombinant *S. cerevisiae* strains is inefficient due to several limitations such as the low uptake rate of pentose sugars and suboptimal heterologous pentose utilization pathways.

1.Cloning and characterization of pentose specific transporters

Pentoses can only enter yeast cells through hexose transporters, which have two orders of magnitude lower affinity towards pentoses than towards glucose. It has also been shown that inefficient pentose uptake is the limiting step in some xylose metabolizing yeast strains. Here we report the cloning and characterization of one arabinose-specific and two xylose-specific transporters from various pentose-utilizing species. These transporters were identified from a total of 18 putative pentose transporters. They were functionally expressed and properly localized in *S. cerevisiae* as indicated by HPLC analysis and immunofluorescence microscopy, respectively. Introduction of these pentose-specific transporters should facilitate pentose sugar utilization *in S. cerevisiae* by improving pentose uptake. More efficient utilization of pentose sugars will lower the cost for lignocellulosic ethanol production.

2.Combinatorial optimization of pentose utilization pathways in *Saccharomyces cerevisiae*

A lot of research has been carried out to overcome the limitations in pentose utilization in *S. cerevisiae*. However, it is challenging to come up with a single strategy which can simultaneously address all major limitations. Here we report the combinatorial optimization of a highly efficient pentose utilization pathway in *S. cerevisiae*. Specifically, we chose the fungal three-step xylose utilization pathway as our target pathway, and performed optimization of the pathway through shuffling of pathway enzymes or promoters. For the protein library based optimization, 10 to 20 different enzyme homologues from various fungal species with different catalytic efficiency and cofactor preference have been cloned for each pathway enzyme, and were assembled into a library of xylose utilizing pathways using our recently developed DNA assembler method. For the promoter library based optimization, promoter mutants with varied strength were created by error-prone PCR and randomly assembled into the pathway to generate a library of xylose utilizing pathways. The resultant libraries exhibited a high efficiency for correct assembly of the complete multi-gene pathway and good diversity of different homologues within the same catalytic step. Using the libraries generated, clones with better combination of enzyme homologues can be selected by faster cell growth. This method can also be applied to directly search for the pathway with a better fit in industrial yeast strains, which may have different metabolic flux patterns compared to the laboratory *S. cerevisiae* strains.

See more of this Session: Meet the Faculty Candidate Poster Session
See more of this Group/Topical: Education

# (3an) Improving Our Quality of Life with Colloidal Interactions: From Consumer Products to Protein Drugs

**Aaron P. R. Eberle**, Center for Neutron Science, NIST, Gaithersburg, MD

Colloidal dispersions exist in nearly every area of our daily lives from paint applied to protect our homes to therapeutic proteins. Universally, their material properties and functionality are defined by the subtle interactions between particles. This in turn leads to local structure, macroscopic phase behavior, and mechanical properties. Although the simplest of colloids, the mono-disperse hard sphere, is well understood there are significant gaps in our understanding of more complex systems that are often polydisperse and interact via a pair-potential.

One focal point of my research is to progress our understanding of how complex colloidal interactions and polydispersity relate to equilibrium structural states and phases, and states out of equilibrium induced by flow. This presentation summarizes two separate research projects include my work during my postdoc: i) "The phase behavior and shear induced structural reordering of a nanoparticle dispersion with adhesive hard sphere interactions," completed at the *University of Delaware*; ii) "Stability and rheology of a model protein (Lysozyme) with long-range repulsion and short-range attraction," that I started during my *National Research Council* postdoctoral appointment at the *NIST Center for Neutron Research*. By using a platform that leverages both experiments and theory I present unique advances that can be universally applied to the formulation and process optimization of complex colloidal systems.

See more of this Session: Meet the Faculty Candidate Poster Session
See more of this Group/Topical: Education

# (3ao) Fundamental Investigations In Polymer Science with High-Tech Applications From Penetrant Diffusion to Tissue Engineering

**Adam K. Ekenseair**, Department of Bioengineering, Rice University, Houston, TX

While the majority of effort in tissue regeneration research has been focused on implantable scaffolds, there are many applications that would be better served with injectable, *in situ* forming materials capable of co-delivering cells and growth factors to optimize tissue regeneration. Injectable scaffolds are minimally invasive and can easily fill complex tissue defects or voids often found in applications such as craniofacial bone regeneration after trauma, tumor resection, or birth defects. Thermogelling polymers, such as poly(N-isopropylacrylamide), which pass through a lower critical solution temperature upon injection into the body, are promising candidates as scaffold backbones. Concomitant chemical crosslinking during thermogellation further enhances the stability and mechanical properties of such materials; however these networks must be made biodegradable on an appropriate timescale for tissue regeneration. This is a major issue for injectable materials, since many commonly used biodegradable polymers, such as polycaprolactone and poly(lactic acid), are not water soluble. Polyamidoamines, an emerging class of polymers, offer a biocompatible, biodegradable, and water-soluble option for the creation of such crosslinked networks; and their efficacy is herein evaluated.

To this end, a novel two-component system was developed. Poly(NiPAAm-co-glycidyl methacrylate) was synthesized by a free radical pathway to introduce functional epoxy rings into the thermogelling polymer. Diamine crosslinking macromers were formed from piperazine and methylene bisacrylamide with varying molecular weights through a step polymerization mechanism. The secondary amines present at the ends of these polyamidoamine macromers reacted readily with epoxy rings, even across phases. Thus, chemically and thermally gelling hydrogel scaffolds were formed and the effects of polymer wt %, time between solution mixing and injection, degree of crosslinking, and crosslinker length on the formation, equilibrium swelling, degradation, and mechanical properties of the scaffold were evaluated. This system is shown to be a promising candidate for injectable tissue engineering scaffolds with little or no syneresis upon gellation and relevant degradation rates.

See more of this Session: Meet the Faculty Candidate Poster Session
See more of this Group/Topical: Education

# (3ap) Structured Polymers for Energy Generation and Storage

**Hany B. Eitouni**, Seeo, Inc, Hayward, CA

Reduction of both greenhouse gas emissions and U.S. dependence on foreign oil will require the widespread adoption of electric vehicles, renewable power generation, and smart electrical grids. The current bottleneck to extensive deployment of both electric vehicles and smart electrical grids is the battery. Adoption by major utility companies of renewable energy plants, such as those based on photovoltaics, is hindered for two reasons. Stabilization of the intermittent power generation requires inexpensive long-life battery systems, which are yet to be developed. Additionally, power derived from photovoltaics is still too expensive to compete with electricity derived from fossil fuels. Deploying these applications will call for entirely new energy generation and storage innovations. In my poster presentation I will highlight examples both from my graduate research and from my accomplishments at Seeo Inc., the start-up company I co-founded. I will discuss how electrochemistry can be used to control assembly of block copolymers and, reciprocally, how structured nanomaterials can be used to control electrochemical device function, particularly in the case of stable high energy batteries. Building on my past work, I will describe my plans to create a research program aimed at obtaining a fundamental understanding of the structure-property-function relationships in nano-structured polymers as they apply to devices for advanced energy generation and storage.

During my graduate work, I synthesized and studied organometallic block copolymers, wherein one block was composed of alternating ferrocene and dialkylsilane units in the main chain. Similarly to typical organic block copolymers, they underwent self-assembly to form microphase-separated ordered structures, and I characterized the phase behavior and thermodynamic interactions of the poly(ferrocenyldimethylsilane) system with multiple copolymers. The thermodynamic interactions in poly(ferrocenyldimethylsilane) diblock copolymers were systematically adjusted by oxidizing the ferrocene moieties with silver salts and examined using SAXS and depolarized light scattering. The polymers retained microphase separated ordered structures upon oxidation and showed systematic changes in the location of the order-disorder transition as a function of extent of oxidation. By controlling the redox properties of the ferrocene moiety in the backbone of the polymer, I developed a method for controlling the self-assembled microstructure and hence bulk material properties. Furthermore, using electrochemical techniques, a novel means of controlling the order-disorder transition of block copolymers was discovered. By applying very small electrochemical potentials to disordered solutions of the organometallic block copolymers, oriented ordered grains were formed near one electrode, the result of electrochemical reactions. After reversing the electrical bias on the system, the ordered grains disappeared and new oriented ordered regions were formed at the opposite electrode. Our work established a fundamentally new mechanism for controlling the self-assembly of block copolymers.

More recently as Director of Materials Development at Seeo Inc., I have focused on developing commercially viable lithium battery systems based on solid-state polymeric electrolytes. Solid-state batteries have long been a goal of rechargeable battery development. Today's conventional batteries rely exclusively on organic liquid electrolytes, whose volatility and reactivity are responsible for the most significant limitations to battery life and safety. Liquid electrolytes continuously react with electrodes at maximum cell voltages and upon cycling, causing reduced lifetime through impedance rise and capacity fade. In contrast, polymer electrolytes limit the mobility of reactive electrolyte species, forming solid interfaces that can remain stable over long periods of time and over many charge/discharge cycles. Remarkably, this drastic enhancement in interfacial stability can yield extremely low rates of degradation and capacity fade, even in cells employing otherwise reactive high energy electrodes. Before the widespread commercialization of lithium ion batteries, solid polymer electrolyte technology was considered to be among the most promising routes to enabling lithium-based batteries. However, despite strong government and private-sector funding, academic and commercial efforts to develop batteries based on solid polymer electrolytes consistently failed. At the heart of each of these failures lay the inability to overcome a single fundamental constraint. Ion transport in polymers is coupled to the motion of the surrounding molecules that are solvating the ions. The presence of mobile molecules is thus essential to allow for a conductive medium; however, the same mobility of molecules is detrimental to the polymer's structural integrity. In all cases, polymer electrolytes with high ionic conductivity failed due to mechanical reasons while mechanically robust ion conductors were severely limited in conductivity. At Seeo, we have employed advanced self-assembled polymer nanostructures to break free from this principal constraint. We are commercializing our proprietary polymeric electrolytes in lithium-based batteries for electric vehicle and grid-tied storage applications. I believe my four-year experience leading an industrial research and development team gives me a unique understanding of how to create an academic research program with capabilities of exploratory science, intellectual property generation, and translation to practical application.

# (3aq) Biofabrication of Bacteria-Based Biohybrid Devices and Self-Folded 3D Hydrogels

**Rohan Fernandes**, Chemical and Biomolecular Engineering, Johns Hopkins University, Baltimore, MD

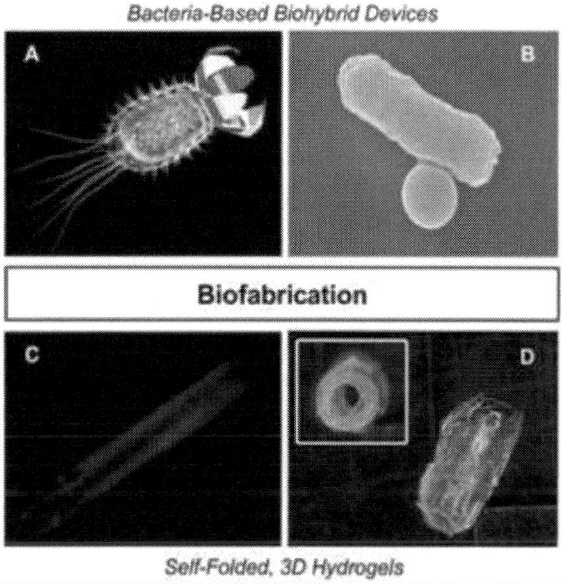


**Biofabrication of bacteria-based biohybrid devices and self-folded, 3D hydrogels.** Schematic (A) of a bacterium attached to a nanostructure. SEM (B) of a single bacterium attached to a functionalized sub-micron bead. Fluorescence image (C) and brightfield image (D) of self-folded, 3D hydrogel sheets.

Biofabricationinvolves the use of biological or bioactive elements as building blocks tomanufacture advanced biological models, therapeutic products and non-medicalbiological systems. In this work, we describe the use of biofabrication inconstructing bacteria-based biohybrid devices for diagnostics and therapeuticsand in creating self-folded, three-dimensional hydrogels for cell and tissueengineering. Bacteria are ideal candidates for use in diagnostics and therapeuticson account of their small sizes, their ability to respond to stimuli, convertchemical energy to motion, swim and grow naturally in inaccessible regions ofthe body as well as their ability to be genetically engineered to produceproducts of interest or exhibit desired characteristics. Recent advances inmicro/nanofabrication technologies have resulted in the generation ofparticles, structures and devices for use in diagnostics and therapeutics. Thiswork describes the interfacing of bacteria with micro/nano particles,structures or devices to construct biohybrid devices that combine theadvantages of both bacteria and the devices. The utility of the bacteria-basedbiohybrid devices in diagnostic and therapeutic applications will be presented.In nature, cells and tissues exist in precise 3D shapes and geometries thatdictate their function. A few techniques exist for patterning cells and tissuesin 3D. Of these, hydrogels are particularly attractive on account of their highwater content, ease of diffusion of oxygen, nutrients, waste and solublefactors and they can be synthesized to be biocompatible. This work willdescribe the creation of self-folded 3D hydrogels and their utility in cell andtissue engineering.

See more of this Session: Meet the Faculty Candidate Poster Session
See more of this Group/Topical: Education

# (3ar) Emerging Technologies In Biomass Exploitation As a Renewable Source of Energy and Material

**Marcus Foston**, Department of Chemistry and Biochemistry, Georgia Institute of Technology, Atlanta, GA

The overarching goal of my proposed research program is to *develop ways to both chemically and biologically exploit and utilize renewable biomass resources.* The long-term focus of my research program is two-fold, I plan to: (1) develop, improve and evaluate methods for the characterization of biomass, biomass conversion, and bio-derived material synthesis; and (2) generate biomass-derived monomeric and polymeric materials for specific energy, electronic, biomedical, and high-performance applications. Potential projects include applications for membranes, drug delivery and with dynamic nuclear polarization NMR and isotopically enriched biomass. During my previous studies, I conducted research employing kinetically controlled cyclodepolymerization of linear siloxane precursors in dilute solution to generate cyclic oligomers in the effort to synthesize complex topological and multi-component polymeric material, such as polyrotaxanes. Recently, my research as part of DOE's BioEnergy Science Center has focused on the chemistry, dynamics, and mechanism of lignocellulosic recalcitrance and deconstruction in an effort to form biofuels with the specific subtask of the application of advanced solution- and solid-state NMR and magnetic resonance imaging techniques.

See more of this Session: Meet the Faculty Candidate Poster Session
See more of this Group/Topical: Education

# (3as) Nano Materials for Microelectronics and Energy Sciences

**Domingo Ferrer**, Microelectronics Research Center, University of Texas at Austin, Austin, TX and Sanjay Banerjee, University of Texas at Austin, Austin, TX

Our research thrust aims at clarifying the mechanisms dictating the synthesis of size-, shape- and composition-controlled atom-sized systems for microelectronics and energy sciences. Our efforts bridge chemical and electrical engineering, materials science and physics. Nanofabrication techniques are pursued for applications involving the demonstration of memory spin-based devices that are scalable to their ultimate physical limits. Diverse routes undertaken to fabricate spin-transfer-torque random access memories (STT RAM) are presented. Monolayer assemblies of FePt nanostructures have been fabricated with different geometries, including nanorods, oval-shaped particles and spherical particles for rational control of their magnetic properties. Graphene transistors are also explored in order to realize low-power and high mobility devices. Large-area monolayer graphene films are grown by chemical vapor deposition, using C2H2 as precursor and Co films as catalyst. The number of graphene layers can be controlled by choosing an appropriate Co film thickness. Electrical transport measurements reveal mobility values that are comparable to graphene grown by CVD on other transition metals. Atomic layer deposition (ALD) is a key process for the microelectronics application studied by our group. Thin films of BeO and MgO are grown for devices involving III-V transistors and magnetic tunnel junctions, respectively. In addition to this, our research group endeavors will serve to answer critical issues in the design and development of low cost, highly efficient nanomaterials that can ease widespread use of clean energy technologies such as solar cells. The students of my research group will be skilled on the fabrication, characterization and application of inorganic nanomaterials to meet the research goals needed to ensure that viable fossil fuel alternatives are developed.

See more of this Session: Meet the Faculty Candidate Poster Session
See more of this Group/Topical: Education

# (3at) Elucidating Reaction Mechanisms and Identifying Bottlenecks In Catalytic Processes

**David W. Flaherty**, Department of Chemical Engineering, University of California at Berkeley, Berkeley, CA

Kinetic and mechanistic investigations of chemical reactions can be clarified by deconstructing the series of elementary steps and individually interrogating their dependence on the prevalent reaction conditions and catalyst structure and composition. The use of decidedly monofunctional systems is critical as an initial step, and consequently results are obtained on single crystal model systems of noble metals, bimetallics, and metal carbides as well as from noble metal clusters supported on inert supports. Complexity can be incrementally restored by the inclusion of materials containing additional chemical functions. Utilizing transient and steady-state kinetic measurements in combination with in situ spectroscopy it is possible to identify "bottlenecks" in the reaction sequences which have consequences for absolute and relative turnover rates. Selective decomposition and formation of heavy hydrocarbons and light oxygenates are investigated to ascertain the role of surface structure and composition as it pertains to stabilizing individual intermediates with the catalytic cycle.

See more of this Session: Meet the Faculty Candidate Poster Session
See more of this Group/Topical: Education

# (3au) Electrostatics In Chemical and Material Processing

**Keith M. Forward**, Department of Chemical Engineering, Massachusetts Institute of Technology, Cambridge, MA

My past research has utilized electrostatics principles to produce unique materials. I have investigated the fundamentals of producing nanomaterials from electrohydrodynamic processes such as electrospinning and electrospraying from a free liquid surface. In this process, electrostatic forces repel surface charges to form Taylor Cones, resulting in electrostatically driven jets or droplets. The electrostatic forces cause the charged liquid to accelerate toward a grounded collector which enhances evaporation and eventually leads to solidification of the jets and droplets, producing nanomaterials with high surface areas and porosities. These materials can be applied in a wide range of fields such as textiles, filtration, tissue engineering, drug delivery systems, pharmaceuticals, nanocomposites, and alternative-energy generation systems including solar cells, fuel cells, and energy storage devices. Correlations have been developed to predict productivity and fiber diameter based upon solution properties, operating parameters, and system configuration.

In addition, my work has focused on the commonly observed, yet poorly understood, electrostatic phenomena of contact charging also known as triboelectrification. In this process, two initially neutral surfaces come in contact with each other and a subsequent transfer charge occurs. Once the two surfaces are separated, an equal but opposite polarity charge remains on the charge surface. The polarity of the surface charge and amount of charge transfer has been empirically studied, however it remains highly debated whether or not ions and/or electrons are responsible for charge transfer. This contact charging produces problems for manufacturing chemicals and materials, particularly in the case of granular materials. Due to the surface area to volume ratio, large amounts of charge can accumulate in granular materials (both in multi-component and single-component systems). A kinetic gas model was employed to determine the frequency of collisions between granular particles. It showed an unbalanced accumulation of the charge transfer species (positive, negative ions, and/or electrons) on the relatively smaller size particles. This model explains the observed behavior of single component systems in which the relatively smaller size particles charged negatively and the larger particles charged positively. This result suggests a negative transfer species (i.e. negative ions and/or electrons) between surfaces.

See more of this Session: Meet the Faculty Candidate Poster Session
See more of this Group/Topical: Education

# (3av) Tailoring Morphologies and Properties of Soft Materials: Insight From Modeling and Simulation

**Jie Feng**, Department of Polymer Engineering, University of Akron, Akron, OH

In my Ph. D. and postdoctoral research, soft materials have been studied using modeling and simulation ranging from all-atom to mesoscopic scales. The results of my research provide a way of understanding the experimental phenomena, predicting the characteristics of soft materials, and tailoring their molecular structures towards optimal system properties. In my poster presentation, several examples of my past and current projects will be presented, such as (1) binding strength and conformation of peptides on metal surfaces investigated using all-atom and coarse-grained simulation; (2) mechanical properties of polymer/CNT composites evaluated using all-atom force field simulation, (3) directed self-assembly of block copolymers using Monte Carlo simulation and time dependent Ginzburg-Landau theory, and (4) hybrid force field and model for the study of interfacial properties. Future challenges and my research plans will be also discussed in the poster session.

See more of this Session: Meet the Faculty Candidate Poster Session
See more of this Group/Topical: Education

# (3aw) Use of Conducting Polymers As Biomaterials for Neural Tissue Engineering Interfaces

**Leandro Forciniti**, ISurTec, New Hope, MN

I graduated with a doctorate in chemical engineering from the University of Texas, Austin. My dissertation work involved investigating the use of the conducting polymer polypyrrole in neural applications. Ultimately, my goal was to make progress in the field of biomaterials such that in the future better therapeutic systems for nerve regeneration could be made. This involved both investigating fundamental questions of how neural cells interact with biomaterials and using that knowledge to synthesize and characterize new biomaterials. I am currently a NSF post-doctoral fellow working on super hydrophobic coatings for biomaterials at ISurTec. As a young faculty member, I would like to work on both investigating the fundamental biology of brain cancers and on investigating the use of conducting polymers for treatments of brain diseases including cancers. I believe my background makes me well suited for tackling both areas. As part of my research I would use multi-scale computational modeling to determine the effect of mechanical forces on brain cancer growth. As an experimental component to this research I would look at determining the effect of extracellular matrix stiffness and ligand density on brain tumorogenesis. While developing a better understanding of brain cancers, I would also investigate the use of conducting polymers as drug delivery devices for treatment. As to not limit the impact of my research, this branch of my research would emphasize the creation of new conducting polymers for a multitude of different brain diseases including the treatment of brain cancers.

See more of this Session: Meet the Faculty Candidate Poster Session
See more of this Group/Topical: Education

# (3ax) Homogeneous/Heterogeneous Chemistry In Energy Conversion

**Claude Franklin Goldsmith**, Inorganic Chemistry, Fritz Haber Institute of the Max Planck Society, Berlin, Germany

Hydrocarbon fuels will remain the energy storage medium of choice for the foreseeable future. As the global demand for these fuels increases, sustainable technologies for both the production and consumption of these fuels will become critical. A suit of reforming processes can be used to synthesize liquid fuels from various feed stocks, such as biomass or natural gas. Examples of intermediate processes include short-contact time partial oxidation, oxidative coupling, hydrothermal reforming, and Fischer-Tropsch synthesis. Advanced conversion technologies – such as homogeneous-charge compression ignition, catalytic combustion, and solid oxide fuel cells – could generate power with increased efficiency.

In order to make these technologies both sustainable and commercially viable, we must understand how the chemical bonds in these fuels are manipulated at the atomic level. Unfortunately, the chemistry of energy conversion is staggeringly complex. Most of these technologies involve high temperatures and high pressures. Under these conditions, homogeneous gas-phase chemistry is coupled with heterogeneous surface-phase chemistry. A detailed mechanism for combustion of comparatively small hydrocarbons requires hundreds of species and thousands of reactions, and the inclusion of surface chemistry dramatically increases the complexity.

My research will help to elucidate the complex chemistry of energy conversion. I will combine various computational methods with experimental techniques to create and validate kinetic mechanisms in homogeneous/ heterogeneous coupled systems. Gas-phase mechanisms will be generated using an automatic reaction mechanism generator, RMG. Thermodynamically consistent surface mechanisms will be generated using scaling relationships, BEP relations, and other methods. Spatial profile reactors will provide experimental data under industrially relevant conditions. The simplified flow field, wide range of operating conditions, and sub-mm resolution of these instruments make them ideal candidates for evaluating microkinetic models. Sensitivity, flux, and reaction-path analyses will highlight which minima and saddle points on a potential energy surface are most important, and these stationary points will be refined with computational quantum chemistry (e.g. DFT) and statistical mechanics. New experiments will be performed to highlight certain reaction channels. This process is iterated until the mechanism is suitably accurate. Finally, mechanism reduction techniques will be applied to create a skeletal mechanism that can be combined with CFD codes for industrial applications.

See more of this Session: Meet the Faculty Candidate Poster Session
See more of this Group/Topical: Education

# (3ay) Understanding and Predicting Chemical Reactivity In Solid Acid Catalysis

**Rajamani Gounder**, Department of Chemical and Biomolecular Engineering, University of California at Berkeley, Berkeley, CA

Chemical reactions catalyzed by Brønsted acids involve cationic transition states and the separation of charge into ion-pairs, which become more stable as acid sites become stronger. Acid sites in aluminosilicate zeolites are weaker and less diverse in composition and strength than in mesoporous or liquid acids, but confinement effects can selectively stabilize transition states relative to the relevant reactants and lead to turnover rates that depend sensitively on the nanostructure of the confining voids. Efforts to examine the individual consequences of acid strength and confinement for catalysis have been limited, however, because these effects are inextricably linked in rate constants and because a microporous structure can contain a diverse range of local environments with significant consequences for turnover rates.

Here, I will discuss mechanistic studies of reactions of alkanes (cracking, dehydrogenation), alkenes (hydrogenation, alkylation, oligomerization), and alkanols (dehydration) on solid acids and protocols to characterize the number, strength and location of acid sites within microporous voids. Fundamental relations between the structures of reactants and catalysts and those of ion-pairs formed at transition states have been established using Born-Haber thermochemical cycles, Marcus theory for charge-transfer reactions and De Donder relations for non-equilibrium thermodynamics. These insights have enabled precise predictions of rate and equilibrium constants for elementary steps and, in turn, of overall catalytic rates and selectivities and of site requirements as reactants, catalysts and reaction paths change. These principles apply generally to acid catalysis and thus provide a methodology also useful in my future research plans to address issues relevant in the production of fuels and chemicals from renewable and fossil-based resources. This research program will integrate fundamental mechanistic studies, methods to characterize active site structure and techniques for the targeted synthesis of catalytic materials.

See more of this Session: Meet the Faculty Candidate Poster Session
See more of this Group/Topical: Education

# (3az) Multi- Scale Functional and Responsive Soft Biomaterials for Tissue Engineering and Drug Delivery Applications

**Murat Guvendiren**, Bioengineering, University of Pennsylvania, Philadelphia, PA

The main focus of my research is the development, characterization and processing of multi-functional and multi-responsive soft biomaterials with well-controlled "dynamic" properties for biological applications with a particular interest in tissue engineering and drug delivery. My past training has provided a foundation for this future work. During my Ph.D. I studied self-assembly, adhesion and interfacial interactions of polymers and soft materials. In particular, I designed and synthesized marine mussel mimetic hydrogels, and investigated their adhesive properties with soft (tissue) and hard substrates (e.g. Ti) in aqueous environments. During my 1st year in postdoctoral studies, I developed skills on nano/micro fabrication and patterning techniques. Specifically, I worked on photopolymerization, nano/micro 2D/3D patterning, elastic instabilities, stimuli responsive soft surfaces and composite hydrogels. For the past 3 years, I have been working on developing polymeric biomaterials for tissue regeneration and drug delivery applications. Some of the specific examples include using surface instabilities to control stem cell shape and differentiation, and investigating the stem cell response to substrate stiffening (in HA gels) in terms of spreading, traction forces (via dynamic traction force microscopy), migration, proliferation and differentiation.

My long-term goal is to establish an interdisciplinary experimental research group on soft functional materials. In this poster session, I would like to show relevant examples of my past research and mainly focused on three projects that will serve as the foundation for initial work in my laboratory:

1. Biomimetic Nanopatterns to Fast-Screen Stem Cell Properties and Function
2. Next Generation Soft Hydrogels with Dynamic Properties for Tissue Regeneration
3. Smart Hydrogel-based Biomaterials for Mechano-Induced Drug Delivery

# (3b) An Empirical Boundary Layer Based Model to Predict Contamination in Multiproduct Refined Petroleum Pipelines

**Anirudh R. Patrachari** and Arland H. Johannes, School of Chemical Engineering, Oklahoma State University, Stillwater, OK

Theoretical convective diffusion models predict interfacial mixing volumes in refined petroleum pipelines using the square root of the travel/pipe length. This is an idealized case obtained by assuming the universal velocity profile and it has been shown to not being able to predict transmix volumes to the required degree of accuracy. This lack of prediction accuracy has been attributed to the presence of pipe fittings, entrance effects and other unaccounted factors. Viable methods have not been developed to account for these factors and hence, empirical models are a popular choice for industrial applications. The empirical models available in literature suggest a variety of exponent values to replace the ideal square root relationship. Values ranging from 0.38 to 0.62 can be commonly found in literature. This work intends to suggest a revised empirical relation for flow through straight pipes, with the power law exponent varying based on the physical properties of the system under consideration. The work also offers a phenomenological context to the development of the empirical model. It is argued that the viscous sub-layer has a major role in the formation of transmix and can be employed as a parameter to estimate the extent of contamination. The effects of bends and other pipe fittings have been accounted for by employing the technique of equivalent lengths that is conventionally know to apply to pressure drop predictions in pipe flows. A comparison of the predictions against literature data validates the improved accuracy of the present work.

See more of this Session: Meet the Faculty Candidate Poster Session
See more of this Group/Topical: Education

# (3ba) Green Strategies for Sol-Gel Immobilization of Biomolecules and Carbon Nanotubes

**Gautam Gupta**, Center for Integrated Nanotechnolgy, Los Alamos, Los Alamos, NM

Immobilization of enzymes, cells, carbon nanotubes and organelles has been relevant for numerous applications including energy harvesting and sensing. One of the important class of molecules are enzymes that are key to new processes because they are environmentally friendly, expensive, fragile, reduce hazardous waste, and one the best catalyst known to mankind, thus the soft immobilization strategies are highly desired. One of the key immobilization techniques used is Sol-gel technique. Sol is generally referred to as a collection of colloidal silica particles ranging from 1-1000 nm in a liquid phase, and sol-gel is obtained by enhancing the interactions between these colloidal particles resulting in the formation of a gel like material.
Several methods have been explored during last few decades to efficiently immobilize biomolecules in sol-gel mainly because of the advantage of room-temperature synthesis and enhanced stability of immobilized enzymes. The common silica precursors include tetra alkoxy silanes, sodium silicate, silaffin polypeptides and even hydrophobic silica precursors. Of all these methods, the most common sol-gel synthesis revolve around a two-step process in which, an alkoxy silane precursor is hydrolyzed using an acid/base catalyst in a water/ alcohol environment followed by removal of toxic solvent and further followed by addition of biomolecules to the system.
I will show a CVD approach that has been developed by our group to bulk immobilization of a wide variety of molecules including enzymes, organelles, cells, sensitive lipid bilayer based systems and carbon nanotubes. Typically, the process is a simple one-step process that involves the exposure of an aqueous solution to the vapors of tetramethoxy silane at room temperature for a desired amount of time. This process follows key principals of green chemistry including: Prevention (No Waste materials), Economy (All chemicals are utilized completely), Less hazardous chemical Synthesis (No acid and base catalyst), Safer solvents (Aqueous route), Energy Efficiency (Ambient Temperature and pressure), and Safe Chemistry (and minimal accident issues and hazards)

See more of this Session: Meet the Faculty Candidate Poster Session

# (3bb) Microfluidic Methods for Nanofiber and Nanoparticle Synthesis

**Venkata R. Gundabala**, School of Physics, Georgia Institute of Technology, Atlanta, GA

This poster will highlight my research interests in the areas of Microfluidics and Functional Coatings. In the poster I will also summarize my past research experience during my PhD and Postdoctoral work involving studies on droplet generation using Microfluidics and on Waterborne polymer nanocomposite coatings. Nanofibers and nanocomposite fibers find wide-ranging applications as tissue templates, filtration media, reinforcements in composite developments, and medical prostheses. Micro and nano particles find wide-ranging applications in encapsulation of food additives, targeted drug delivery, photonic applications, etc. My past research in Microfluidics has shown the ease with which one can change from droplet generation mode to a whipping jet mode in a microfluidic channel. Based on that, my current interest lies in developing novel microfluidic techniques for steady generation of monodisperse emulsion droplets as well as solid particles and particles with core-shell morphologies. In addition, I will focus on implementing electrospinning in microfluidic devices, with an aim to expand the parameter space of control over the fiber properties. My other research interests include investigating the applicability of nanocomposite materials as functional coatings through synthesis and film formation studies.

See more of this Session: Meet the Faculty Candidate Poster Session
See more of this Group/Topical: Education

# (3bc) Designing Biomaterial Scaffolds for Tissue Regeneration and Cellular Assays

**Gargi Ghosh**, Department of Chemical and Biological Engineering, University of Wisconsin-Madison, Madison, WI

The flow of signal within a cell upon receiving cue from extracellular stimuli can be represented by biochemical reactions. My research interests are focused at the interface of Engineering, Chemistry and Biology to dissect chemical signaling pathways and characterize how quantitative changes in signal transduction can alter the cell fate. Specifically, I am interested in rational designing of biomaterial scaffolds for applications in tissue regeneration, three dimensional cell culture, and cellular arrays/ assays. The research will span the fundamental studies to better understand cell-material interactions including wettability, surface topography, surface chemistry and stiffness of bioactive surfaces in regulating cell fate to more targeted clinical applications including development of 3D cell culture matrices to investigate interaction between stromal cells and cancer cells as well as development of cellular arrays/assays for quantifying receptor mediated cell behavior. My Ph.D research (involving immobilization of human umbilical vein endothelial cells (HUVECs) onto cellulose triacetate membrane to develop cell based biosensor for screening cancer) as well as postdoctoral research (involving development of hydrogel based three dimensional protein array for high-throughput profiling of epidermal growth factor receptor (EGFR) in cancer cellular extracts) which provide me complementary skills to develop a biomaterial based approach to address health related issues will be highlighted and my plans to develop an independent research program will be discussed.

See more of this Session: Meet the Faculty Candidate Poster Session
See more of this Group/Topical: Education

# (3bd) Ultrafast Dynamics of Interfacial Soft Matter

**Valeria Garbin**, Chemical and Biomolecular Engineering, University of Pennsylvania, Philadelphia, PA

Self-assembly at fluid interfaces is at the basis of natural and industrial processes: the interfacial assembly of amphiphiles provides the architecture of biomembranes, and is exploited in pharmaceutical and cosmetic manufacturing; nanoparticles can form films with tunable electromagnetic properties, and catalyze reactions at the liquid/liquid interface. In spite of their relevance for applications, the dynamic response to external flow and applied deformation of these "soft interfaces" remains largely unexplored because of the inherent challenges in accessing relevant regimes in experiment.

My research will focus on characterizing the response of soft interfaces to highly dynamic deformations and on understanding their nonlinear behavior when driven out of equilibrium. Rapid, leading order deformations of complex interfacial materials can indeed produce a wide range of nonlinearities–including shear-thinning, inhomogeneity in the microstructure, and mechanical failure. In my doctoral and postdoctoral research I have gained extensive experience in precise dynamic measurements using advanced imaging techniques (high-speed imaging, diffractive optics, X-ray scattering, microinterferometry) and micromanipulation techniques (optical tweezers, microfluidics, ultrasound), to study problems in biomedical fluid mechanics and colloidal science. Combining techniques in the forefront of optics, acoustics and interfacial science will allow me to develop a research program on the nonlinear dynamics of soft interfaces.

In my initial research, I will probe the mechanical response of novel structures presenting soft interfaces (e.g. nanoparticle capsules and membranes) using large-amplitude and fast deformations, for applications to responsive and functional materials. I will also study nonlinear phenomena arising in biomembranes subject to stress (e.g. molecular rearrangements and pore formation), which remain poorly understood despite their central role in many biomedical processes.

See more of this Session: Meet the Faculty Candidate Poster Session
See more of this Group/Topical: Education

# (3be) Polymer Electrolytes for Energy and Water Technologies

**Daniel T. Hallinan Jr.**, Department of Chemical and Biomolecular Engineering, University of California, Berkeley, Berkeley, CA

Fundamental investigations on polymer electrolytes require understanding the electrochemical driving forces associated with ions in polymers. Electrode interfaces are an essential part of any electrochemical system. As such, interactions between polymers and electrodes (inorganics) are relevant. A key part of my faculty research will therefore focus on polymer-inorganic composites for energy technologies. For example, in battery electrodes a polymer electrolyte can serve multiple functions as not only the ion conductor but also the binder of the active inorganic material. It can mitigate otherwise detrimental volume changes of the active material and improve safety by removing the need for flammable solvents of liquid electrolyte. However, side-reactions between the polymer and inorganic components can degrade battery performance. In order to develop compatible polymer-inorganic systems, the interactions between the two must be better understood. Molecular scale techniques such as infrared and Raman spectroscopies, nanoscale techniques such as hard and soft x-ray scattering and electron imaging, and interface sensitive techniques such as reflection mode scattering and X-ray Photoelectron Spectroscopy will prove useful for studying polymer-inorganic interactions. My postdoctoral experience has provided access to several sophisticated x-ray techniques at Lawrence Berkeley National Laboratory.

In addition to interfacial effects in electrochemical systems, ion transport through the bulk polymer electrolyte can be important. The interplay among cations, anions, and polymer functional groups affects ionic conductivity and, thereby, device performance. *In situ* electrochemical techniques, for example with reflection infrared spectroscopy, are needed to probe both the thermodynamic interactions in polymer electrolytes and the ion transport, as well as polymer mobility. My extensive work with infrared spectroscopy during my PhD combined with development of *in situ* electrochemical techniques during my post doctoral fellowship, will provide a solid foundation for developing such techniques as a faculty member.

Ion transport in polymer electrolytes is also relevant for water desalination. As I have extensive experience studying water in polymers, I would like to develop a project for membrane purification and desalination of water. Access to clean, fresh water is an important human need that requires a low-energy solution. I expect research funding in this field to grow in the coming years.

See more of this Session: Meet the Faculty Candidate Poster Session
See more of this Group/Topical: Education

# (3bf) Computational Materials Design for Catalysis and Energy Materials

**Jeong Woo Han**, Department of Nuclear Science and Engineering, Massachusetts Institute of Technology, Cambridge, MA

During the past decade the theoretical description of science and engineering phenomena has undergone a dramatic development. Today's advances in computational capabilities and algorithms make it possible to design new materials with properties tailored to specific applications within the detail and accuracy required for computational results to compare with experiments. We have guided experimental efforts, using the information of new materials at the atomic and molecular level obtained by the state-of-the-art theoretical methods such as density functional theory (DFT) calculations. Our focus here is on the discovery of materials with desired properties for a wide range of important applications in catalysts and energy materials. Specific examples include as follows;

(1) Synthesis and separation of enantiopure chemicals is of enormous importance in the pharmaceutical industry. We studied several classes of solid surfaces that are chiral to aid the development of new materials for applications in chiral chemical processing. These materials include chiral minerals such as quartz, highly stepped metal surfaces that are intrinsically chiral, and atomically flat metals that are decorated with chrial molecules such as amino acids.

(2) Transition metal carbides have received considerable attention as promising catalysts due to their electronic and catalytic properties similar to the noble metals, but with more resistance to sulfur poisons. Alkali promoters are known to enhance the efficiency of the catalytic activity of these materials, but the detailed mechanisms remain elusive. We examined how alkali promoters bind on these surfaces, providing a useful foundation for the catalytic selectivity in the chemical reactions such as Fischer-Tropsch synthesis.

(3) Solid oxide fuel cells (SOFCs) have been attractive given its fuel flexibility and high conversion efficiencies. The slow rate of oxygen reduction at the cathode is, however, the main barrier to achieve higher power output in SOFCs at intermediate temperatures. Recently, lattice strain was shown to have a significant impact on facilitating oxygen ion transport, vacancy formation, and surface reactions for SOFC-related materials. We mechanistically investigated strain

effects on the surface chemistry, electronic structure, and oxygen incorporation, showing how the lattice strain should be controlled to attain fast oxygen reduction kinetics on the cathode materials.

Our computational modeling has elucidated fundamental understanding of the correlation between materials structure and properties. Theoretical guidance based on this can play a valuable role on novel chemical processes and important technological advances by directing experimental efforts to the most promising of many possible materials that can be considered.

Postdoctoral Advisor: Professor Bilge Yildiz, Department of Nuclear Science and Engineering, Massachusetts Institute of Technology, Cambridge, MA

Doctoral Thesis Advisor: Professor David S. Sholl, School of Chemical and Biomolecular Engineering, Georgia Institute of Technology, Atlanta, GA

Publications:


1. **Jeong Woo Han** and Bilge Yildiz, *Enhanced one dimensional mobility of oxygen on strained $LaCoO_3$ (001) surface*, J. Mater. Chem. **Advance Article** (2011) DOI:10.1039/C1JM12830B.
2. Zhuhua Cai, Yener Kuru, **Jeong Woo Han**, Yan Chen, and Bilge Yildiz, *Surface Electronic Structure Transitions at High Temperature on Perovskite Oxides: The Case of Strained $La_{0.8}Sr_{0.2}CoO_3$ Thin Films*, **under revision** (2011).
3. **Jeong Woo Han**, Joanna N. James, and David S. Sholl, *Chemical speciation of adsorbed glycine on metal surfaces*, J. Chem. Phys. **135** (2011) 032127.
4. Helia Jalili†, **Jeong Woo Han†**, Yener Kuru, Zhuhua Cai, and Bilge Yildiz, *New Insights into the Strain Coupling to Surface Chemistry, Electronic Structure, and Reactivity of $La_{0.7}Sr_{0.3}MnO_3$*, J. Phys. Chem. Lett. **2** (2011) 801-807. †Equally contributed
5. **Jeong Woo Han**, Liwei Li, and David S. Sholl, *Density Functional Theory Study of H and CO Adsorption on Alkali-Promoted $Mo_2C$ Surfaces*, J. Phys. Chem. C **115** (2011) 6870-6876.
6. **Jeong Woo Han**, Helia Jalili, Yener Kuru, Zhuhua Cai, and Bilge Yildiz, *Strain Effects on the Surface Chemistry of $La_{0.7}Sr_{0.3}MnO_3$*, ECS Trans. **35** (2011) 2097-2104.
7. Heather L. Tierney, **Jeong Woo Han**, April D. Jewell, Erin V. Iski,

Ashleigh E. Baber, David S. Sholl and E. Charles H. Sykes, *Chirality and Rotation of Asymmetric Surface-Bound Thioethers*, J. Phys. Chem. C **115** (2011) 897-901. Part of the "Alfons Baiker Festschrift"

8. **Jeong Woo Han** and David S. Sholl, *Enantiospecific adsorption of amino acids on hydroxylated quartz (10-10)*, Phys. Chem. Chem. Phys. **12** (2010) 8024-8032.
9. Darin O. Bellisario, **Jeong Woo Han**, Heather L. Tierney, Ashleigh E. Baber, David S. Sholl, and E. Charles H. Sykes, *Importance of Kinetics in Surface Alloying: A Comparison of the Diffusion Pathways of Pd and Ag Atoms on Cu(111)*, J. Phys. Chem. C **113** (2009) 12863-12869.
10. **Jeong Woo Han** and David S. Sholl, *Enantiospecific Adsorption of Amino Acids on Hydroxylated Quartz (0001)*, Langmuir **25** (2009) 10737-10745.
11. **Jeong Woo Han**, John R. Kitchin, and David S. Sholl, *Step decoration of chiral metal surfaces*, J. Chem. Phys. **130** (2009) 124710.
12. **Jeong Woo Han**, Joanna N. James, and David S. Sholl, *First principles calculations of methylamine and methanol adsorption on hydroxylated quartz (0001)*, Surf. Sci. **602** (2008) 2478-2485.

See more of this Session: Meet the Faculty Candidate Poster Session
See more of this Group/Topical: Education

# (3bg) Modeling, Optimization and Analysis of Energy Systems Under Uncertainty

**Bri-Mathias S. Hodge**, National Renewable Energy Laboratory, Golden, CO

In order to plan the future energy system, expected technological changes must be accounted for; however, there is significant uncertainty at both the planning and operational levels. Some of the uncertainties involved at the planning level include: the technologies that will be integrated and the degree to which they will be utilized, the localization of plants, the necessary accompanying transmission and distribution infrastructure and the structure of the market and support systems present in the future electricity system. At the operational level, the uncertainty is mainly a result of the variability inherent in many renewable energy technologies. While the flexibility needed to handle the relatively small amount of uncertainty currently present in the system is sufficient to handle lower variable generation penetration rates, fundamental changes to power system planning and operations will be needed in order to accommodate larger penetration rates. These complex problems cannot be solved by considering the technologies alone; models of the system must be able to capture the effects of social, economic and policy decisions in order to capture all of the relevant phenomena that may affect the state of the system. The representation of these new systems requires the concurrent use of a number of tools taken from the process systems engineering, operations research, electrical engineering and statistical fields to characterize system states at various levels of time, geographic dispersion and model fidelity, as dictated by the questions to be answered. In summary, I propose to develop, extend and apply classical and cutting-edge analytical techniques to the study of large-scale and complex problems in the socio-economic technology energy system space where consideration of both engineering and policy aspects is required and sources of system uncertainty must be considered.

See more of this Session: Meet the Faculty Candidate Poster Session
See more of this Group/Topical: Education

# (3bi) Application of Nanoporous Materials In Crystallization, Membrane Separations, and Catalysis

**Benjamin Hamilton**[1], Michael D. Ward[2], Marc A. Hillmyer[3] and Daniel F. Shantz[1], (1)Artie McFerrin Department of Chemical Engineering, Texas A&M University, College Station, TX, (2)Molecular Design Institute, Department of Chemistry, New York University, New York, NY, (3)Chemistry, University of Minnesota, Minneapolis, MN

My research interests reside at the nexus of nucleation phenomena, interfacial science, and the performance of catalytic and separations materials. Recent studies on crystal growth phenomenon stress the importance of the competition between the energetically favorable internal environment of a crystal and the energetically unfavorable interface between a crystal and its surroundings in determining crystallization outcome. These studies indicate manipulation of the crystallization environment – by varying pressure, temperature, solvents, additives, nucleation surface, size confinement, and other parameters – can alter crystallization outcomes in a rational manner. The results of these studies dovetail with those from colloids studies, which highlight the importance of interfacial energies in determining the properties of colloidal systems. I intend to exploit my understanding of nucleation phenomenon and my multidisciplinary background to improve the performance of crystalline catalysts and membranes, and to reduce membrane fouling owing to crystals and aggregates.

This poster describes the effects of size confinement, temperature, and tailor-made additives on the nucleation, polymorphism and stability of molecular crystals in nanometer size confinement. One of the sources of size confinement, nanoporous polymer monoliths generated by selectively etching well-aligned cylindrical cores from di- and tri-block polymers, exhibited versatility that inspired me to broaden my interests for my postdoctoral work to include applications of nanoporous materials in separations and catalysis. Consequently, the poster also describes the fabrication of nanoporous ceramic/organic dendrimer hybrids and their use in membrane and catalysis applications. Collectively, this poster highlights my core expertise in crystallization and supporting background in polymer science, synthetic chemistry, fabrication and applications of nanoporous materials, membrane separations, and catalysis.

**Thesis:** Hamilton, B.D. The Influence of Nanoscale Size Confinement on the Phase Behavior of Molecular Organic Crystals. Ph.D. Dissertation, University of Minnesota, Minneapolis, MN, 2009.

**Thesis Advisors:** Michael D. Ward, NYU; Marc A. Hillmyer, University of Minnesota

**Post-Doctoral Research Advisor:** Daniel F. Shantz, Texas A&M University

See more of this Session: Meet the Faculty Candidate Poster Session
See more of this Group/Topical: Education

# (3bj) Biomolecular Interfaces: Understanding and Design of Zwitterionic Materials for Biomedical Applications

**Yi He** and Shaoyi Jiang, Chemical Engineering, University of Washington, Seattle, WA

Research in biomaterials has shown significant growth in recent years. Poly-ethylene glycol (PEG) is one of most widely used biomaterials. However, PEG is known to be not suitable for long-term applications since it is subject to oxidation. My proposed research will take a different approach by focusing on developing zwitterionic polymer based biomaterials through a combination of molecular modeling tools and experimental methods. The major goal of my research program is to explore the unique feature of zwitterionic biomaterials and to apply the underlying design principles to various applications, including biolubrication, cartilage tissue repair, and drug delivery.

See more of this Session: Meet the Faculty Candidate Poster Session
See more of this Group/Topical: Education

# (3bk) Shrinking the Gap Between Simulations and Experiments

**Beverly Brooks Hinojosa**, Chemical Engineering, University of Florida, Gainesville, FL

With the continuous advancements in computational power and efficiency, the system sizes that may be simulated using high accuracy quantum mechanical calculations have significantly expanded. I have used density function theory (DFT) to provide key insights into two areas of interest in modern day chemical engineering, electronics and surface chemistry. I have focused on collaborating with experimentalists in an effort to advance collectively the fundamental understanding of various materials.

In the area of surface science, I collaborated with Dr. Jason Weaver's experimental group in the investigation of adsorption of simple molecules on polar and non-polar surfaces of $SrTiO_3$. Specifically, I examined the role of surface termination and coverage on the favorability of molecular and dissociative adsorption of $NH_3$ and $H_2O$. These two molecules serve as simple analogs of a broad range of alcohol and amine reactions that can occur on the oxide surface. The goal of the computational study is to map the effect of local surface structure on the adsorption and reactive behavior observed in ultrahigh vacuum surface science experiments.

My other area of interest is electronics. In particular, I focused on understanding the role atomic structure plays within the pyrochlore systems ($A_2B_2O_7$) , which have received significant attention for use in high-permittivity dielectrics, capacitors, and high-frequency filter applications. The highly flexible structure allows for incorporation of a wide selection of A and B cations. There is a need to understand the effect of local atomic structure and dynamics on the macroscopic properties to enable rational design within the large range of possible chemical compositions. I focused my efforts on understanding the dielectric properties, in complex bismuth based pyrochlores, such as $Bi_{3/2}M^{2+}Nb_{3/2}O_7$ and $Bi_{3/2}M^{2+}Ta_{3/2}O_7$, where $M^{2+}$ = Zn, Mg. To de-convolute the cooperative effects leading to dielectric relaxation in these more technologically relevant bismuth pyrochlores, I explained the development of atomic displacements within bismuth titanate ($Bi_2B^{4+}{}_2O_7$). I examined the role of chemical substitutions on the local geometric and electronic structure within BZN, BZT, BMN, and BMT. I worked directly with experimentalists to connected my DFT results with available experimental data to provide further explanation of the role atomic substitutions and structure on the material properties.

These two areas of interest have provided an opportunity to simultaneously understand the underlying principles of condensed matter physics and surface science and to collaborate directly with experimentalists within these areas.

See more of this Session: Meet the Faculty Candidate Poster Session
See more of this Group/Topical: Education

# (3bl) Development and Characterization of UHMWPE Fiber-Reinforced Hydrogels for Soft Tissue Replacement Applications

**Julianne L. Holloway**[1], Saadiq El-Amin[2], Natalie Kelly[2], Suzanne A. Maher[2], Mark R. VanLandingham[3], Anthony M. Lowman[1] and Giuseppe R. Palmese[4], (1) Chemical and Biological Engineering, Drexel University, Philadelphia, PA, (2) Hospital of Special Surgery, New York City, NY, (3)U.S. Army Research Laboratory, Aberdeen, MD, (4)Drexel University, Philadelphia, PA

Meniscal tears are the most common orthopedic injuries to the human body. However, poor treatment options exist due to the complex anisotropic property distribution within the meniscus. A fiber-reinforced hydrogel-based synthetic meniscus allows tailoring of the mechanical properties and molding of the implant to the size and shape of the native cartilage. Physically cross-linked poly(vinyl alcohol) (PVA)-ultra high molecular weight polyethylene (UHMWPE) fiber composites were characterized in compression (0.1 MPa - 0.8 MPa) and tension (0.1 MPa - 250 MPa) showing fine control over mechanical properties based on PVA concentration, UHMWPE fiber volume fraction, and hydrogel freeze-thaw cycles. Morphology and crystallinity analysis of PVA hydrogels showed both increases in crystallinity and structural ripening, phase separation, with freeze-thaw cycles. Indicating that both factors increase hydrogel mechanical properties with cycling. Swelling studies were performed in an osmotic solution to replicate the swelling pressure present in the knee. Minimal swelling was observed for hydrogels with a PVA concentration of 30-35wt%, independently of hydrogel freeze-thaw cycles. This allows for independent tailoring of hydrogel modulus and pore structure using freeze-thaw cycles and swelling behavior using polymer concentration to match a wide range of properties needed for various soft tissue applications. Chemical grafting and oxygen plasma treatments were performed on UHMWPE fibers to improve fiber-hydrogel interfacial properties. Interfacial shear strength (IFSS) was characterized using a single fiber pull-out test. Improvements to IFSS occurred from 11 kPa without any treatment to above 200 kPa following treatment. Lastly, *ex vivo* gait simulations were performed on PVA-UHMWPE fiber implants using a human cadaver knee model. Preliminary results show improvement in contact pressures compared to a meniscectomy. In conclusion, PVA-UHMWPE fiber composites can be tailored over a wide range of mechanical properties, porosity, swelling behavior, and fiber-hydrogel adhesion. Indicating the potential to produce an implant that can reproduce the anisotropic nature of the meniscus. This research can also be applied to other soft tissue applications, including the ACL and the annulus fibrosus.

See more of this Session: Meet the Faculty Candidate Poster Session
See more of this Group/Topical: Education

# (3bm) Nanofibers for Tissue Engineering and Drug Delivery

**Jong Kyu Hong**, School of Chemical Engineering, Oklahoma State University, Stillwater, OK

Electrospinning is a widely established polymer processing technology which allows generation of fibers (in nanometer to micrometer size) that can be collected to form non-woven structures. By choosing a suitable polymer and appropriate solvent system, fiber size can be controlled. Since the technology allows the possibility of tailoring the mechanical properties and biological properties, there has been a significant effort to adapt the technology in tissue regeneration and drug delivery. This abstract focuses on recent developments in adapting this technology for tissue regeneration applications. In particular, different configurations of nozzles and collector plates are summarized from the view of cell seeding and distribution. With the possibility of obtaining structures with large pore sizes cell infiltration to layers below the surface is possible. Further developments in obtaining thick layers of tissues and thin layered membranes are discussed. In addition, possibility of developing a simple 3D system using electrosprayed fibers is described.

See more of this Session: Meet the Faculty Candidate Poster Session
See more of this Group/Topical: Education

# (3bn) Intersections In Electrochemisty and Microbiology

**Lewis Hsu**, Sonny Astani Department of Civil and Environmental Engineering, University of Southern California, Los Angeles, CA

In my doctoral work, I have focused on application of bioelectrochemical systems towards remediation of heavy metals. This work has utilized techniques in molecular biology and microbiology to understand transformations in bioelectrochemical reactors. Coupling this approach with standard electrochemical tests yields more insight into these reactors, and thus a more complete understanding of the redox phenomena associated with the microbiology in these reactor systems.

Several projects highlighting this intersection of electrochemistry and microbiology will be highlighted with the following projects: (1) Remediation of chromium by *Shewanellae* at the cathode of a microbial fuel cell, (2) Probing the mechanisms of electron transfer in *Shewanella oneidensis* MR-1, and (3) Microbial community analysis of a chromate reducing biocathode.

Development of this electrochemisty-microbiology intersection in bioelectrochemical systems into a independent research plan will also be discussed.

See more of this Session: Meet the Faculty Candidate Poster Session
See more of this Group/Topical: Education

# (3bo) Advanced Dynamic Optimization and Control for Large-Scale Systems

**Rui Huang**, Chemical Engineering, Carnegie Mellon University, Pittsburgh, PA

Air Separation Unit (ASU) is an energy intensive process. It is economically attractive to apply advanced control techniques to the ASU. Current advances in dynamic optimization algorithms enable us to consider more complicated model and to improve controller's performance. This work addresses the following problems:

(1) **Reducing computational delay:** Implementing nonlinear model predictive control (NMPC) based on first principle dynamic model exhibit many advantages against traditional linear MPC based on empirical model. However, computational delay associated with solving large-scale optimization problem degrade the performance of the controller. Advanced step NMPC algorithm [1] is utilized to implement NMPC on the ASU problem and the computational delay is reduced by 200 times.

(2) **Offset-free in the presence of plant model mismatch:** Offset-free behavior which requires the output of the process to be exactly at the set point is an important requirement in control practice. An NMPC framework based on various state estimators is proposed and implemented on the ASU to achieve offset free even in the presence of plant model mismatch.

(3) **Robust NMPC:** It is important to guarantee robust stability of NMPC in the presence of large uncertainties. An multi-scenario formulation is proposed to guarantee the robust stability of NMPC algorithm. It is implemented on the ASU process.

(4) **Economic-oriented control:** With the first-principle dynamic model, it is preferred to directly solved the NMPC problem to maximize the economic profit of the process. An infinite-horizon formulation with guarantee stability is proposed to solve the economic-objective control problem for the ASU.

It is worth pointing out that the proposed methods are not restricted to the ASU problem. On the contrary, they are expected to work for any large-scale processes.

**Reference**:

[1] V. M. Zavala and L.T. Biegler. The advanced step NMC controller, optimality, stability and robustness. *Automatica 2009, 45: 86.*

See more of this Session: Meet the Faculty Candidate Poster Session
See more of this Group/Topical: Education

# (3bp) Bio-Inspired, Self-Healing Organic-Inorganic Hybrids for Bone Tissue Engineering

**Jijun Huang**[1], Grace Lau[2], Lihong He[3], Manuel Houmard[2], Phillip B. Messersmith[4] and Antoni Tomsia[5], (1)Materials Sciences Division, Lawrence Berkeley National Laboratory, Berkeley, CA, (2)Materials Science Division, Lawrence Berkeley National Lab, Berkeley, CA, (3)Biomedical Engineering, Northwestern University, Evanston, IL, (4)Northwestern University, Evanston, IL, (5)Lawrence Berkeley National Laboratory, Berkeley, CA

Bio-inspired, self-healing organic-inorganic hybrids for bone tissue engineering

Jijun Huang**[1]**, Grace Lau**[1]**, Manuel Houmard**[1]**, Lihong He**[2]**, Phillip B. Messersmith**[2]**, and Antoni P. Tomsia**[1]**

[1]Materials Sciences Division, Lawrence Berkeley National Laboratory, Berkeley, CA 94720

[2]Biomedical Engineering Department, Northwestern University, Evanston, IL 60208

Repair and replacement of skeletal tissue by synthetic materials remains one of the great challenges in tissue engineering. Unlike nature bone and other tissue, the bone and dental implants do not have the function of self-healing. In order to achieve dramatic improvements in in vivo performance of composites for skeletal tissue repair, new ways of approaching composite design and fabrication are needed. We recently employed a biomimetic or bio-inspired approach to develop self-healing bioactive composites by combining freeze casting technology with infiltration and in situ polymerization. This new approach allows us to be able to fabricate a series of lamellar-structured biocomposites of $TiO_2$ or ZnO with self-healing polymers based on poly(methyl methacrylate). The formed biocomposites show considerably improved compressive strength and fracture toughness in addition to their intrinsic self-healing capability. In this work, we will demonstrate our design strategy, fabrication technology, and improved mechanical properties of the biomimetic composites.

# (3bq) Nanoengineered Materials for Energy and Electronic Applications

**Carlos Hangarter**, Metallurgy Division, NIST, Gaithersburg, MD

Nanostructured materials, such as nanoparticles, nanowires, and nanoporous scaffolds, are attractive materials for electronic and energy applications because of their potential for device miniaturization and unique properties which arise from surface/interface dominated behavior and size dependent effects. These building blocks can be further complexed with hybridization or axial and radial interfaces, to create composites, multilayered, core/shell, and nanopeapod structures that give rise to novel properties for catalysis, communications, data storage, logic operations, sensing, and energy conversion/storage.

My research interests include advanced materials and processing technology to synthesize nanostructures with an embedded design and intrinsic functionality. In my poster I will report results from my graduate work; conducting polymers nanowire sensors with tunable selectivity and electronic behavior; magneto-transport properties of multilayed and core-shell ferromagnetic nanowire devices. Additionally, I will present recent progress on both dealloyed PtNi thin films for oxygen reduction reaction and photovoltaic properties of three dimensionally structured chalcogenides. Finally, I will describe future directions and my approach to cultivating an independent research group with a unique perspective in nanoengineered materials.

See more of this Session: Meet the Faculty Candidate Poster Session
See more of this Group/Topical: Education

# (3br) Modeling and Design Optimization for Development of High Performance Microfluidic Devices

**Mranal Jain**, Cain Dept. of Chemical Engineering, Louisiana State University, Baton Rouge, LA

My post-doctoral/ doctoral research experience includes modeling of electrokinetic transport processes and design optimization of microfluidic devices. Microfluidic devices are portable devices capable of automating conventional analytical laboratory tests with short analysis time, low reagent/ sample consumption. Such devices require fluid and particle manipulation to achieve various functionalities. My research focus is to gain fundamental understanding of transport phenomena at smaller length scales and apply it towards design of efficient microfluidic devices. My background in process systems engineering research (masters program) provides me with a unique perspective towards the microfluidics research problems. I believe that the performance of microfluidic devices can be significantly improved by application of optimization and control techniques.

In accordance with my background, my research interest lies at the interface of microfluidics and process systems engineering. I would like to contribute in following research areas:

- Computational transport phenomena in bio-microfluidic devices
- Design optimization for microfluidic devices using surface and topology optimization
- Micromixing techniques and their performance analysis
- Classical and non-linear electrokinetics
- Application of process control techniques in microfluidics

In this poster, I will present summary of my research results. I will also discuss about various research challenges in above areas, their significance and possible approaches to tackle them.

See more of this Session: Meet the Faculty Candidate Poster Session
See more of this Group/Topical: Education

# (3bs) Core|Shell Optical Nanoantenna

**Vladan Jankovic** and Jane Change, Chemical Engineering, University of California, Los Angeles, Los Angeles, CA

## Au|$SiO_2$|Yb:Er:$Y_2O_3$ core|shelloptical nanoantenna: experiment & simulation

**V. Jankovic, J. Chang**

*Department of Chemical Engineering, University of California, LosAngeles, Engineering V-1011,*

**ABSTRACT**

The conversion of electromagnetic (EM) energy from freepropagating radiation to localized energy and vice versa in the radio frequency(RF) and microwave domains is accomplished with the use of antennas. Opticalantennas are analogous to their RF and microwave counterparts, but there arecrucial differences in their physical properties and scaling behavior becausemetal is a highly dispersive material with finite conductivity at opticalfrequencies. Optical antennasare not driven by galvanic transmission lines like RF antennas, instead, localizedoscillators such as atomic emitters are brought close to the feed point of theantennas, and electronic oscillations are driven capacitatively.

In this work, Au nanoparticles of different shapes (spheres, rods and stars)were used as antenna elements, $Er^{3+}$ ions in an $Y_2O_3$host matrix were used as atomic emitter antenna driving elements while the capacitative gap between the antenna element and the atomicemitter was controlled by deposition of an ultra-thin $SiO_2$ innershell between the Au nanoparticle and the Yb:Er:$Y_2O_3$ outer shell. Au nanospheres were grown by nucleating Au seeds using asodium borohydride in the presence cetyltrimethylammoniumbromide (CTAB) micelles and aging theresulting Au seeds (<2nm) over a few days thus producing monodispersespheres of radius 10nm and an extinction peak around 520nm. Au nanorods were grown by adding a small amount of freshlyprepared seed solution to a growth solution of Au(I) ions, and CTAB where theCTAB micelle preferentially adsorbs on the [100] Au face and directs the growthof Au nanorods in the [111] direction. Au nanorod aspect ratios from 2 to 5 were achieved by varyingthe concentration of the reducing agent, ascorbic acid, resulting in plasmon resonances that ranged from 650nm to 850nm, idealfor coupling with rare earth ion based emitter materials. Au nanostars with extinction peaks ranging from 550nm to 650nmwere synthesized using the same protocol used for the nanorodsexcept that the order of the precursor addition was changed. A 4-5nm silicaspacer layer was deposited through a controlled TEOS hydrolyzationreaction and was shown to be effective in preventing quenching yet enablingenergy coupling between the Au nanorod and the RE-iondoped oxides. Spatially and compositionally controlled Yb:Er:$Y_2O_3$outer shells were deposited using both wet chemistry methods and radical enhanced atomic layer deposition(RE-ALD).

Upconversion (UC) spectral, powerdependence and radiative lifetime measurements with 532nm,750nm 980 nm and 1064nm laser excitation were used to assess the coupling of theAu optical antenna to the emitter ions as a function of antenna shape, spacerlayer thickness and spectral and spatial mode overlap efficiency. Preliminaryoptical characterization showed a 2X earlier onset of upconversionwith 980nm excitation for Yb:Er:$Y_2O_3$coupled to an Au nanorod antenna compared to pure (uncoupled)Yb:Er:$Y_2O_3$ nanoparticles.

Powerdependence measurements with 980nm excitation showed a >5 slope indicating amulti-photon absorption induced luminescence process for the Au-coupled erbiumand a <2 slope for the uncoupled erbium, indicating a two photon absorption(expected for erbium with 980nm excitation). These optical antenna core|shell particles have potential application in bio-imagingand light trapping for solar and sensor applications.

See more of this Session: Meet the Faculty Candidate Poster Session
See more of this Group/Topical: Education

# (3bt) Nanostructured Functional Materials for (Bio)Sensing and Energy-Related Applications

**Wenzhao Jia**, Department of Chemical, Materials & Biomolecular Engineering, University of Connecticut, Storrs, CT

The development of nanostructured functional materials with unique properties is of great interest in current research of materials science and chemical engineering, and advances the improvement of many applications, especially chemosensors and biosensors aiming at the rapid and cost-effective detection of various targets with improved sensitivity and selectivity, and more efficient catalysts for higher energy conversion. Therefore, this research focuses on the fabrication of novel nanostructured materials, including noble metal functionalized nanofibers and metal oxide based nanostructures, and further explores their enhanced performances in sensing or energy-related applications. Specifically, on one hand, noble metals were deposited on the surface of electrospun nanofibrous template through appropriate physical, chemical, or electrochemical approach. The resultant composite nanomaterials, inheriting not only the excellent conductivity and superior catalytic ability of noble metal, but also the large surface area and the high porosity of electrospun nanofibers, have been successfully applied in both vapor/gas phase detection and liquid phase detection for various analytes with dramatically improved sensing performances. On the other hand, as promising substitutes for noble metals in sensing application, metal oxides with different nanostructures were facilely synthesized via direct thermal oxidation, hydrothermal reaction, and galvanic replacement reaction. The excellent electrocatalytic properties of the as-prepared nanostructured metal oxides were investigated and demonstrated in the detection of hydrogen peroxide which is the most widely used model compound.  The methodology for the fabrication of novel nanomaterials in this work can be extended to the development of other novel multifunctional nanomaterials for a wider range of applications.

See more of this Session: Meet the Faculty Candidate Poster Session
See more of this Group/Topical: Education

# (3bu) Biomaterials for Modulating Dendritic Cell-Derived Immune Responses

**Kye Il Joo**, Chemical Engineering and Materials Science, University of Southern California, Los Angeles, CA

My research focus has been on the design and characterization of polymeric biomaterials for modulating the immune responses. It has also been devoted to studies on development of novel nanoparticle delivery system such as viral vectors and functional liposomes for translational medicine.

## (1) Polymeric Biomaterials Containing Cytokines for Modulating Dendritic Cell-Derived Immune Responses

Development of an effective vaccine and immunotherapy technology for cancer involves eliciting a potent tumor antigen-specific immune response by introduction of tumor antigens and generating systemic immunity against the tumors. Dendritic cells (DCs) are the most powerful antigen-presenting cells, capable of initiating immune response by stimulating both T cells and B cells, and much effort has therefore been devoted to methods for delivering appropriate antigens to DCs to develop effective vaccines. Immunization by adoptive transfer of autologous DCs that are loaded with antigens in vitro is, however, labor-intensive and requires significant regulatory concerns. In addition, most transplanted DCs remain non-functional, and few home to the lymph nodes where the subsequent activation of T-cell immunity is induced. More cost-effective and efficient protocols involve delivering antigen to host DCs in vivo by direct injection of vaccine vectors to program DCs in situ. Thus, the ability to deliver antigens to host DCs in vivo is a key step towards generating a potent antitumor immunity.

In this study, we demonstrate a general strategy to enhance antigen-uptake by DCs in vivo. It depends on recruiting and housing host DCs at the immunization injection site using functionalized biomaterials incorporating inflammatory cytokines. This approach offers the sustained and localized release of the cytokine granulocyte-macrophage colony-stimulating factor (GM-CSF), which has been identified as a potent stimulator of DC recruitment and proliferation, from the subcutaneously injected hydrogels / implanted PLGA scaffold. The engineered biomaterials provide a controllable subcutaneous microenvironment where host DCs could be recruited and further educated in situ by following injection of immunogens (antigens) at the recruited site. The porous PLGA scaffold & Thermo-sensitive mPEG-PLGA hydrogels capable of solution-gel transition through changes in temperature enable greater efficiency of drug/cytokine loading and also exhibit a sustained and localized release of GM-CSF in the target site. The enhanced antigen-specific immune response was observed when viral vector-based vaccine carriers were administered at the site implanted with these engineered hydrogel biomaterials.

### (2) Targeted Gene Delivery with Viral Vectors

One of the most challenging problems in virus-mediated gene therapy is how the gene can be specifically delivered into the target tissues or cells. Our lab has developed a novel and efficient strategy that allows the retro/lentiviral vectors to enter specifically into the target cell. This innovative design is based on engineering retro/lentiviral vectors by altering the viral envelope glycoprotein (Env), the protein that is responsible for binding the virus to cell surface receptors and for mediating entry. Compared with conventional methods that manipulate the delicate coupling interactions of the binding and fusion domains of Env glycoprotein, which usually causes decreased viral infectivity, our engineering approach involved the incorporation of a targeting antibody and pH-dependent fusogenic protein as two distinct molecules on the viral surface. A major advantage of this scheme over others where the viral protein is engineered with a foreign binding component is that the fusion protein maintains its full biological activity so that viral titer is not killed for increased specificity. This targeting methodology is flexible and can be extended to other forms of cell type-specific recognition with the different targeting antibody (or other cell-binding protein) to mediate targeting. The flexibility and broadness of this method will facilitate the application of targeted gene delivery for therapy and research.

Adeno-associated virus (AAV) vectors have attracted considerable interest because of its great premise as a vector for human gene therapy. However, their applicability is limited due to the restricted range of cells that they can efficiently transduce. Therefore, it is desirable to redirect AAV vector tropism that specifically transduces selected target cells in vivo. In this study, we have developed a new method for site-specific modification of AAV2 in order to generate a targeted AAV2 vector. This approach involves the insertion of a genetically encoded aldehyde tag onto AAV2 capsids, which can be metabolically modified to generate an aldehyde group as a unique chemical handle for further site-specific introduction of targeting ligands without causing a significant loss of viral titer. We demonstrated that this aldehyde tag enables covalent attachment of hydrazide-functionalized molecules including antibodies and peptides to AAV2 vector in a site-specific manner. It showed that antibody conjugation to AAV2 could significantly enhance viral transduction in both permissive and non-permissive cell lines expressing alternative cell-surface receptors. In addition, RGD-peptide conjugated AAV2 exhibited tumor targeting in vitro. These results demonstrated a general and efficient means for targeting AAV vectors.

See more of this Session: Meet the Faculty Candidate Poster Session
See more of this Group/Topical: Education

# (3bv) Exploiting Synthetic Materials to Study and Direct Immune Response

**Christopher M. Jewell**, Departments of Bioengineering, Materials Science and Engineering, and The Ragon Institute of MGH, MIT, and Harvard, Massachusetts Institute of Technology, Cambridge, MA

Over the past several decades, synthetic materials have contributed dramatically to new prophylactic and therapeutic treatments for a range of illness and disease, but drug delivery – the time, concentration, location, and specificity with which biological cargo is delivered – continues to be a major obstacle for medicine. My research interests have focused on designing and exploiting synthetic materials to achieve increased levels of control over the delivery of biological cargo. During this poster session I will highlight examples from my graduate and postdoctoral research which demonstrate the utility of synthetic biomaterials for enhancing the delivery of drugs, proteins, and nucleic acids *in vitro* and *in vivo*, and more recently, in the context of translational research focused on vaccination and immunotherapy. Building on this past work, I will describe my plans to nucleate a distinct, independent research program at the interface of immunology and materials engineering.

I completed doctoral studies at the University of Wisconsin – Madison under the guidance of Professor David Lynn in the department of Chemical and Biological Engineering. One major research area centered on development of degradable multilayered films that permit spatial or temporal control over the delivery of DNA and proteins from the surfaces of biomedical devices such as intravascular stents. In collaboration with the Abbott research group at UW, I initiated a parallel project aimed at exploiting the properties of a redox-active ferrocene-containing lipid to reversibly control DNA interactions using chemical and electrochemical cues. In October 2009 I accepted a postdoctoral fellowship from the Ragon Institute of MGH, MIT, and Harvard, initiating postdoctoral research with Professor Darrell Irvine at MIT. My postdoctoral work focuses on enhancing vaccination by exploiting the properties of biomaterials to increase the persistence of vaccine components in lymph nodes – the key site of immune response generation – and to engineer specific immunological profiles by controlled release of small molecules (immunomodulators) known to bias immune response.

A central theme of my research is application of basic science to develop new technologies, which is illustrated by my past projects in drug delivery that have generated 4 patent filings and 21 peer-reviewed publications. Immunology is an area flush with opportunities to apply biomaterials in the study of basic questions and to develop new strategies for improving human health. My expertise in biomaterials design, immunology, and vaccinology uniquely positions me to capitalize on these opportunities. The overarching goal of my research lab will be to exploit biomaterials to generate immune responses with specific, tunable characteristics. This concept has two research thrusts: 1) basic investigations aimed at understanding interactions between synthetic materials and key immune cells, and 2) development of biomaterial vaccines to combat infectious diseases and autoimmune disorders.

**Related References:**


(1) **C. M. Jewell***, J. Jung*, R. P. Carney, F. Stellacci, and D. J. Irvine, "Oligonucleotide delivery by cell-penetrating "striped" nanoparticles." **2011** – In revision.

(2) **C. M. Jewell**, S. C. Bustamante López, and D. J. Irvine, "In situ engineering of the lymph node microenvironment via intranodal injection of adjuvant-releasing polymer particles." *Proc. Natl. Acad. Sci. USA* **2011** – In press.

(3) C. M. Metallo, P. A. Gameiro, E. L. Bell, K. R. Mattaini, J. Yang, K. Hiller, **C. M. Jewell**, Z. R. Johnson, D. J. Irvine, L. Guarente, J. K. Kelleher, M. G. Vander Heiden, O. Iliopoulos, and G. Stephanopoulos, "Reductive glutamine metabolism by IDH1 mediates lipogenesis under hypoxia." *Nature* **2011** – In press.

(4) D. J. Irvine and **C. M. Jewell**, "Vaccine and Immunotherapy Delivery.", *Comprehensive Biomaterials*, Ed. P. Ducheyne, Elsevier Press, **2011** – In press.

(5) R. M. Flessner, **C. M. Jewell**, D. G. Anderson, and D. M. Lynn, "Degradable Polyelectrolyte Multilayers that Promote the Release of siRNA." *Langmuir* **2011**, *27*, 7868-7876.

(6) A. S. Breitbach, A.H. Broderick, **C. M. Jewell**, S. Gunasekaran, Q. Lin, D. M. Lynn, and H. E. Blackwell, "Surface-mediated release of a synthetic small-molecule modulator of bacterial quorum sensing: Gradual release enhances activity." *Chemical Communications* **2011**, *47*, 370-372.

(7) B. Sun, R. M. Flessner, E. M. Saurer, **C. M. Jewell**, N. J. Fredin, and D. M. Lynn, "Characterization of pH-induced changes in the morphology of polyelectrolyte multilayers assembled from poly(allylamine) and low molecular

weight poly(acrylic acid)." *Journal of Colloid and Interface Science* **2011**, *355*, 431-441.

(8) N. J. Fredin, R. M. Flessner, **C. M. Jewell**, S. L. Bechler, M. E. Buck, and D. M. Lynn, "Characterization of nanoscale transformations in polyelectrolyte multilayers fabricated from plasmid DNA using laser scanning confocal microscopy in combination with atomic force microscopy." *Microscopy Research and Technique* **2010**, 10.1002/jemt.20830.

(9) E. M. Saurer, **C. M. Jewell**, J. M. Kuchenreuther, D. M. Lynn, "Assembly of Erodible, DNA-Containing Thin Films on the Surfaces of Polymer Microparticles: Toward a Layer-by-Layer Approach to the Delivery of DNA to Antigen-Presenting Cells." *Acta Biomaterialia* **2009**, *5*, 913-924.

(10) **C. M. Jewell**, M. E. Hays, Y. Kondo, N. L. Abbott, and D. M. Lynn, "Chemical Activation of Lipoplexes Formed from DNA and a Redox-Active, Ferrocene-Containing Cationic Lipid." *Bioconjugate Chemistry* **2008**, *19*, 2120-2128.

(11) **C. M. Jewell** and D. M. Lynn, "Surface-Mediated Delivery of DNA: Cationic Polymers Take Charge." *Current Opinion in Colloid and Interface Science* **2008**, *13*, 395-402.

(12) **C. M. Jewell** and D. M. Lynn, "Multilayered Polyelectrolyte Assemblies as Platforms for the Delivery of DNA and Other Nucleic Acid-Based Therapeutics." *Advanced Drug Delivery Reviews* **2008**, *60*, 979-999.

(13) C. L. Pizzey, **C. M. Jewell**, M. E. Hays, D. M. Lynn, N. L. Abbott, Y. Kondo, S. Golan, and Y. Talmon, "Characterization of the Nanostructure of Complexes formed by a Redox-Active Cationic Lipid and DNA." *Journal of Physical Chemistry B* **2008**, *112*, 5849-5857.

(14) M. E. Hays, **C. M. Jewell**, Y. Kondo, D. M. Lynn, N. L. Abbott, "Lipoplexes Formed by DNA and Ferrocenyl Lipids: Effect of Lipid Oxidation State on Size, Internal Dynamics and Zeta-Potential." *Biophysical Journal* **2007**, *93*, 4414-4424.

(15) J. Zhang, S. I. Montañez, **C. M. Jewell**, D. M. Lynn "Multilayered Films Fabricated from Plasmid DNA and a Side-Chain Functionalized Poly(β-amino ester): Surface-Type Erosion and Sequential Release of Multiple Plasmid Constructs from Surfaces." *Langmuir* **2007**, *23*, 11139-11146.

(16) B. Sun, **C. M. Jewell**, N. J. Fredin, D. M. Lynn, "Assembly of Multilayered Films Using Well-Defined, End-Labeled Poly(acrylic acid): Influence of Molecular Weight on Exponential Growth in a Synthetic Weak Polyelectrolyte System." *Langmuir* **2007**, *23*, 8452-8459.

(17) M. E. Hays, **C. M. Jewell**, D. M. Lynn, N. L. Abbott, "Reversible Condensation of DNA using a Redox-Active Surfactant" *Langmuir* **2007**, *23*, 5609-5614.

(18) **C. M. Jewell***, S. M. Fuchs*, R. M. Flessner, R. T. Raines, D. M. Lynn, "Multilayered Films Fabricated from an Oligoarginine-Conjugated Protein Promote Efficient Surface-Mediated Protein Transduction." *Biomacromolecules* **2007**, *8*, 857-863.

(19) **C. M. Jewell**, J. Zhang, N. J. Fredin, M. R. Wolff, T. A. Hacker, D. M. Lynn, "Release of Plasmid DNA from Intravascular Stents Coated with Ultrathin Multilayered Polyelectrolyte Films." *Biomacromolecules* **2006**, *7*, 2483-2491.

(20) **C. M. Jewell**, M. E. Hayes, Y. Kondo, N. L. Abbott, D. M. Lynn, "Ferrocene-Containing Cationic Lipids for the Delivery of DNA: Oxidation State Determines Transfection Activity." *Journal of Controlled Release* **2006**, *112*, 129-138.

(21) N. L. Abbott, **C. M. Jewell**, M. E. Hayes, Y. Kondo, D. M. Lynn, "Ferrocene-Containing Cationic Lipids: Influence of Redox States on Cell Transfection." *Journal of the American Chemical Society* **2005**, *127*, 11576-11577.

(22) **C.M. Jewell**, J. Zhang, N. J. Fredin, D.M. Lynn, "Multilayered Polyelectrolyte Films Promote the Direct and Localized Delivery of DNA to Cells." *Journal of Controlled Release* **2005**, *106*, 214-223.

**These authors contributed equally*

See more of this Session: Meet the Faculty Candidate Poster Session
See more of this Group/Topical: Education

# (3bw) Production of Renewable Fuels From Lignocelluosic Biomass by Catalytic Technologies

**Jungho Jae**, Chemical Engineering, University of Massachusetts, Amherst, Amherst, MA and George W. Huber, Chemical Engineering, University of Massachuseets - Amherst, Amherst, MA

Diminishing petroleum resources combined with concerns about global warming and dependence on fossil fuels are leading our society to search for renewable sources of energy. In this respect, lignocellulosic biomass has a tremendous potential as a renewable energy source, if we develop economical processes for converting biomass into useful fuels and chemicals [1]. My Ph.D. research involves the design of new catalysts and reactors for biomass conversion. Specifically, I have focused on catalytic fast pyrolysis of lignocellulosic biomass.

Catalytic Fast Pyrolysis (CFP) is a promising technology for the direct production of gasoline range aromatics from lignocellulosic biomass in a single catalytic reactor [2]. The most challenging part of this process is controlling the complex homogeneous and heterogeneous reaction chemistry to maximize aromatic yield and suppress the production of coke [3]. To gain fundamental understanding of this process, we studied reaction chemistry for CFP of glucose (i. e. cellulose model compound) with isotopic labeling of $^{13}C$ glucose and ex-situ FT-IR technique in a pyroprobe coupled with GC-MS. It was found that glucose is completely decomposed in one second and converted into dehydrated products, including anhydrosugars and furans. The dehydrated products then enter into the zeolite catalyst pore where they are converted into aromatics, CO, $CO_2$, and water through a series of dehydration, decarbonylation, decarboxylation, oligomerization, and dehydrogenation reactions. The isotopic labeling studies revealed that aromatic formation reaction proceeds through a common intermediate or "hydrocarbon pool" composed of these dehydrated products [4]. This is a shape selective reaction where the product selectivity is related to the catalyst properties. The shape selectivity of zeolite catalysts for CFP have systematically been studied with different zeolites including small pore zeolites (ZK-5 and SAPO-34), medium pore zeolites (Ferrierite, ZSM-23, MCM-22, SSZ-20, ZSM-11, ZSM-5, IM-5, and TNU-9), and large pore zeolites

1. *Depolymeriation of lignocellulosic Biomass into Fuel Precursors: Maximizing Carbon Efficiency by Combining Hydrolysis with Pyrolysis*, Jungho Jae, Geoffrey A. Tompsett, Yu-Chuan Lin, Torren R. Carlson, Jiacheng Shen, Taiying Zhang, Bin Yang, Charles E. Wyman, W. Curtis Conner and George W. Huber, *Energy Environ. Sci.*, **3** (2010) 358 – 365

2. *Production of Green Aromatics and Olefins by Catalytic Fast Pyrolysis of Wood Sawdust,* Torren R. Carlson, Yu-Ting Cheng, Jungho Jae and George W. Huber, *Energy Environ. Sci.*, **4** (2011) 145-161

3. *Catalytic Fast Pyrolysis of Glucose with HZSM-5: The combined homogeneous and heterogeneous reactions*, Torren R. Carlson, Jungho Jae, Yu-Chuan Lin, Geoffrey A. Tompsett, and George W. Huber, *J. Catal.*, **270** (2010) 110-124

4. *Mechanistic Insights from Isotopic Studies of Glucose Conversion to Aromatics over ZSM-5*, Torren R. Carlson, Jungho Jae, and George W. Huber, *ChemCatChem.*, **1** (2009) 107-110

5. *Investigation into the Shape Selectivity of Zeolite Catalysts for Biomass Conversion,* Jungho Jae, Geoffrey A. Tompsett, Andrew J. Foster, Scott M. Auerbach, Raul F. Lobo, and George W. Huber, *J. Catal.*, **279** (2011) 257-268

See more of this Session: Meet the Faculty Candidate Poster Session
See more of this Group/Topical: Education

# (3bx) Novel Methods for Simulating the Self-Assembly of Complex Particles

**Eric Jankowski**, Chemical Engineering, University of Michigan, Ann Arbor, MI

The self-assembly of nanoparticles is a ubiquitous phenomenon that underlies problems of crucial importance in medicine and biology and which can be exploited for the fabrication of customized materials and smart devices. The link that connects colloid crystal formation to virus capsid assembly to enzyme catalysis to ion transport in fuel cells is the minimization of free energy between interacting thermal particles with complicated shapes. Given the recent proliferation of synthetic techniques that enable the fabrication and characterization of mondisperse anisotropic particles on nanometer length scales we have unprecedented opportunities to utilize these building blocks to solve practical problems. My research program focuses on the design and implementation of simulation techniques for studying complex self-assembling systems, providing a conduit between theory and experiment.

Of particular utility in the design of building blocks for robust assembly of target patterns are screening methods that can identify potential barriers to assembly. I demonstrate the efficacy of a pathway-based approach I have developed for the prediction of strong self-assembly candidates for models of patchy anisotropic colloids, CdTe/CdS tetrahedra, and sticky spherical clusters. This technique depends upon the calculation of vibrational and rotational entropic contributions of clusters to partition functions and I discuss the details of these algorithms.

Dissipative systems such as ATP-fueled enzymes or self-propelled particles are examples of complicated systems that can require special modeling considerations that have only recently become straightforward to implement. Using a massively multithreaded approach on consumer graphics hardware I study a model of self-propelled colloids and virus capsomer analogues. In doing so, behaviors are observed that are inaccessible on node-based computing architectures without considerable investments in capital and development time. Control over structural morphology is demonstrated via tuning of thermodynamic parameters and the development of new theory to describe these systems is discussed.

See more of this Session: Meet the Faculty Candidate Poster Session
See more of this Group/Topical: Education

# (3by) Self-Assembled Microcontainers for Three-Dimensional Control Over Chemicals In Cellular Microenvironments

**Yevgeniy V. Kalinin**, Chemical and Biomolecular Engineering, Johns Hopkins University, Baltimore, MD

Self-assembled microcontainersfor three-dimensional control over chemicals in cellular microenvironments

**Directing the chemotactic self-organization of *E. coli* in the shape of a helix.** A) A conceptual schematic of the desired chemotactic self-organization. *E. coli* cells (green cylinders) self-organize to a helical pattern based on an underlying helical chemical pattern of an attractant that is diffusing out of the container. B) Time-lapse images of green fluorescent *E. coli* self-organizing from a uniform distribution to a helical pattern. C1,C2) Phase contrast images of regions marked as C1 and C2 in panel B showing *E. coli* bacteria in the vicinity of the container.

The ability to create well-defined and controlled three-dimensional cellularmicroenvironments has been an area of active research over the last few years. Thisposter will show how self-assembled three-dimensional patterned hollow microcontainers can aid in creating and controlling shapeand composition of chemical patterns on the microscale.It will be shown how to control the shape and duration of the chemical patternsreleased by the containers and how these chemical patterns can guide spatialself-organization of motile cells (see the figure on the right). Alternatively,by confining living cells to the inside of the microcontainersit is possible to only selectively allow small molecules to permeate the cells.The poster will discuss the possibilities the containers offer fortransplantation of insulin-secreting pancreatic beta cells since the cells canbe shielded from the immune system of their new host by the microcontainers.

See more of this Session: Meet the Faculty Candidate Poster Session
See more of this Group/Topical: Education

# (3bz) The Tail Wagging the Dog: Insights Into Catalysis In R67 Dihydrofolate

**Ganesh Kamath**, Computer Science and Mathematics Division, Oak Ridge National Laboratory, Oak Ridge, TN

Enzymes are phenomenal catalysts. Different enzyme folds catalyzing same chemical reaction could provide vital insights into the enzyme mechanism. Hydride transfer reaction catalyzed by E. coli dihydrofolate reductase (EcDHFR) has been previously investigated. Plasmid encoded R67 DHFR, sharing no structure or sequence homology with EcDHFR, also catalyzes hydride transfer reaction between dihydrofolate (DHF) and its cofactor nicotinamide adenine dinucleotide phosphate (NADPH). R67 DHFR is a homotetramer and shows numerous characteristics of a primitive enzyme including promiscuity in binding of substrate/cofactor, formation of non-productive complexes, and the absence of a conserved acid in its active-site. Furthermore, R67's active-site is a pore, which is mostly accessible by bulk solvent. This present study uses a computational approach to characterize the mechanism of hydride transfer. Not surprisingly, NADPH remains fixed in one half of the active site pore using numerous interactions with R67. Also stacking between the nicotinamide ring of cofactor and the pteridine ring of substrate, DHF, at the hourglass center of the pore, holds the reactants in place. However, large movements of the para-aminobenzoylglutamate tail of DHF occur in the other half of the pore due to ion-pair switching between symmetry related K32 residues from two subunits. The tail movement at the edge of the active-site, coupled with the fixed position of the pteridine ring in the center of the pore, leads to puckering of the pteridine ring and promotes transition state formation. Flexibility coupled to R67 function is unusual as it contrasts with the paradigm that enzymes use increased rigidity to facilitate attainment of their transition states. A comparison with chromosomal DHFR indicates a number of similarities, including puckering of the nicotinamide ring and changes in the DHF tail angle, accomplished by different elements of the dissimilar protein folds.

**Future research areas:**

## [A] High throughput screening and development of ionic liquids as a highly efficient catalysis medium for bio-ethanol production

Cellulosic ethanol from abundant biomass has attracted significant attention as a possible low-cost renewable energy source to reduce the nation's dependence on foreign oil. This goal is hindered by the lack of technology enabling the release of fermentation sugars from biomass and a need to improve the catalytic efficiency of naturally occurring cellulases. Normally, the catalytic efficiency of an enzyme can be altered through mutations or the use of different substrates. A third option is the alteration of the dynamics of the enzyme based on solvent coupling. While water is considered the biological solvent, non aqueous enzyme dynamics remains unexplored and limited to volatile organics. Ionic liquids are novel materials which are environmentally friendly and have low vapor pressure and exits as liquids at room temperature. These ionic liquids have been proposed as a medium for overcoming the recalcitrance of biomass by improving the dissolution of cellulose and improving the activity of cellulases. Ionic liquids with different functional groups can provide control over the local chemical micro-environment. We propose a systematic study based on computational methods to investigate neat and mixtures of solvents for developing highly efficient medium for catalysis and improved dissolution rates of biomass. The computational methods would involve both classical and first principles calculations to determine the thermodynamic and transport properties of neat ionic liquids and their interactions with cellulose and cellulase enzyme to quantify the solvent effect on dynamics of the ternary complex. The refinement of non-polarizable models and also development of explicit anharmonic and harmonic contributions to models/ polarizable models will be considered based on the performance of the existing models. The objective function used to screen the existing ionic liquids would be the maximization of the catalytic efficiency and stability of the cellulase enzyme in these ionic liquids while macroscopic properties of the ionic liquids would be used as additional constraints. The fundamental chemical insights obtained will guide the synthesis of novel ionic liquids by variation of functional groups to match the requirements.

## [B] Evaluation and ranking of free energy of adsorption as a rational design of ionic liquids for direct exfoliation of nanoscale graphene from graphite.

Research on graphenes—monolayers of carbon atoms arranged in a honeycomb lattice—is proceeding at a relentless pace as scientists of both experimental and theoretical bents seek to explore and exploit its superlative attributes including giant intrinsic charge mobility, and record-setting thermal conductivity, stiffness, and strength. Of course, fully exploiting the remarkable properties of graphene requires reliable, large-scale production methods which are non-oxidative and introduce few defects, criteria not fully satisfied by any known approach. A major advance in this direction was the recent discovery by our group that ionic liquid (IL)-assisted exfoliation of graphite leads to isolation of few- and single-layer graphene sheets with yields two orders of magnitude higher than with earlier approaches using "surface energy-matched" solvents like N-methyl-pyrrolidone. The overarching goal of this research is to develop a molecular-level view of the IL/graphene interface to elucidate and better control key chemical interactions responsible for efficient exfoliation. Our approach takes full advantage of the modularity of IL solvents which offer an unprecedented opportunity to control the chemistry presented at the interface. A vital component of this research will involve computational calculations based on model nanographenes, with in silicofeedback informing experimental results. If successful, our findings will afford valuable insight, laying the foundation for the first rational route to preparing practical quantities of isolated, processable graphene sheets based upon properly designed ILs, facilitating the fabrication of graphene-based devices and catalyzing graphene expansion into other areas including batteries, supercapacitors, separation technologies, photovoltaics, membranes, catalysis, and environmental remediation.

**Research Interests**

**Non-aqueous enzymology using novel media such as ionic liquids**

The use of non-aqueous enzymology for the enhancement of catalytic activity of cellulase enzyme, overcome recalcitrance problem in biomass and increase the production and quality of the intermediates required for bioethanol production.

**Protein dynamics and folding**

Development of accurate non polarizable computational models and methods to understand the physics and mechanism of biomacromolecules structure and function, energetic materials and nanomaterials.

**Drug Docking and Discovery**

The understanding of protein motions, folding and catalytic activity in the presence of bound ligands/substrates through classical and semi-empirical computational models and neutron /x-ray scattering would help design novel drugs to combat diseases and also provide invitro methods to engineer novel materials to provide cost efficient energy.

The use of high throughput screening using theoretical methods and computation for materials ranging from special liquids to peptides and complementing the results with suitable experimentation forms the basis of this research.

**Development of metal oxides for chemical warfare agents detection.**

The use of computational methods to understand behaviors of hazardous chemicals and mechanisms for adsorption thereby enabling us to develop novel surfaces for chemical warfare agent's detection.

I have a diverse professional and academic background in chemical engineering and bioscience. My experience at Wayne State University where I finished my PhD was in molecular modeling and determination of thermodynamic and transport properties of organic and energetic materials under Dr. Jeff. Potoff. This modeling capability was further advanced to biomacromolecules, when I did post doctoral studies at University of Maryland. At Oak Ridge National Laboratory, these theoretical models were used to understand catalytic behavior and dynamics in variety of Rossman fold and cellulase enzymes.

The focus of my research is use of theoretical methods to evaluate and rank materials ranging from ionic liquids to graphene and peptides and being a contributor towards energy problems and combating diseases. Ionic liquids as media for biomass are considered to increase the solubility of biomass thereby overcoming the recalcitrance problem. Moreover the enhancement of catalytic activity of cellulase enzyme acting on the cellulose/biomass in the presence of aqueous/novel media is sought. I consider this research to be in focus and very appropriate with your Department's mission and also in line with the country's mission of developing a sustainable renewable source of energy

See more of this Session: Meet the Faculty Candidate Poster Session
See more of this Group/Topical: Education

# (3c) Electrochemistry for Energy: Air-Breathing Enzymatic Electrodes for Batteries and Fuel Cells

**Joshua Gallaway**, Energy Institute and Chemical Engineering, City College of New York, New York, NY

The enzyme laccase can reduce oxygen in ambient air, making it an ideal cathode catalyst for an air-breathing battery/fuel cell hybrid in the spirit of a zinc-air battery, but on a smaller, more designable scale. Laccase is manufactured from the common fungi *Trametes versicolor*. As enzymes are obtained from cell lysis and chromatography, they may be produced on a mass scale from farmed bacteria or fungi. Micro power sources based on catalysts with little cost limitation will allow them to be deployed ubiquitously, to power highly specialized miniature devices.

The chief obstacle to using enzyme catalysts in man-made power sources is that their structure is suited to operation in a cellular environment. In certain areas this is a benefit, such as functionality at physiological temperature and pH. However, electron exchange in biological systems is handled by small diffusional redox molecules, transporting electrons via diffusion. In contrast, batteries and fuel cells rely on electron exchange at a heterogeneous interface (an electrode).

A composite soft electrode was designed to catalyze oxygen reduction with laccase and simultaneously conduct electrons with incorporated metal coordination compounds. Metal complex molecular structure was engineered to increase electron flow, and an optimized molecular structure was found. Catalytic rate was further increased by improving the synthesis for complex attachment to the composite. This proved that macro-structure was important and achieved unprecedented bioelectrocatalytic current densities of 20 $mA/cm^2$. Prototype biofuel cells from this design using both hydrogen and methanol as anode fuels were successful, with $mW/cm^2$ performance.

Going forward, enzyme catalysis must take place on gas-phase oxygen in ambient air. The goal is continuous operation of enzyme-catalyzed air electrodes with power, stability, and lifetime appropriate for practical use.

# (3ca) Nanoconfined Organic Molecules and Polymers: Fundamentals and Scalable Device Applications

**Dun-Yen Kang**, School of Chemical & Biomolecular Engineering, Georgia Institute of Technology, Atlanta, GA

Recent work on the optical, electronic, and heat-transport applications of 1D-nanostructured organic molecules and polymers strongly highlights the enhancement in their properties over 2D thin films of the same materials. Nanoporous materials could provide nanoconfinement for organic molecules or polymers with controllable channel dimensions, orientation, and interface chemistry. On the other hand, scalable techniques for fabricating thin films and membranes of nanoporous materials have been developed. Hence, the creation, scientific investigation, and application of photon/electron/phonon-conductive organic guest molecules or polymers in 1D nanoconfined geometries within nanoporous thin films could revolutionize the functionality of these guest entities while possessing excellent processibility covering a wide range of length scales ($10^{-6}$-$10^{2}$ m). A host of new device platforms for "green" energy, photoelectricity, biomimetic, and biomedical devices can be created via the "marriage" of functional organic molecules/polymers and 1D nanoporous membranes.

The realization of these exciting possibilities requires a new level of understanding of the structure and properties of these complex hybrid nanostructures, as well as the capability to engineer and optimize their performance. My interdisciplinary research program will therefore be dedicated to the science and engineering of nanostructured host-guest devices comprising functional molecules and polymers confined in 1D-nanoporous thin films. Our proposed novel device platform, and our formulation of reliable fabrication-structure-property relations in such devices, will together allow us to pursue technological applications that include: (1) thermoelectric generators containing thermoelectric organics for harvesting waste heat from portable electrical devices, vehicles, or power plants; (2) artificial antenna systems with dye-molecules mimicking green plants for storage of photonic energy in an enthalpic form; and (3) biosensors with conducting functional polymers for cancer clinic testing.

I have acquired a strong research background in advanced nanoporous materials synthesis and characterization, surface chemistry and functionalization in nanoconfined environments, study of molecular transport phenomena in nanostructured materials, and fabrication of nanotubular thin films and membranes. This interdisciplinary expertise will strongly support the development of the proposed research program.

Doctoral Thesis Advisors: Professor Sankar Nair and Professor Christopher Jones, School of Chemical & Biomolecular Engineering, Georgia Institute of Technology, Atlanta, GA

Publications:

[1] D.-Y. Kang, C. W. Jones, and S. Nair, "Modeling Molecular Transport in Composite Membranes with Tubular Fillers" (submitted).

[2] D.-Y. Kang, J. Zang, C. W. Jones, and S. Nair, "Single-Walled Aluminosilicate Nanotubes with Organic-Modified Interiors", *The Journal of Physical Chemistry C,* 115(15), p. 7676-7685 (2011).

[3] D.-Y. Kang, J. Zang, E. R. Wright, A. L. McCanna, C. W. Jones, S. Nair, "Dehydration, Dehydroxylation, and Rehydroxylation of Single-Walled Aluminosilicate Nanotubes", *ACS Nano,* 4(8), p. 4897-4907 (2010).

See more of this Session: Meet the Faculty Candidate Poster Session
See more of this Group/Topical: Education

# (3cb) Engineering Proteins and Peptides for the Investigation and Treatment of Infectious Disease

**Amy J. Karlsson**, Chemical and Biomolecular Engineering, Cornell University, Ithaca, NY

My research interests are focused on engineering proteins and peptides that can be used to better understand fungal infections and study potential antifungal compounds. I plan to use two approaches to design and synthesize proteins and peptides for studying antifungal properties of the compounds and for studying the interactions of fungal pathogens with host cells. The first approach is to use organic synthesis to produce compounds that mimic and improve on the activity of naturally occurring antifungal peptides. Organic synthesis provides the ability to tightly control the peptide sequence and to easily incorporate non-natural amino acid building blocks into the peptides. During my PhD research, I used this approach to synthesize antifungal peptides composed of non-natural, beta-amino acids that mimicked the structure of natural antimicrobial peptides. The designed peptides had both improved antifungal activity and improved specificity for fungal cells compared to natural antimicrobial peptides. The second approach to peptide and protein design and synthesis that I will apply to my research is directed evolution. The protein production machinery in bacteria can be harnessed to perform protein synthesis, and the proteins can be evolved to improve on their natural functions or to have novel functions. Directed evolution can be used without any knowledge of the mechanism of action of a protein or peptide and offers the ability to create large numbers of potentially improved sequences that can be screened for desired functions. My postdoctoral research utilizes directed evolution to engineer high-affinity antibodies that fold in the intracellular environment, providing the opportunity to study the effects of antibody binding to intracellular targets. My experience with both organic synthesis and directed evolution provides a complementary set of techniques that can be used to tailor a protein or peptide engineering approach based on what is known about a particular fungal pathogen. Thus, my future plans are to make use of this molecular toolkit to create proteins and peptides that can be used to address basic questions underlying fungal pathogenesis or that can inhibit key mechanisms in fungal disease.

See more of this Session: Meet the Faculty Candidate Poster Session
See more of this Group/Topical: Education

# (3cc) Rational Design of Nanoparticles for Biological Applications

**Mark J. Kastantin**, Chemical and Biological Engineering, University of Colorado, Boulder, CO

Nanoparticles are increasingly used to interact with biological systems. Accordingly, the number of methods to make nanoparticles with a wide variety of geometries and chemical compositions has risen sharply in the last decade. What is lacking, however, is comprehensive understanding of how nanoparticle properties affect their interaction with proteins and cells in biological systems. Put another way, why would one choose a particular geometry and chemical composition to accomplish a goal (targeted drug delivery, for example)?

Towards the goal of bottom-up nanoparticle design, I propose to use single-particle methods to discern mechanisms of nanoparticle interactions with biological systems such as proteins and cells. This research goal will be integrated with a study of self-assembly fundamentals in order to efficiently create new types of nanoparticles as they become justified by insights into nanoparticle behavior.

My experience in this area stems from my doctoral work developing a modular, micellar platform for the creation of multifunctional nanoparticles for targeted drug delivery. In my post-doctoral research, I use single-molecule methods to make correlations between molecular conformation and the dynamic behaviors of adsorption, desorption and diffusion. These methods have lead to powerful mechanistic insight regarding the behavior of proteins and nucleic acids at interfaces. In combination, expertise in these areas has the potential to identify a set of "design criteria" for nanoparticles in addition to extending the ability of nanoparticles to probe and interact with proteins, cells, and tissue.

See more of this Session: Meet the Faculty Candidate Poster Session
See more of this Group/Topical: Education

# (3cd) DNA Nanoparticles That Penetrate the Human Mucus Barrier

**Anthony J. Kim**, School of Medicine, Johns Hopkins University, Baltimore, MD

Protective mucus coatings typically trap and rapidly remove foreign particles from the eyes, gastrointestinal tract, airways, nasopharynx, and female reproductive tract, thereby strongly limiting opportunities for controlled drug and gene delivery at mucosal surfaces. We prepared polymeric DNA nanoparticles that rapidly penetrate sputum expectorated from the lungs of patients with cystic fibrosis (CF), a disease characterized by hyperviscoelastic mucus secretions. Polymeric DNA nanoparticles capable of overcoming human mucus barriers open significant opportunities for improved gene delivery at mucosal surfaces.

See more of this Session: Meet the Faculty Candidate Poster Session
See more of this Group/Topical: Education

# (3ce) Towards a $

**N.V.S.N. Murthy Konda**, U.S. Department Of Energy, National Energy Technology Laboratory, Morgantown, WV

The need for sustainable energy has never been more than what it is now and is growing year-after-year. While the role of each of the emerging clean and green technologies (such as carbon capture/storage, hydrogen fuel cell cars, biofuels, gasification-based processes, renewables and zero-carbon communities) varies geographically and with time, it is important that all these technologies are developed synergistically and utilized up to their fullest potential. However, cost competitiveness, energy efficiency, environmental performance and reliability of these technologies are yet to be improved significantly. To achieve these, holistic life-cycle analyses on economics, efficiency and emissions are needed during both the design and operational stages. Furthermore, in order to reap the benefits to the maximum extent possible, innovation at and tight-integration amongst multiple length- and time-scales is required thus calling for a multi-scale modeling approaches.

Motivated by these and based on my extensive research experiences in relation to all the aforementioned aspects of energy systems engineering (ESE, including modeling, design, control and optimization of various systems ranging from a particle/molecule to devices to processes to networks), in my future research career, I would like to focus on developing and applying generic/tailor-made methods and tools to foster innovative development and accelerate deployment of clean and green energy technologies. To be specific, my short/medium term focus will be on the following Themes (T):

T1 - Hydrogen and biofuels driven transportation (technologies and infrastructure)

T2 - Carbon capture, transport and sequestration (technologies and infrastructure)

T3 - Clean coal technologies based on gasification (inc. integrated gasification combined cycle)

T4 - Renewable energy sources (including biomass, solar and wind)

T5 - Energy systems of the future (urban energy systems, rural electrification, zero-carbon homes)

T6 - Water and healthcare infrastructures and technologies

See more of this Session: Meet the Faculty Candidate Poster Session
See more of this Group/Topical: Education

# (3cf) Intersurface Forces and Dissolution of Minerals At Contact Points

**Kai Kristiansen**, Department of Chemical Engineering, University of California at Santa Barbara, Santa Barbara, CA

The overall theme of my research is to understand the molecular and nanoparticle behavior at the interface of surfaces. Important parts of advancing in this direction are to develope and implement new experimental devices [1] for interfacial science in general and integrate them into a Surface Forces Apparatus [2] in particular. As the size of particles reaches characteristic length scales, novel physical properties (compare to bulk properties) start to emerge [3]. The carbon nanotubes are one candidate and I will present a study of the interactive forces in aqueous films of carbon nanotubes on mica surfaces at nanometer scale.

Control and understanding of dissolution processes are of critical importance in various areas such as corrosion, chemical mechanical polishing (CMP) and geology. Despite the apparent importance of electrochemical reactions in the dissolution process [4], very little is known about the relationship between surface potentials, overlapping electric double layers at the contact between dissimilar minerals, and dissolution rates; primarily due to the lack of experimental setups, which can visualize and measure dissolution rates *in-situ* between two closely apposed mineral surfaces in real time with atomic scale resolution. We have developed a new electrochemical Surface Forces Apparatus and I will present results of dissolution of amorphous quartz as function of applied potential on an apposing gold surface, which shows dissolution in the range of nanometers per hour [5].

[1] Kristiansen K, McGuiggan P M, Carver G, Meinhart C, and Israelachvili J; "3D Force and Displacement Sensor for AFM and SFA measurements". *Langmuir* **24** (2008) 1541-1549.

[2] Israelachvili J, Min Y, Akbulut M, Alig A, Carver G, Greene G W, Kristiansen K, Meyer E E, Pesika N, Rosenberg K, and Zeng H; "Recent advances in the surface forces apparatus (SFA) technique". *Reports on Progress in Physics* **73** (2010) 036601.

[3] Min Y J, Akbulut M, Kristiansen K, Golan Y, and Israelachvili J; "The role of interparticle and external forces in nanoparticle assembly". *Nature Materials* **7** (2008) 527-538.

[4] Greene G W, Kristiansen K, Meyer E E, Boles J, and Israelachvili J; "Role of electrochemical reactions in pressure solution", *Geochimica et Cormochimica Acta* **73** (2009) 2862-2874.

[5] Kristiansen K, Valtiner M, Greene G W, Boles J, and Israelachvili J; "Pressure solution – the importance of the electrochemical surface potentials", submitted to *Geochimica et Cosmochimica Acta*.

See more of this Session: Meet the Faculty Candidate Poster Session
See more of this Group/Topical: Education

# (3cg) Fluid Mechanics of Complex Fluids: From Advanced Materials to Biomedical Applications

**Amit Kumar**, Chemical and Biological Engineering, University of Wisconsin-Madison, Madison, WI

Complex fluids are ubiquitous in nature and technological settings. Some examples of great interest include nanoparticle suspensions, blood, and swimming micro-organisms among others. The key feature of these systems is that their bulk properties are determined by their underlying microstructure, which in turn is strongly influenced by the complex hydrodynamic interactions between the constituent particles. A particular focus of my research has been to develop the state of the art numerical algorithms to accurately and efficiently compute the hydrodynamic interactions between the suspending particles, thus enabling predictive simulations on these systems. In this poster, I will present two specific examples from my research showing the power of these predictive simulations. In the first example, I will discuss the complex orientation behavior observed in sheared suspensions of non-spherical dicolloidal particles. A dicolloidal particle is modeled as two fused spheres of varying radii and center to center separation. A novel theory relating the orientation behavior to the observed normal stress difference in the suspension will be presented. These simulations were carried out with a Particle-Mesh-Ewald Stokesian Dynamics technique developed for an arbitrarily shaped rigid particle. In the second example, I will discuss studies related to the blood flow in the microcirculation. Microcirculation refers to the part of the circulatory system with the smallest blood vessels. As a model problem, I will present results for the flow induced segregation in suspensions of mixtures of deformable capsules with different rigidities. Flow induced segregation based on particle rigidity has potential applications in the separation or detection of diseased red blood cells from their normal counterpart. These simulations were carried out with an accelerated implementation of the boundary integral method valid for an arbitrary geometry. In addition to these problems, several other future directions will be discussed with relevance to biomedical and advanced materials research.

See more of this Session: Meet the Faculty Candidate Poster Session
See more of this Group/Topical: Education

# (3ch) Passivating Alkali Metals During Pyrolysis of Biomass to Obtain Higher Yield of Sugars

**Najeeb Kuzhiyil**, Center for Sustainable Environmental Technologies, Iowa State University, Ames, IA

Sugars are excellent feedstocks for producing fuels and chemicals using fermentation and chemical technologies. My research is focused on producing sugars from biomass and cellulosic wastes. By depolymerizing cellulose and hemicellulose of biomass using fast pyrolysis, which is rapid heating in the absence of oxygen, anhydrosugars such as levoglucosan (LG) can be produced. However, inherently contained alkali in biomass catalyzes the glucose ring fragmentation reaction instead of the depolymerization reaction during pyrolysis resulting in very low yield of anhydrosugars. By passivating alkali, the reaction mechanism can be driven through the depolymerization pathway to increase the yield of anhydrosugars. The present study invented a method to passivate alkali using phosphoric and sulfuric acid infusion thereby increasing the yield of anhydrosugars significantly. A correlation between the amount of alkali contained in biomass feedstocks and the amount of acid to be infused to obtain maximum yield of levoglucosan was developed that can be used for commercial production of sugars from biomass. The recovery of these sugars from the pyrolysis reactor was found to be difficult because of the polymerization and charring of LG at the operating temperature of 500°C. Analytical pyrolysis of acid infused biomass feedstocks at various temperatures indicated that a temperature range of 350-400° C can be the best for recovering the LG formed in the reactor. Methods are under development to maximize the recovery of pyrolytic sugars from a fluid bed and auger reactors using low temperature pyrolysis and increased sweep gas flow.

See more of this Session: Meet the Faculty Candidate Poster Session
See more of this Group/Topical: Education

# (3ci) Immuno-Liposome Nanoparticles for Single Cell Array

**Kwang Joo Kwak**, Chemical and Biomolecular Engineering, The Ohio State University, Columbus, OH

**Immuno-Liposome Nanoparticles for Single Cell Array**

Kwang Joo Kwak[1,2]

[1]NSF Nanoscale Science and EngineeringCenter for Affordable Nanoengineering of Polymeric Biomedical Device(NSEC-CANPBD)

[2]Department of Chemical and BiomolecularEngineering

The Ohio StateUniversity, Columbus, Ohio 43210

*Contact author-E-mail: kwak.63@osu.edu

Recent advances in biomimetic lipidmembranes provide opportunities to develop applications such as biosensing, gene/drugdelivery, and cellular recognition. My research interests are focused on tetheredimmuno-liposome nanoparticles (t-ILNs) that can be conjugated on a planarsurface through a self-assembled monolayer of a thiolipid and modified with celladhesion molecules and therapeutic targeting antibodies. The multi-antibodyconjugated t-ILNs have been investigated to be able to isolate extremely rarecells such as cancer stem cells from the whole human blood. The t-ILNs arraybased on a centrifugal force has been demonstrated to improve sensitivity andspecificity of stem cell isolation by increasing the focal adhesion between celland the ILN surface and preventing non-specific binding on the interfaces. Novel t-ILN array based cell isolationhas higher potential for many other important biomedical applications such asprognosis of cancer patients and monitoring targeted therapy.

See more of this Session: Meet the Faculty Candidate Poster Session
See more of this Group/Topical: Education

# (3cj) Fabrication of High Added-Value Crystalline Products and Nanostructured Materials

**R. Lakerveld**, Department of Chemical Engineering, Massachusetts Institute of Technology, Cambridge, MA

Crystalline materials are a key component of a vast numberof high added-value products and emerging technologies. The product qualityrequirements of crystalline materials are increasingly stringent in order tocomply with demands for more sustainable fabrication processes and enhancedfunctional performance of the final product.

Fabrication processes of crystalline materialsspan several length scales as illustrated in Figure 1. At the smallest scale,elementary building blocks (e.g. ions, organic molecules, proteins, DNA tiles)define the chemical identity of the final product. These building blocks areassembled via reversible and non-covalent interaction forces into clusters witha specific structural identity. The efficiency for further processing and thefunctionality of the final product are determined by this structural identity. Forexample, the precise confirmation of an active pharmaceutical ingredient in acrystal lattice can have dramatic effects on the shape of the crystals to be furtherprocessed or on the bioavailability of the final product. Furthermore, emergingnanostructured materials derives their exciting functional properties, inaddition to the chemical composition, from periodic or non-periodic structuralfeatures. In the final step of the fabrication process, crystalline clustersare grown to a final state while preserving the chemical and structuralidentity of the existing clusters. During this stage, final product propertiessuch as the crystal size distribution evolve.

Current fabrication processes for crystallinematerials fail to meet future demands for sustainability and enhancedproperties of the final product. The fundamental reason is that actuation onthe scale of individual building blocks is poorly exploited and instead only macroscopicvariables such as temperature, pressure, and composition are being used. These macroscopicactuators are able to influence the growth phase of the fabrication process,but offer limited flexibility to control the formation of the individual crystallineclusters.

My long-term research goal is to drasticallyimprove the performance of fabrication processes for crystalline materials by directingthe assembly of crystalline clusters with actuation at the nano- and microscalecombined with a subsequent dedicated growth phase.

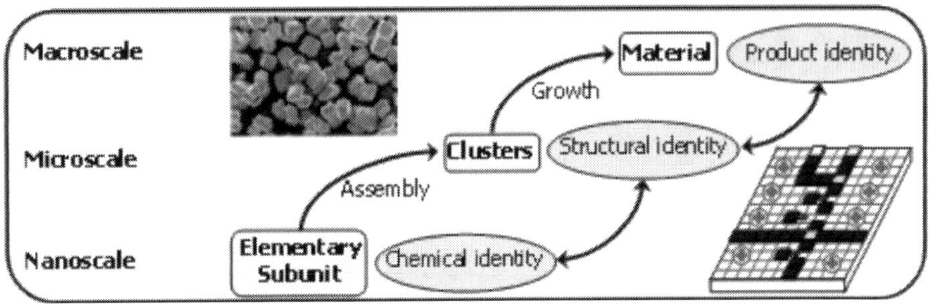


**Figure 1:** Multi-scale fabrication process of crystallinematerials.

See more of this Session: Meet the Faculty Candidate Poster Session
See more of this Group/Topical: Education

# (3ck) The Effect of Interfacial Interactions of Polymers On Friction and Adhesion

**Lucas J. Landherr**, Polymers Division, National Institute of Standards and Technology, Gaithersburg, MD

Friction and adhesion are phenomena that have grown in interest as a result of developments in research into both solid and fluid interfaces. Achievements in the field of nanoscale devices have increased demand for improved lubricants on that length-scale, as the dominance of surface effects makes friction and adhesion of particular concern. These factors also have important contributions in fluid flow, as advancements in superhydrophobic surfaces have allowed for significant levels of slip and drag reduction. In both areas of research, the influence of polymers on interfacial interactions has the potential for reducing friction and adhesion, thus resulting in improved conditions for a wide range of applications.

For this poster presentation, I will discuss efforts to apply polymeric solutions to both fields of research to create conditions of low friction and adhesion. Specifically in terms of solid friction, thin films of polydimethylsiloxane (PDMS) networks tethered to self-assembled monolayers (SAM) were developed to function as optimal microscale lubricants. These PDMS-SAM hybrids were found to exhibit extremely low friction coefficients, as low as 0.0012, which is the lowest ever recorded for a dry lubricant. The friction properties were determined to result from low shear stress and dramatically small contribution of adhesion to the friction force. In investigating fluid friction, polymeric flow over superhydrophobic surfaces was studied in an attempt to enhance the drag reduction and slip phenomena that can occur in each separate component. Interfacial rheometry and microfluidic measurements were conducted to determine the slip length, fluid profile, and friction properties. Overall, I will discuss results and analysis in the context of previous friction and adhesion studies, as well as potential future work on drag reduction and lubrication experiments and applications.

See more of this Session: Meet the Faculty Candidate Poster Session
See more of this Group/Topical: Education

# (3cl) Hierarchical and Nanostructured Zeolite Materials for Computer Microprocessors and Biomass Conversion

**Christopher M. Lew**[1], Yushan Yan[2] and Michael Tsapatsis[1], (1)Chemical Engineering & Materials Science, University of Minnesota, Minneapolis, MN, (2) Chemical Engineering, University of Delaware, Newark, DE

Zeolite thin films and membranes have been widely investigated over the last 15 years, mainly as separation membranes and membrane reactors. Their uniform microporosity and the ability to fine-tune their chemical composition and physical architecture lead to important macroscopic properties that can be exploited in a growing number of applications. The diversity in their use is demonstrated in this work through the exploration of zeolites as low-dielectric constant (low-*k*) films for computer microprocessors and as catalysts for biomass conversion.

Next-generation microprocessors require smaller feature sizes, but the slow development of new low-*k* films with better material properties is reducing progress toward improved computing performance. Pure-silica-zeolites (PSZ) are a strong candidate for replacement low-*k* materials because their intrinsic porosity and crystallinity give them a low *k* value and strong mechanical properties. Several other characteristics, including hydrophobicity, wet-etch chemical resistance, pore size distribution, and thermal conductivity were evaluated and optimized for low-*k* applications. In order to obtain the desired aforementioned properties, PSZ films were carefully engineered through different synthesis and modification techniques, including pre- and post-synthesis functionalization techniques and "on-wafer" crystallization synthesis methods at ambient pressures. Successful device integration of a viable new low-*k* alternative requires that all of these properties meet minimum specifications. As a result of the methods developed in this work, PSZ low-*k* films are a leading low-*k* candidate and considerably outperform other alternative materials.

As an essential catalyst in the oil refining industry, zeolites may also play a key role in the conversion of biomass to fuels and commodity chemicals. Studies are currently being performed to investigate the interaction between the kinetics and molecular dimensions of Sn-containing zeolites and biomass-derived sugars. New and novel nanostructured zeolite catalysts are also being prepared through the use of hierarchical synthesis strategies and innovative organic structure-directing agents. These zeolites with designed architectures may be useful as multifunctional catalysts or as thin films with unique optical and structural properties.

See more of this Session: Meet the Faculty Candidate Poster Session
See more of this Group/Topical: Education

# (3cm) Ultra-Thin Porous Film Coating Via Molecular Layer Deposition

**Xinhua Liang**, Chemical and Biological Engineering, University of Colorado at Boulder, Boulder, CO

Porous thin films/coatings have wide applications in separations, corrosion protective coatings, gas storage, controlled drug delivery, and catalyst encapsulation for controlling reaction selectivity and maintaining catalytic activity. Ultra-thin porous films/coatings are especially important for some applications. Inorganic films/coatings with few defects, such as zeolite membranes and mesoporous films, have been prepared by hydrothermal synthesis and sol-gel method. While these films/coatings have well defined uniform porous structures, it is difficult to control their thickness with the precision of angstroms. In this paper, we demonstrate a general method to create porous inorganic films/coatings with precisely controlled thickness down to several angstroms from dense organic/ inorganic hybrid aluminum alkoxide (alucone) films grown by molecular layer deposition (MLD). The organic constituents in the film can be removed either by mild water etching at room temperature, or calcination in air at elevated temperatures. Because of the layer by layer growth process for MLD, the deposited polymer films have regular structure and the oxidation of MLD polymer films produces uniform interconnected highly porous structures with high surface area. The porous structure has both micropores and mesopores with BET surface area as high as 1200 $m^2/g$. The porous alumina films were used to stabilize highly dispersed Pt nanoparticles (< 2 nm) supported on high surface silica gel particles. The Pt catalyst with a porous ultra-thin alumina layer was much more stable to calcination in air, even at 800 °C. We expect this porous thin film/coating growth technique, which offers both precise control of the thickness down to several angstroms and well defined porous structure, to have wide applications in catalyst functionalization, controlled drug delivery and nano-filtration.

See more of this Session: Meet the Faculty Candidate Poster Session
See more of this Group/Topical: Education

# (3cn) Design, Control and Estimation of New Energy Systems: Application to Energy Efficient IGCC Plants with Carbon Capture

**Fernando V. Lima**, Department of Chemical Engineering and Materials Science, University of Minnesota, Minneapolis, MN

In the recent years, incentives on the efficient and sustainable use of fossil fuels and on the use of renewable sources of energy have risen again in the United States and worldwide. These incentives are largely motivated by high energy costs dictated by oil prices in the international market and the strong environmental concerns associated with carbon dioxide ($CO_2$) emissions to the atmosphere that are released during the combustion of fossil fuels and are linked to global climate change. The use of process systems engineering tools, such as process modeling, control, design and optimization, may be the key to allow the implementation of the next generation of technologies that enable the efficient and sustainable use of the current major energy sources. However, process systems engineering approaches for new energy systems are still scarce. My research will contribute with novel applications of existing process systems tools and the development of new tools for the improvement of new energy systems. In this poster, I will focus on the application of systems tools to coal-based Integrated Gasification Combined Cycle (IGCC) plants with carbon capture for electricity generation, but other systems of interest include biorefineries, coal and biomass-based gasification plants for the production of power, chemicals and liquid fuels.

Regarding the application of interest, IGCC plants have emerged recently as a more efficient technology for coal-based electricity generation when compared to conventional pulverized coal-fired power plants. IGCC units also allow the mitigation of $CO_2$ emissions as they enable $CO_2$ capture at different positions of the process flowsheet. Designing new energy efficient coal-based plants for power generation that minimize the amount of $CO_2$ emissions is one of the great current engineering challenges, especially now that a few IGCC demonstration projects with carbon capture are under development in the U.S. [1].

Regarding the tools of interest, my research will explore the use of process operability concepts [2, 3] to design new energy efficient process units. Process operability both permits the verification of a current process design limitations and enables the consideration of control objectives at the design stage. This fact is important because once the design is fixed, process control will not be able to overcome its inherent limitations. Individual units will be studied first, followed by their energy integration. Process integration increases the overall plant efficiency, but results in several dynamics and control challenges such as nonlinearities and time-scale separation in the process model, and the requirement of transitions due to the change in targets and operating conditions. The implementation of model-based controllers and state estimation algorithms [4] will be investigated to facilitate the operation of new energy systems in a constrained, fluctuating and integrated environment.

As an example of the application of process systems engineering tools to a unit in the IGCC flowsheet, I will present my current research results on membrane reactors for carbon capture in IGCC units [5].

References

[1] Anthony N. Zinn. CCPI and ICCS demonstration projects. In 2010 NETL $CO_2$ Capture Technology Meeting, Pittsburgh, PA, September 2010.

[2] Fernando V. Lima and Christos Georgakis. Input-output operability of control systems: the steady-state case. J. Proc. Cont., 20:769-776, July 2010.

[3] Fernando V. Lima, Zhenya Jia, Marianthi Ierapetritou, and Christos Georgakis. Similarities and differences between the concepts of operability and flexibility: the steady-state case. AIChE J., 56:702-716, March 2010.

[4] Fernando V. Lima and James B. Rawlings. Nonlinear stochastic modeling to improve state estimation in process monitoring and control. AIChE J., 57:996-1007, April 2011.

[5] Fernando V. Lima, Prodromos Daoutidis, Michael Tsapatsis, and John J. Marano. Modeling and Optimization of Membrane Reactors for Carbon Capture in IGCC Units. To be Presented at AIChE Annual Meeting, Minneapolis, MN, October 2011.

See more of this Session: Meet the Faculty Candidate Poster Session
See more of this Group/Topical: Education

# (3co) Structure, Rheology and Dielectric Properties of Nanostructured Amphiphile/Ionic Liquid Mixtures

**Carlos R. Lpez-Barrn** and Norman J. Wagner, Chemical Engineering, University of Delaware, Newark, DE

Room temperature ionic liquids (RTIL) have received a great deal of attention due to their remarkable properties including high thermal and chemical stability, negligible vapor pressure, high ionic conductivity and wide electrochemical window. Nanoscale self-assembly in RTIL extends the usefulness of these compounds by providing structure and viscoelasticity. During the last year I have begun a systematic exploration of self-assembly of amphiphilic compounds (including ionic surfactants and block copolymers) in ionic liquids. I am particularly interested in the relation between the structure, the thermodynamics and the rheological and dielectric properties of amphiphile/ionic liquid systems. To study different aspects of these relationships, we use a number of experimental techniques, including SALS, SANS and flow-SANS, shear rheology, flow-birefringence and rheo-dielectrics measurements, optical polarizing microscopy and electronic microscopy as well as titration calorimetry. During the poster session I will present some of our most remarkable findings. For instance, we observed spontaneous formation of vesicles and formation of a very stable bicontinuous sponge phase in solutions of didodecyldimethylammonium bromide (DDAB) in ethylammonium nitrate (EAN), which are not observed in DDAB/ water solutions. Another exciting finding is the formation of very long, entangled, viscoelastic and highly conductive wormlike micelles formed by a Pluronic triblock copolymer in EAN. Additionally, I am exploring two of the applications of these systems: (1) synthesis of highly ordered templated mesoporous silicates and (2) intra and inter-micelle photopolymerization of functionalized amphiphiles to generate nanostructured ruber-like materials.

See more of this Session: Meet the Faculty Candidate Poster Session
See more of this Group/Topical: Education

# (3cp) Multiscale Simulation to Advance Micellar Drug Delivery

**Sharon M. Loverde**, Chemical & Biomolecular Engineering, University of Pennsylvania, Philadelphia, PA

Multiscale simulation is an emerging tool to bridge relevant time and length scales in self-assembly and interfacial processes in soft matter systems. Utilizing different techniques, such as Dissipative Particle Dynamics and coarse-grained molecular dynamics, I have examined phenemona ranging from the ns to ms timescales: for example—the break-up of mixed worm like micelles, solubility of the anti-cancer drug paclitaxel in different micellar morphologies, and finally, the interaction of micelles with lipid membranes. Utilization of these novel coarse-grained methodologies that remain closely tied with experimental measurements is key to the development of a new level of molecular engineering design for applications such as self-assembly and delivery.

See more of this Session: Meet the Faculty Candidate Poster Session
See more of this Group/Topical: Education

# (3cq) Biomass Depolymerization Using Biphasic H2O-CO2 Mixtures

**Jeremy S. Luterbacher**[1], Jefferson W. Tester[1] and Larry P. Walker[2], (1) Department of Chemical and Biomolecular Engineering, Cornell University, Ithaca, NY, (2)Department of Biological and Environmental Engineering, Cornell University, Ithaca, NY

Sustainably producing concentrated solutions of monosaccharides is a key bottleneck in the conversion of lignocellulosic biomass to biofuels or bioproducts. Most pretreatment and enzymatic hydrolysis processes are run at low-solid concentration (<10 wt%) and use chemical catalysts while high-solids enzymatic hydrolysis reactions are almost always performed with air-dried pretreatment mixtures.

Initial work during my doctoral research focused on exploring the use of biphasic $H_2O$-$CO_2$ mixtures as an alternate medium for high-solids (up to 40 wt%) pretreatment. Early studies were done in a small (25 ml) unstirred reactor using a single temperature stage. More recently, two-temperature stage pretreatment was introduced and optimized in a larger 1 L stirred reactor to take advantage of the biomass depolymerization temperature dependent reaction sequence (a short high-temperature stage at 200-210°C was followed by a longer low-temperature stage at 160-170°C). Optimally pretreated substrates (210°C, 16 min for hardwood and 210°C, 1 min for switchgrass) were used as feedstock in high-solids enzymatic hydrolysis reactions that gave glucose yields above 80% for both switchgrass and hardwood after 48 hrs of hydrolysis. Therefore, without additional chemical catalysts or any drying, two-temperature stage $H_2O$-$CO_2$ pretreatment coupled with high solids enzymatic hydrolysis can produce monosaccharide solutions of 185 gr/L and 148 gr/L for mixed hardwood and switchgrass, respectively. This suggests that $H_2O$-$CO_2$ pretreatment is an attractive alternative to chemically catalyzed processes such as dilute acid pretreatment.

Parallel to these studies, efforts are underway to better understand the relationship between the effects of pretreatment and the enzymatic depolymerization mechanisms of cellulosic substrates. A fluorescence confocal microscopy method was developed for observing and measuring the binding of cellulases *in situ*, for following the temporal morphological changes of cellulosic materials during their depolymerization by a commercial cellulase mixture. The Spezyme CP cellulase cocktail was supplemented with a small fraction of fluorescently labeled *T. Reseii* CBH I, which served as a reporter to track cellulase diffusion and binding onto the internal physical structure of various substrates (bacterial microcrystalline cellulose, cellulose filter paper and pretreated hardwood and switchgrass). Kinetic models were developed and parameters were extracted using the variation in fluorescence intensity of the substrate and the bound enzyme over time and were successfully used to predict the extent of similar reactions in solution.

See more of this Session: Meet the Faculty Candidate Poster Session
See more of this Group/Topical: Education

# (3cr) Scientific Challenges to Develop a Nonaqueous Secondary Li-Air Battery

**Bryan D. McCloskey**, Science and Technology, IBM Almaden Research Laboratory, San Jose, CA

The 2008 IBM Almaden Research Grand Challenge competition launched a project focused on developing rechargeable Li-air batteries, whose theoretical specific energy (11,700 Wh/kg of Li) potentially offers a significant improvement over the 160 Wh/kg specific energy found in state-of-the-art Li-ion batteries. Despite this enormous potential specific energy, considerable scientific challenges remain to successfully develop a secondary Li-air battery. This poster will address some of these challenges, with a particular emphasis on two current areas of Li-air research at IBM: electrolyte stability and electrical passivation of the cathode due to electrochemical product formation.

To better understand electrolyte stability in the presence of the working electrochemical cathode reaction (*i.e.*, $Li^+ + O_2 + 2e^- \rightarrow Li_2O_2$), quantitative Differential Electrochemical Mass Spectrometry (DEMS) was used to analyze Li-$O_2$ cells employing various electrolytes and porous carbon paper as the cathode. In conjunction with the gas-phase DEMS analysis, electrodeposits formed during discharge were characterized using *ex-situ* analytical techniques. This poster will outline results which indicate that the reversibility of Li-$O_2$ electrochemistry is strongly dependent on the choice of solvent employed in the cell.

In addition, lithium-oxygen discharge electrodeposits formed on smooth, glassy carbon electrodes were characterized by both electrochemical and surface analytical techniques. Ferrocene was used as an *in-situ* electrochemical redox probe to correlate cell performance with electrical passivation as a result of discharge product formation. These studies revealed that cell failure during discharge was coincident with exponentially decreasing charge transfer at the cathode surface. *Ex-situ* analysis of the electrodeposit using focused ion beam micromachining, combined with conductive atomic force microscopy, yielded results consistent with the *in-situ* findings.

See more of this Session: Meet the Faculty Candidate Poster Session
See more of this Group/Topical: Education

# (3cs) Modeling Advanced Materials for Green Chemistry and Energy Related Applications

**Thomas A. Manz**, School of Chemical and Biomolecular Engineering, Georgia Institute of Technology, Atlanta, GA

There are significant opportunities for using computational chemistry methods to study advanced materials for green chemistry, biosensing, and energy related applications. Examples of such opportunities include: (a) the development of more solective transition metal complexes for catalytic chemical transformations, (2) the study of graphene for biosensing applications, (3) the study of single-molecule magnets and half-metals for sensors and information storage, and (4) the study of ferro-electric materials for use as dielectrics in super-capacitors that store large amounts of electricity. My research group will use computational chemistry methods like density functional theory to study a variety of problems related to these research areas.

I have experience in the application of density functional theory to both periodic and non-periodic systems. My post-doctoral research involved the development of an improved method for computing net atomic charges, bond orders, and atomic spin moments in both periodic and non-periodic systems. Systems studied during my post-doctoral research included: (a) magnetite -- a highly correlated material whose bulk is a half-metal, (b) a non-collinear single-molecule magnet, (c) metal organic frameworks, and (d) H elimination reactions on Cu and Pt surfaces. My Ph.D. dissertation involved the development of quantitative structure activity relationships for olefin polymerization catalyzed by Ti and Zr complexes with mixed cyclopentadienyl/aryloxide ligation. I performed density functional theory calculations and compared these to data collected by experimental collaborators in order to develop quantitative models of the relationships between catalyst structure and reactivity.

Post-doctoral advisor: Professor David Sholl, School of Chemical and Biomolecular Engineering, Georgia Institute of Technology

Ph.D. committee: Professors James Caruthers, Kendall Thomson, and Nicholas Delgass, School of Chemical Engineering, Purdue University; Professor Mahdi Abu-Omar, Chemistry Department, Purdue University

Selected Publications:

(1) Manz, T. A.; Sholl, D. S. "Chemically Meaningful Atomic Charges that Reproduce the Electrostatic Potential in Periodic and Nonperiodic Materials," J. Chem. Theory Comput., **6** (2010) 2455 - 2468.

(2) Manz, T. A.; Phomphrai, K.; Medvedev, G.; Krishnamurthy, B. B.; Sharma, S.; Haq, J.; Novstrup, K. A.; Thomson, K. T.; Delgass, W. N.; Caruthers, J. M.; Abu-Omar, M. M. "Structure-activity correlation in titanium single-site olefin polymerization catalysts containing mixed cyclopentadienyl/aryloxide ligation," J. Am. Chem. Soc. **129** (2007) 3776-3777.

(3) Manz, T. A.; Sholl, D. S. "A Dimensionless Reaction Coordinate for Quantifying the Lateness of Transition States," J. Comput. Chem. **31** (2010) 1528-1541.

(4) Manz, T. A.; Sharma, S.; Phomphrai, K.; Novstrup, K. A.; Fenwick, A. E.; Fanwick, P. E.; Medvedev, G. A.; Abu-Omar, M. M.; Delgass, W. N.; Thomson, K. T.; Caruthers, J. M. "Quantitative Effects of Ion Pairing and Sterics on Chain Propagation Kinetics for 1-Hexene Polymerization Catalyzed by Mixed Cp'/ArO Complexes," Organometallics, **27** (2008) 5504-5520.

(5) Watanabe, T.; Manz, T. A.; Sholl, D. S. "Accurate Treatment of Electrostatics during Molecular Adsorption in Nanoporous Crystals without Assigning Point Charges to Framework Atoms," J. Phys. Chem. C. **115** (2011) 4824 - 4836.

See more of this Session: Meet the Faculty Candidate Poster Session
See more of this Group/Topical: Education

# (3ct) Polymer Networks: Modeling and Emerging Applications

**Hassan Masoud**, George W. Woodruff School of Mechanical Engineering, Georgia Institute of Technology, Atlanta, GA

We develop a particle-based computational model to explicitly simulate the mechanics and transport properties of polymer gels, i.e. cross-linked polymer networks immersed in Newtonian fluids. Our hybrid numerical approach consists of a bond-bending lattice spring model to capture the micromechanics of random networks of interconnected elastic filaments and the dissipative particle dynamics to explicitly model the viscous fluid and diffusive objects. We use our mesoscale model to examine how permeability and diffusivity of gels change when they undergo mechanical deformations and establish the dependence of the transport properties on the variation of the network structure under stretching and shear loads. We also employ the model to examine mechanisms for controlled release of nanoparticles enclosed within microgel capsules. We show that the capsule swelling results in a steady particle release, with a relatively slow release rate that is set by the diffusion through the network. We also demonstrate that the fluid flow though the capsule's membrane during deswelling leads to a rapid particle release. This release, however, is limited as network mesh size decreases when capsule shrinks. To enhance the rapid release from the deswelling capsule, we introduce solid microstructures, e.g. rigid microrods, inside the capsule. The length of the rods is comparable to the capsule internal diameter and the rod radius is comparable to the size of nanoparticles. The rods stretch the membrane of deswelled capsule and promote formation of larger pores, which allow rapid and massive release of nanoparticles. Our findings reveal a new approach for regulating the release from micro-carriers by controllably changing the pore size. Furthermore, our three dimensional fully-coupled model of the cross-linked polymer networks provides a foundation for future studies of a broad range of problems in engineering, medicine, and biology that involve active and responsive polymer gels.

Bio:

I am a Ph.D. candidate in Mechanical Engineering at the Georgia Institute of Technology, and expect to receive my degree by June 2012. I received my B.S. degree with Highest Honors in Aerospace Engineering from the Sharif University of Technology (Tehran, Iran) in 2006. Subsequently, I joined the State University of New York at Buffalo where I earned my M.S. degree in Mechanical Engineering in 2009.

I have developed very broad research interests in fluid mechanics as well as soft matter physics and biomimetic design. During my Masters studies, I derived exact analytical solutions to study the flow field and particle deposition inside an evaporating colloidal drop. The focus of my Ph.D. research has been on the application of computational methods to a wide variety of problems involving interaction between deformable/rigid objects and their surrounding fluid. The problems I have considered range from the aerodynamics of flexible flapping wings and transport in random polymer networks to dynamics of fluid-filled microcapsules and colloidal flow in geometrically patterned microchannels. My high quality research has so far resulted in a book chapter and more than seven invited and regular articles published in leading archival journals and/or presented at major international conferences of the field. My works have also been highlighted in the *National Science Foundation News* and *Georgia Tech Research Horizon Magazine*, selected as a finalist for the *Padden Award for Excellence in Polymer Physics Research*, and featured in the *Virtual Journal of Nanoscale Science & Technology*. For more detailed information, please see my enclosed summary of research accomplishments. In the future, I plan to apply my expertise in fluid mechanics, mathematics, and soft matter physics to solve state-of-the-art problems in engineering and biology. Specifically, I am interested in computational design of adaptive and responsive soft materials.

In parallel with performing research, I have participated in writing several grant proposals for NSF, DOE and AFOSR, and served as a peer-reviewer for more than a dozen professional journals in my fields of interest. I have also paid equal attention to improvement of my teaching ability by taking teaching practicum course, delivering several lectures for two graduate courses, and teaching recitation and lab sections for three undergraduate courses. Therefore, given my academic background, I feel comfortable teaching undergraduate courses in Thermal/Fluid Science and graduate courses in Fluid Mechanics, Conduction and Convection Heat Transfer, Combustion, and Advanced Numerical Methods. Moreover, I have developed my mentoring skills by supervising several undergraduate research students at Georgia Tech.

See more of this Session: Meet the Faculty Candidate Poster Session
See more of this Group/Topical: Education

# (3cu) Multiscale Modeling of Biophysical and Biochemical Aspects of Viral Life Cycles

**Eric R. May**, Chemistry & Biophysics, University of Michigan, Ann Arbor, MI

Understanding viral life cycles is essential for the development of antiviral therapies, but also for the development of nanotechnologies based upon viral capsids and protein cages. My current and previous research have laid a foundation for understanding key components of viral systems (single viral proteins, virus capsids, membranes), but moving forward I want to integrate these components into a more holistic understanding of viral processes. I am interested in determining material properties of biological systems and understanding how these properties affect the phase and shape behavior of the system. As well as studying the coarse properties of these systems, I also want to understand the specific (local) interactions that play a crucial role in imparting bulk characteristics to the system. Identification of these "key" interactions can lead to testable hypotheses (i.e. mutagenic studies) about the significance of the identified interactions. This information can lead to targeted therapies, design principles and control mechanisms (i.e. pH triggered uncoating) for nanotechnologies. In this presentation I shall cover results related to membrane and virus material properties, viral protein structural predictions, protein-protein interaction predictions and free energy calculations on transition pathways. Taken individually, these different studies each probe an important aspect of viral systems and many of the methods employed extended across multiple spatial and temporal scales. Taken as a whole, these different components and methodologies provide a multiscale approach for studying the complex processes associated with viral life cycles.

See more of this Session: Meet the Faculty Candidate Poster Session
See more of this Group/Topical: Education

# (3cv) Multiscale Modeling of Liquid Repellency and Self Assembly Process

**Ateeque Malani**, Department of Chemical Engineering, Massachusetts Institute of Technology, Cambridge, MA

Interfacial phenomena arise in number of industrially important situations such as liquid repellency, multiphase flow, particulate dispersants, condensation etc., in which the structure of surface plays a key role. Similarly, self assembly lies at the core of many chemically important processes, such as synthesis of silica gel, zeolites, ordered mesoporous materials and Metal organic frameworks, which gives rise to different morphology and porosity in the resulting final structure. My research goal is to design and control the structure, morphology, chemistry and porosity of these materials by studying the phenomena at molecular scale to obtain the basic insight about the chemical and thermodynamic factors that affect these processes. I will highlight my past research effort in these areas and discuss how these studies will be helpful in the design of clean energy and separation technologies.

Surface roughness and chemistry are central to the engineering of materials used in water- and oil repellent applications. The success of these material is determined by their wetting behavior which is assessed by monitoring static and dynamic contact angle phenomena. The primary objective is to construct a methodological framework to study surface properties using molecular simulations and prepare repellent surfaces through experiments using input from molecular modeling. Towards this goal I have developed novel method to study three phase static and dynamic contact angle phenomena. As proof of concept, the methodology has been validated extensively using a simple Lennard-Jones (LJ) fluid in contact with a face-centered cubic LJ crystal surface. Excellent agreement is observed between the static (equilibrium) contact angles obtained by the new method and those reported in the literature. In the experimental part, I am using electrostatic fiber formation technique "electro-spinning" to prepare nonwoven fibrous materials using different polymers. The results obtained from the molecular modeling and experimental study will be discussed.

During my previous postdoctoral research work, I have studied the self assembly of silica-gel formation using multiscale modeling. Understanding mechanism of silica polymerization is an important problem in material science. Here, I have developed a new model of silica polymerization and have studied this using reactive Monte Carlo simulations. In comparison to previous models and simulations, this new model enables study of polymerization reaction at low density and room temperatures. The network evolution from simulation is in excellent agreement with experimental observations. The analysis of simulation results suggest that polymerization occurs in three stages starts with initial oligomerization followed by ring formation, then cluster aggregation and finally gelation. This detailed mechanism provides the opportunity for tailoring the morphology, structure and porosity of the final resulting structure by altering processing conditions.

See more of this Session: Meet the Faculty Candidate Poster Session
See more of this Group/Topical: Education

# (3cw) Solar Thermochemistry for Sustainable Fuel and Food Production and for Industrial CO2 Capture and Sequestration

**Ronald Michalsky**[1], Peter Pfromm[1], Bryon Parman[2] and Vincent Amanor-Boadu[2], (1)Department of Chemical Engineering, Kansas State University, Manhattan, KS, (2)Department of Agricultural Economics, Kansas State University, Manhattan, KS

Solar thermochemistry for sustainable fuel and foodproduction and for industrial $CO_2$ capture and sequestration

RonaldMichalsky [1, 2], Peter H. Pfromm [2],Bryon Parman [3, 4], Vincent Amanor-Boadu [4]

[1] IGERTassociate in biorefining, [2]Department of Chemical Engineering,

[3] IGERTtrainee in biorefining, [4] Department ofAgricultural Economics,

KansasState University, Manhattan, Kansas, USA

Solarthermochemistry has the potential to overcome the storage issue of solar energyby capturing solar energy in form of chemical energy. The synthesis of ammoniais targeted here since ammonia is essential for modern agriculture, recognizingthat by some estimates food production will have to increase perhaps 70% by2050, and that additional ammonia may still be needed for bio-energy crops.Ammonia is also being proposed by others for hydrogen storage and transport, oras a direct fuel for modified diesel engines. The Haber-Bosch process is the stateof the art industrial process to produce ammonia. Spending 2% of the world'senergy budget as natural gas for ammonia, the process consumes more than onepound of natural gas to produce one pound of ammonia, releasing around threepounds of carbon dioxide.

The poster reportson thermodynamic theory and experimental proof-of-concept of a solarthermochemical cycle to produce ammonia from water and air at atmosphericpressure. A solar concentrator was used. A detailed and sophisticatedeconomical analysis indicating profitability of the concept at full scale (over1000 metric tons ammonia per day) was performed by an NSF IGERT Ph.D. studentteam consisting of Mr. Michalsky and Mr. Bryon Parman and is also shown here.

The potential ofsolar thermochemistry in carbon dioxide capture and sequestration isdemonstrated with two concepts for future work: using flue gas from coal-firedpower plants for low-cost production of polycrystalline silicon from silicondioxide, and secondly substituting coal with hydrogen obtained from solar water-splittingto produce steel in a solar concentrator.

Ronald Michalsky has received an M.S. (Diplom-Ingenieur F.H.)in bioprocess engineering from the Technical University Mittelhessen, Germany. His M.S. thesis work resulted in three peer-reviewed journal publications. Since2008 he is a Ph.D. candidate in chemical engineering and currently an NSF IGERTassociate in biorefining at Kansas State University where he is pursuing thefundamental thermodynamics and experimental proof of concept for solarthermochemical production of ammonia from steam and air at ambient pressure. Afirst publication on the Ph.D. work is in peer review at the time of thiswriting. Mr. Michalsky has presented on his M.S. and Ph.D. work at nationalconferences including the AIChE Annual Meetings. Mr. Michalsky is mentoringundergraduate research in the Chemical Engineering curriculum and is mentoringNSF REU students in the summer of 2011.

See more of this Session: Meet the Faculty Candidate Poster Session
See more of this Group/Topical: Education

# (3cx) Metal Nanoparticles for Advanced Technologies: A High-Throughput Approach to Study Structural Degrees of Freedom

**William D. Michalak**, Chemical Engineering, Carnegie Mellon University, Pittsburgh, PA

Metal nanoparticles on structured supports are used in many technological applications such as biosensing, energy harvesting, catalysis, and electronics. In every case, the functions and properties of the metallic nanostructures depend on their composition and structure (i.e. size, shape, and spatial distribution). Challenges to using metal nanoparticles in these applications are the difficulties of optimizing the structure-property functionality over a large structural domain. To accelerate optimization, it is desirable to use high-throughput/combinatorial methods; however, the most common applications of high-throughput methods in the physical and material sciences focus almost exclusively on compositional space. Therefore, a practical approach to create and study metal nanostructures in a high-throughput manner would greatly accelerate the understanding of their structure-sensitive properties.

In my graduate studies at Carnegie Mellon University with Profs. Andrew Gellman and James Miller, I discovered a novel approach to create a morphological gradient of metal particles in a repeatable and controlled manner, which extends directly to high-throughput fabrication and characterization of structure-sensitive properties because it allows for simultaneous fabrication over a relatively large spatial domain. The method is based on inducing precursor thin films to dewet from a substrate through spinodal dewetting. Spinodal dewetting allows the creation of particles that have well-defined structural properties by adjusting a few variables: initial film thickness, annealing temperature and annealing time. The morphologies of the particles were characterized using scanning tunneling and atomic force microscopies, and hydrodynamic stability and integral geometry theories to confirm the dewetting mechanism. In addition, the hydrodynamic instability theory provides a connection to the thermophysical properties of the system. The dewetting approach is general to any metal/ceramic support system and provides an alternative, inexpensive, and robust means to rapidly create metal nanostructures. It shows promise for large scale production of the metal structures, as well as understanding basic material properties.

I am currently working with Prof. Gabor Somorjai at the University of California, Berkeley as a post-doctoral researcher. I continue to study the novel functionalities of metal nanoparticles as a function of size and shape using state-of-the-art experimental techniques. My initial research goals as a faculty member include the rational design of renewable-energy and clean-energy systems. I will focus on molecular-level understanding and high-throughput methodologies to accelerate development. The experience I gained during my graduate work, and the experience I continue to gain as a postdoc, position me to make significant contributions in a faculty position.

See more of this Session: Meet the Faculty Candidate Poster Session
See more of this Group/Topical: Education

# (3cz) Process Optimization and Economic Analysis of the Production of Biocrude and Other Lipidic Materials by Sewage Sludge Microorganisms In Wastewater Treatment Plant Biorefineries

**Andro H. Mondala**, Dave C. Swalm School of Chemical Engineering, Mississippi State University, Mississippi State, MS

Municipal wastewater treatment plant activated and anaerobic digester sludges present an abundant source of lipidic materials in its raw form or after enhancement via manipulation of cultivation conditions. Initial studies found that the enhancement process, which involves the batch cultivation of activated sludge in a high carbon-to-nitrogen (70:1) ratio wastewater medium with a high sugar load (60 g/L), increased its lipid content by 70 % compared with raw activated sludge. Biodiesel yield from the enhanced activated sludge also reached 10 % CDW compared to 2 – 3 % in raw sludge. However, the conversion of substrate to lipids was relatively lower than those observed in most pure cultures of oleaginous microorganisms. Optimization of specific environmental parameters of the batch fermentation process was conducted, particularly the culture pH, since changes in culture pH levels are known to significantly affect the composition of the microbial community in activated sludge. The effect of the culture pH was investigated by testing different fermentation runs with controlled pH levels of 4, 7, and 10 and compared with previous results wherein pH levels were uncontrolled. Subsequently, different feeding regimes and fermentation modes (discontinuous fed-batch, semi-continuous feeding, continuous) were also tested to improve lipid yields and conversion. The enhancement process was also applied to anaerobic digester sludges in order to asses the potential of extracting lipids and other high value products from all sludge streams and reduce the output of waste sludges from wastewater treatment facilities. The lipid extracts obtained from raw and enhanced sludges were then fractionated using polar and nonpolar solvents and characterized using gas chromatographic methods to determine the identities and profiles of the lipidic materials and determine their potential applications. Process and economic analysis of the process was conducted with the aid of the SuperPro® simulation software in order to design and evaluate the proposed wastewater treatment plant biorefinery concept in an industrial scale.

See more of this Session: Meet the Faculty Candidate Poster Session
See more of this Group/Topical: Education

# (3d) Image-Based Fluid Dynamics Modeling for Biomedicine and Beyond

**Roman S. Voronov**, Chemical and Biomolecular Engineering, University of Pennsylvania, Philadelphia, PA

**Research Interests:**
Image-Based Modeling is an emerging multi-disciplinary field that finds numerous applications in many diverse industries ranging from Biomedicine to Enhanced Oil Recovery. By using state-of-the-art imaging techniques in conjunction with high performance fluid dynamics computer modeling the image-based approach enables researchers to gain insights into systems that otherwise would have been inaccessible. For example, using imaging techniques, such as 3D Computed Tomography or Confocal Microscopy, allows one to obtain exact geometries of complex structures of interest, such as that of living porous media in-vivo or that of inorganic porous media (such as rock formation). Perhaps the main advantage of this modeling approach is that the implementation of the obtained structures in simple robust computer simulations gives practical results that are based on actual experimental data and do not suffer from mathematical complexity or from over-simplifications that are often necessary in other modeling techniques.

**Future Direction:**
As faculty I would like to continue applying Image-Based Modeling in different areas of Biomedicine, as well as in other industries. In particular I would like to unite the knowledge that I obtained from the two topics that I worked on during my Ph.D. and during my postdoctoral appointment: Bone Tissue Engineering and Blood Thrombosis, respectively. I believe that the two are closely related, because Bone Tissue Engineering grafts must interact with the patients' blood post-implantation. Proper blood flow through the artificial tissue must be ensured in order to eliminate waste products from the construct, to mechanically stimulate the bone cells, and to deliver oxygen and nutrients to them. Image-Based Modeling presents itself as the perfect tool for exploring this system, because a) growth of the of the bone tissue can be captured by imaging it at different time points and b) effects of the dynamically changing porous structure on the hemo-chemical and mechanical processes within the construct can be readily calculated via high performance computing.

Besides Biomedicine my interests include flows through inorganic porous media (in particular multi-phase flow through porous rock as it is related to enhanced oil recovery), drug delivery in the brain, modeling industrial chemical processes, financial optimization and entrepreneurship. My overall philosophy is to use simple and well-developed simulation methods in order to get practical and realistic results based on actual physical data. Therefore, in the future, I foresee myself collaborating closely with experimentalists and/or having an experimental component in my own laboratory.

**Successful Proposals:** NIH F32, American Heart Association Postdoctoral Training Grant, Teragrid Supercomputing Resources.

**Post-doctoral Advisors:** Scott L. Diamond & Skipp Brass, Chemical and Biomolecular Engineering/School of Medicine, *University of Pennsylvania*

**Ph.D. Co-Advisors:** Dimitrios V. Papavassiliou & Vassilios I. Sikavitsas, Chemical, Biological and Material Engineering, *University of Oklahoma*

**Research Experience:**
My academic career path has been a blend of many fields of science and engineering. My formal training was in simulation, which I have learned on many different physical scales: Density Functional Theory (*femto*), Molecular Dynamics (*nano*), and Lattice Boltzmann Fluid Dynamics (*meso*). Yet, projects that I worked on were always close collaborations with experimentalists: Friction Reducing Surfaces for the NAVY, Bone Tissue Engineering, Enhanced Oil Recovery and Blood Coagulation/Thrombosis. As a result of these collaborations I have acquired invaluable and versatile experience in many research areas: material science, biology, medicine, biochemistry, biomedical imaging, and engineering design, just to name some. Moreover, as a result of intensive training at various supercomputing centers (OU Supercomputing Center for Education & Research - OSCER, Texas Advanced Computing Center - TACC and Lawrence Berkeley National Laboratory – LBNL), I became proficient in high performance computer modeling.

**Teaching Experience:**
Aside from my research career, I also have extensive teaching experience. I spent a semester teaching basic science to highschool students as a part of a NSF fellowship; I TAed and guest-lectured undergraduate lecture and laboratory courses in Chemical Engineering both at the University of Oklahoma and at the University of Texas (for which I won a Campus Wide Best TA Award); I was repeatedly chosen by my department to be the "Pro/II Expert" (duties included lecturing undergrads, holding office hours and developing course materials); and,

lastly, I mentored graduate students in my group.

**Postdoctoral Project:** *"Intra-Thrombus Chemo-Transport and Local Stress Mechanics Under Flow."*

**PhD Dissertation:** *"Fluid Shear Stress and Nutrient Transport Effects Via Lattice Boltzmann and Lagrangian Scalar Tracking Simulations of Cell Culture Media Perfusion through Artificial Bone Tissue Engineering Constructs Imaged With MicroCT."*

**Selected Publications:**

**R. S. Voronov**, S. VanGordon, V.I. Sikavitsas and D.V. Papavassiliou, **2010**. *"Computational Modeling of Flow Induced Shear Stresses Within 3D Salt-Leached Porous Scaffolds Imaged via Micro-CT"*. J. of Biomechanics, 43 (7), 1279-1286.

**R. S. Voronov**, S. VanGordon, V.I. Sikavitsas and D.V. Papavassiliou, **2010**. *"Efficient Lagrangian Scalar Tracking Method for Reactive Local Mass Transport Simulation through Porous Media"*. Int. J. for Numerical Methods in Fluids. doi: 10.1002/fld.2369

**R. S. Voronov**, S. VanGordon, V.I. Sikavitsas and D.V. Papavassiliou, **2010**. *"Distribution of Flow-Induced Stresses in Highly Porous Media"* Appl. Phys. Lett. 97, 024101.

**R.S. Voronov**, D. Papavassiliou and L. Lee, **2006**: "*Boundary Slip and Wetting Properties of Interfaces: Correlation of the Contact Angle with the Slip Length.*" Journal of Chemical Physics, 124, 204701.

*Reviews*:
**R.S. Voronov**, D.V. Papavassiliou and L.L. Lee, **2008**: "*Review of Fluid Slip over Superhydrophobic Surfaces and Its Dependence on the Contact Angle.*" Ind. Eng. Chem. Res., 47 (8), 2455-2477. doi: 10.1021/ie0712941

See more of this Session: Meet the Faculty Candidate Poster Session
See more of this Group/Topical: Education

# (3da) Engineering Immunological Functions with Biomaterials

**James J. Moon**, Biological Engineering, Materials Science & Engineering, and The David H. Koch Institute for Integrative Cancer Research, Massachusetts Institute of Technology, Cambridge, MA

Synthetic materials such as biodegradable polymericparticles and hydrogels have revolutionized the traditional fields of drugdelivery and biopharmaceuticals. Yet, application of these new materials andengineering techniques remain largely unexplored in immunological studies. Myresearch goal is to develop synthetic biomaterials that can regulateimmunological functions and use this system to design new vaccines andimmunotherapies. During this presentation, I will highlight my past researchefforts in drug delivery, polymer synthesis, immunology, and cellular andtissue engineering, and I will discuss how these engineering technologies willbe applied to augment immunological functions for development of novel medicaltreatments, including vaccines against infectious diseases andimmunotherapies for cancers.

As currently licensed vaccine adjuvants are inadequate ineliciting protective immunity against cancer or intracellular pathogens such asmalaria or HIV, there is a great interest in new vaccine technologies that canpotently activate humoral and cellular immune responses. Toward this goal, I have developed anovel synthetic vaccine platform based on interbilayer-crosslinkedmultilamellar vesicles (ICMVs) and demonstrated their efficacy in a malariamodel. ICMVs were formed by fusing liposomes intomultilamellar vesicles and subsequent crosslinking of adjacent lipid headgroupsacross lipid bilayers within multilamellar vesicles. These particlesexhibited substantially enhanced protein antigen loading and extended drugrelease kinetics compared with traditional drug delivery vehicles (e.g. liposomes and polymeric particles). Crosslinking of lipid bilayers enhancedstability of particles in serum, allowing efficient delivery of cargo antigens toantigen-presenting cells. ICMVs encapsulating model antigen ovalbumin andFDA-approved adjuvant elicited potent antibody and CD8+ T cell responses *in vivo*, comparable to those for strongviral vector vaccines. With the goal of translating these results to aninfectious disease, the potency of ICMVs was examined in a malaria model. ICMVscarrying a malaria antigen and adjuvant elicited robust humoral immuneresponses with high-avidity antibody titers lasting more than a year. Particleimmunization also led to prominent germinal center formation in lymph nodes,accompanied with increased, partially explaining the long-lasting andhigh-avidity antibody titers elicited by particle immunizations. These results highlight thepotential of synthetic vaccines based on biomaterials in vaccinations against infectiousdiseases.

In addition to particulate systems, scaffolds and hydrogelsare another mode of drug delivery for biological pharmaceuticals. During mygraduate work, I developed synthetic, biomimetic hydrogels, designed to undergoprotease-mediated degradation. Unlike protein-based extracellular matrix gelsisolated from natural sources, synthetic gels based on polymers can beprecisely designed/modified to change their biophysical and biochemicalproperties for manipulation of drug release kinetics and cellular behaviors.With the main goal of inducing and sustaining therapeutic angiogenesis forwound healing and tissue engineering applications, rational design of hydrogelswere pursued: changes in polymer structures, photolithographic micropatterning of biological drugs within hydrogels, andco-culture system with stem cells led to fabrication of hydrogels that wererobustly vascularized with newly formed blood vessels integrated with hosts'vasculature *in vivo*.

## References


1. **Moon JJ**, Suh H, Bershteyn A, Stephan MT, Liu H, Huang B,Sohail M, Luo S, Um SH,Chiu W, and Irvine DJ. Interbilayer-crosslinked multilamellar vesicles forpotent humoral and cellular immune responses, 10, 243-251, 2011, **NatureMaterials**.

- Featured by *Nature Materials: News and Views, 10, 166-68, 2011, and NatureBiotech: Research Highlights, 29, 330, 2011.*

2. StephanMT, **Moon JJ**, Um SH, Bershteyn A, andIrvine DJ. Therapeutic cell engineering with surface-conjugated syntheticnanoparticles, 16, 1035-41, 2010,**NatureMedicine**.

3. **Moon JJ**, Suh H, Luo S, YadavaA, and Irvine DJ. Induction and maintenance of robust immune responses againstPlasmodium vivax malaria using interbilayer-crosslinked multilamellar vesicles.(***In preparation***).

4. **Moon JJ**, Suh H, Yadava A, and Irvine DJ. Vaccinationwith polymeric nanoparticles vaccines incorporating toll-like receptor agonistsgenerates potent immune responses against Plasmodium viva malaria. (***In preparation***).

5. **Moon JJ**, Saik JE, Poche RA,Leslie-Barbick JE, Lee SH, Smith AA,

Dickinson ME,and West JL. Biomimetic hydrogels with pro-angiogenicproperties, 31, 3840-3847, 2010, **Biomaterials**.

6. **Moon JJ**, Hahn MS, Kim I, Nsiah BA, and West JL.Micropatterning of poly (ethylene glycol) diacrylate hydrogels with biomolecules to regulate andguide endothelial morphogenesis, 15, 579-585, 2009, **Tissue Engineering**.

7. Hu Y, Atukorale PU, Lu JJ, **MoonJJ**, Um SH, Cho EC, Wang Y, Chen J, and Irvine DJ. Cytosolicdelivery mediated via electrostatic surface binding of protein, virus, or siRNAcargos to pH-responsive core-shell gel particles, 10, 756-765, 2009, **Biomacromolecules.**

8. Leslie-Barbick JE, **Moon JJ**,and West JL. Covalently-Immobilized Vascular Endothelial Growth Factor PromotesEndothelial Cell Tubulogenesis in Poly(ethyleneglycol) Diacrylate Hydrogels, 20, 1763-1779, 2009, **Journal of BiomaterialsScience.**

9. **Moon JJ**, and West JL. Vascularization of engineered tissues: approaches to promoteangiogenesis in biomaterials, 8, 300-310, 2008, **Current Topics in Medicinal Chemistry**.

10. Lee SH*, **Moon JJ***, and West JL.Three-dimensional micropatterning of bioactive hydrogels via two-photon laserscanning photolithography for guided 3D cell migration, 29, 2962-2968, 2008, **Biomaterials**.

11. Gobin AM*, **Moon JJ***, and West JL. EphrinA I-targeted nanoshells forphotothermal ablation of prostate cancer cells, 3,351-358, 2008, **InternationalJournal of Nanomedicine**.

12. **Moon JJ**, Lee SH, and West JL. Synthetic biomimetic hydrogels incorporated withephrin-A1 for therapeutic angiogenesis, 8, 42-49, 2007, **Biomacromolecules**.

13. Lee SH, **Moon JJ**, Miller JS, and West JL. Poly(ethylene glycol) hydrogels conjugated with acollagenase-sensitive fluorogenic substrate tovisualize collagenase activity during three-dimensional cell migration, 28,3163-3170, 2007, **Biomaterials**.

14. Hahn MS, Taite LJ, **Moon JJ**,Rowland MC, Ruffino KA, and West JL. Photolithographic patterning of polyethylene glycol hydrogels, 27, 2519-2524, 2006, **Biomaterials**.

15. Lee P, LinR, **Moon J,** and Lee LP. Microfluidic alignment of collagen fibers for in vitro cellculture, 8, 35-41, 2006, **BiomedicalMicrodevices.**

16. DeLong SA, **Moon JJ**, and West JL. Covalentlyimmobilized gradients of bFGF on hydrogel scaffoldsfor directed cell migration, 26, 3227-3234, 2005, **Biomaterials**.

17. Lee SH,Miller JS, **Moon JJ**, and West JL. Proteolytically degradable hydrogels with a fluorogenic substrate for studies of cellular proteolytic activity and migration, 21, 1736-1741, 2005, **Biotechnol Progr.**

18. **Moon JJ**, Matsumoto M, Patel S, Lee L, Guan JL, and Li S. Role of cell surface heparansulfate proteoglycans in endothelial cell migration and mechanotransduction,203, 166-176, 2005, **JCell Physiol.**(selected as a cover)

19. Li S, **Moon JJ**, Miao H, Jin G, Chen BP, YuanS, Hu Y, Usami S, and Chien S. Signal transduction in matrix contraction and themigration of vascular smooth muscle cells in three-dimensional matrix, 40,378-388, 2003, **J.Vasc Res.**

*Authorscontributed equally

See more of this Session: Meet the Faculty Candidate Poster Session
See more of this Group/Topical: Education

# (3db) Synthetic Control of Metabolic Pathways to Improve Productivity of Biomass-Based Chemicals and Drugs

**Tae Seok Moon**, Department of Pharmaceutical Chemistry, University of California, San Francisco, San Francisco, CA

The past decade has witnessed the power of synthetic biology in the creation of synthetic devices and systems that help to understand complex biological systems. However, its potential for real-world applications has not been fully exploited. One of its promising applications is the construction of programmed cells that are able to integrate multiple environmental signals and to implement synthetic control over biological processes. My research interests are focused on developing microbes that are able to process multiple input signals and to produce desirable outputs. Specifically, I aim to create synthetic gene circuits in order to control metabolic pathways and improve productivity of biomass-based chemicals and drugs. During my poster session I will present examples of my research projects that have embodied my research ideas. I will also discuss the challenges and potentials of my research plans.

See more of this Session: Meet the Faculty Candidate Poster Session
See more of this Group/Topical: Education

# (3dc) Synthesis and Gas Adsorption Study of Porous MOF Materials

**Bin Mu** and Krista S. Walton, Chemical and Biomolecular Engineering, Georgia Institute of Technology, Atlanta, GA

Metal-organic frameworks (MOFs) or porous coordination polymers (PCPs) have become the focus of intense study over the past decade due to their potential for advancing a variety of applications including air purification, gas storage, adsorption separations, catalysis, gas sensing, drug delivery, and so on. These materials have some distinct advantages over traditional porous materials such as the well-defined structures, uniform pore sizes, chemically functionalized sorption sites, and potential for post-synthetic modification etc. Thus, synthesis and adsorption studies of porous MOFs have increased substantially in recent years. During my PhD study in School of Chemical and Biomolecular Engineering at Georgia Institute of Technology, the major goal of my project is to search, synthesize, and test these novel hybrid porous materials for adsorptive removal of toxic industrial chemicals (TICs) and chemical warfare agents (CWAs), and installing the benchmark for new-generation nuclear, biological, and chemical (NBC) filters. Our work demonstrates that MOFs provide significant improvement over the current NBC filters (an impregnated activated carbon), and are potential advanced materials for improved air purification system which have wide applications involving individual protective device and collective protective system.

See more of this Session: Meet the Faculty Candidate Poster Session
See more of this Group/Topical: Education

# (3dd) Harnessing Heat As a Route for Efficient Photovoltaics

**Prashant Nagpal**, Physical Chemistry and Applied Spectroscopy, Los Alamos National Labratory, Los Alamos, NM

More than 80% of current world energy needs are met through fossil fuels and more than 60% of this energy is released to the environment as "waste heat". An important obstacle in efficiently harnessing the emitted radiation using infrared photocells is the energetically broad distribution of the emitted photons. Similar energy losses limit the photovoltaic power conversion efficiencies to 31%. A less studied alternative for efficient energy conversion of heat into electricity utilizes a secondary emitter which emits a narrow frequency light utilizing the thermal energy, and this light is then converted into electricity using semiconductor photocell. This thermophotovoltaic energy conversion can have power conversion efficiencies as high as 60%. However, progress has been hindered by the choice of efficient emitters and infrared photocells.

Efficient thermophotovoltaic emitters should be able to tailor both the frequency and directionality of the emitted radiation. Radiative thermal emission from hot bodies is broad in energy and is non-directional. Metal nanostructures, patterned as three-dimensional photonic crystals, selectively tailor the light-matter interaction near the photonic bandgap and the increased photonic density of states is used to selectively emit light of particular frequency from a hot metallic nanostructure. Here I will present results from selective tailoring of thermal emission at low temperatures for development of efficient thermophotovoltaic emitters. Another possible route towards tailoring thermal emission is utilizing surface plasmons. While these hybrid photon-electron waves on metal surfaces are typically excited using light or fast electrons, recently we have explored excitation of surface plasmon polaritons using heat. I will discuss recent results on fabricating ultrasmooth patterned surface with nanometer scale precision for low-loss propagation of surface plasmons. These patterned surfaces will then be combined with designed nanostructures for modulating the frequency of emitted radiation, and also the directionality of the emitted glow. Such unprecedented control over hot body radiation opens up new opportunities and challenges for utilizing plasmonics, while also opening up alternate route towards efficient photovoltaic conversion.

See more of this Session: Meet the Faculty Candidate Poster Session

# (3de) Interdisciplinary Biomaterial Research

**Grinia M. Nogueira**, Chemical Engineering, MIT, Cambridge, MA

My research interests are related to biomaterials fabrication and characterization. My motivation for working in this field is the ability to develop tools and materials that could help to save or improve lives.

I strongly believe that collaboration and interdisciplinary research are critical factors required for a successful project. During my PhD in Brazil, my research was primarily focused on fabricating membranes and hydrogels of silk fibroin, a fibrous protein that forms the silkworms' cocoons, for application as biomaterials. In addition to working with silk, I had collaborated with clinical and biological researchers, with a goal to optimize the durability of cardiac valves by reducing their in vivo calcification. I was also selected to perform and conduct a research project with Kimberly Clark-Brasil, which led to a patent on the incorporation of biopolymers in the paper making process to improve properties such as adsorption and mechanical strength of paper. Also, while pursuing my PhD, I was awarded a fellowship that allowed me to come to MIT in the Department of Materials Science and Engineering (DMSE), where I performed research on layer-by-layer assembly and thin film characterization for 6 months.

I worked as post doctorate researcher at the DMSE/MIT for one year, studying spray-layer-by-layer assembly of polymer and nanoparticles in order to create Bragg reflectors. After one year of pure materials science research, I received an offer to work in a collaborative project with the Departments of Chemical Engineering and Biological Engineering at MIT, where I am currently studying and applying microfabrication tools to create cell microenvironments. This project has provided me with a unique post doctorate experience by combining materials science, cell biology, and clinical research with engaged and highly competent scientists.

A research area that I would like to explore in more detail is the use of biopolymers, such as silk fibroin, chitosan, and alginate to construct biomaterials. These biopolymers have unique characteristics that can be explored individually or in combination with other materials. For instance, silk fibroin is well known for its high mechanical strength, chitosan for its antibacterial and blood clotting properties, and alginate for its high cell compatibility. Besides being renewable resources, these biopolymers can be obtained and processed at relatively low cost which could allow for the fabrication of affordable biomaterials in developing countries.

I also plan to collaborate with clinical and biological scientists in order to develop or improve existing biomaterials. Medicine still faces many challenges and unknowns that could be more quickly uncovered in a multidisciplinary collaboration. My research goals include creating tools to investigate biological questions, fabricating affordable biomaterials to allow easier access to health care, and extensive collaboration with clinical and biological fields. I look forward to further discussing these goals in this poster session.

See more of this Session: Meet the Faculty Candidate Poster Session
See more of this Group/Topical: Education

# (3df) Synthetic Polypeptide Macromers: Components of a Hydrogel Toolkit for Modeling Cell-Matrix Interactions

**Abigail M. Oelker**, Chemical Engineering, Massachusetts Institute of Technology, Cambridge, MA

Theoverarching goal of my research is to design hydrogels for biomedicalapplications and as templates with which to investigate cell-matrixinteractions. Cellular functions including DNA synthesis, migration, anddifferentiation are co-regulated by both the stiffness of the extracellularmatrix (ECM) and the presence of soluble signaling molecules (growth factors,cytokines, hormones, etc.): as a postdoctoral scientist working with ProfessorPaula T. Hammond (chemical engineering) and Professor Linda G. Griffith(biological engineering, mechanical engineering) at MIT, I have synthesized andcharacterized a set of synthetic polypeptides with tunable size, composition,and secondary structure as components of a hydrogel toolkit for modeling thesecell-matrix interactions. This system, described further in my abstract submitted to theCell-Biomaterial Interactions Session (08B05), provides a level of control thatallows us to design hydrogels with stiff rod-like segments or flexible coilsegments, thus possibly impacting mechanical and other hydrogel properties. Iam investigating the relationships between polypeptide structure andcomposition, as well as hydrogel formulation (i.e. mass fraction and cross-linkdensity) on the physicochemical properties of the resulting hydrogels. Thesecharacterization data will be used to create a library of hydrogel substrateswith systematically and independently varied stiffness, permeability, andligand presentation for *in vitro* study of the fundamentalparameters of cell-matrix interactions.

After obtaining mybachelor's degree (Lehigh University, B.S. chemical engineering), I developed hydrogel-basedbiomaterials with Professor Mark W. Grinstaff atBoston University. Through my doctoral work (PhD chemistry), I gained expertisein the design, synthesis, and characterization of hydrogels for ophthalmologicapplications such as *in situ* gelling surgical adhesives and permanentcorneal implants. One of my research goals was the development ofapplication-specific hydrogel adhesives to treat a variety of ocular wounds:the main benefit of these materials is that, unlike sutures, the current standardof care, hydrogel adhesives can seal wounds without inflicting secondarytrauma, inducing astigmatism, or providing a pathway for infection. In thisproject, I formed hydrogel adhesives by mixing a peptide-based dendriticmolecule with bi-functional PEG polymers, which reacted under mild conditions(i.e. pH 7.4, room temperature) to rapidly form transparent, elastic hydrogelswith high water content. We studied three types of hydrogels cross-linked byformation of either thiazolidine or more robust pseudoproline bonds finding that the degradation time ofthe hydrogel adhesive could be tuned from 1.5 to 24 weeks depending on the typeof cross-link bond formed. The constituent macromersand resulting hydrogels did not induce *in vitro* cytotoxicity and, whencompared to sutures in an *ex vivo* model, exhibited equal or higherefficacy in closing corneal wounds. These results demonstrate the feasibilityof hydrogel adhesives for corneal wound repair and highlight the tunable natureof hydrogel physicochemical properties supporting our hypothesis thathydrogel adhesives can be designed to suit a variety of wounds ranging fromquick-healing corneal incisions to slow-healing corneal transplants.

In addition, I created a hydrogel implant designed toserve as a space-filling, permanent onlay for repairof corneal defects caused by trauma or disease. This onlay,which consisted of two layers: a thin, stiff polyHEMAhydrogel and a thicker, softer dendrimer hydrogellayer, was designed to support surface epithelialization, provide a protectivebarrier against infection, and promote integration with the surrounding stromaltissue. Due to their extremely hydrophilic nature, hydrogels do not readilysupport the adhesion and spreading of cells and, as such, these materials serveas neutral templates upon which to engineer specific cell-material interactionsat the implant interface. Our results demonstrated that peptide sequences, suchas arginine-glycine-aspartic acid (RGD), a ubiquitous recognition sequence for $\alpha_V\beta_3$integrin binding, and ECM proteins, such as collagen, can be covalentlytethered to hydrogels in order to facilitate the adhesion and spreading ofcorneal cells. Although additional long-term studies are necessary to refinethe surface modification strategy and fabrication protocol, the promisingoptical, mechanical, cell adhesion, and interface properties of the proposedimplant provide motivation for continued investigation into its development asa functional corneal onlay.

While finishing mypostdoctoral research, I will expand my knowledge of polymer chemistry,hydrogel characterization, and analysis of cell-biomaterial interactions. Expertise in these areas is critical for my future research, in which I willdesign synthetic scaffolds for *in vitro* tissue modeling and tissueengineering applications. I believe it is imperative to conduct appliedresearch for the design of translational biomaterials in harmony with basicstudy of the underlying tissue physiology and pathophysiology. For this reason,I will develop a research program that includes three main areas: (1) scaffolddesign and characterization, (2) examination of cell/ scaffold interactions, and(3) creation and study of relevant *in vitro* models of both healthy anddiseased tissue.

See more of this Session: Meet the Faculty Candidate Poster Session
See more of this Group/Topical: Education

# (3dg) Functional Surfaces and Interfaces for Composite and Biomedical Applications

**Amy M. Peterson**, Department of Chemical & Biological Engineering, Drexel University, Philadelphia, PA

Materials that can repair cracks and recover from mechanical failure are desirable. Because remendable materials both repair and prevent the propagation of cracks on the micro scale, they offer the potential for increased durability, safety, and cost efficiency for many applications. The focus of my PhD work was to understand the kinetic and thermodynamic parameters that control thermoreversible Diels-Alder bond formation in different types of healing polymeric systems. Such studies are necessary to optimize the healing ability of Diels-Alder-based systems and also contribute to the understanding of networks containing equilibrium controlled bonds. I have developed three healing systems for epoxy-amine thermosets based on the thermoreversible Diels-Alder reaction of furan and maleimide. In one, crack healing of a traditional epoxy-amine thermoset is induced by thermally reversible cross-linking of a secondary phase. In another, a furan-functionalized epoxy-amine thermoset can be healed with a bismaleimide solution at room temperature and minimal pressure and significant load recovery is possible multiple times in a given location. The third system allows for interfacial healing of glass fiber-reinforced epoxy-amine composites via compatible functionalization of glass fibers and the polymer network.

Interactions at the interface between the body and an implanted device determine the failure or success of the device. Optimizing surface conditions of titanium implants is necessary to increase the lifetime of implants and quality of life for millions of patients. Changes in topology have been shown to effect cell adhesion, shape, and motility. It is important that an implant surface is optimized for desired cell (such as osteoblast) adhesion and growth, while adhesion of undesired cells and proteins is minimized. One way to improve osteoblast adhesion and growth is to add appropriate growth factors to the enviroment. My posdoctoral research focuses on the development of improved biointerfaces with tailorable properties that are scalable and the ability to release growth factor.

See more of this Session: Meet the Faculty Candidate Poster Session
See more of this Group/Topical: Education

# (3dh) Computational Molecular Science: From Biology to Nanotechnology and Beyond

**Amish J. Patel**, Chemical and Biological Engineering, Rensselaer Polytechnic Institute, Troy, NY

Computational science has been invaluable in helping us understand the molecular basis of diverse phenomena across various disciplines, ranging from nanotechnology to biology. Water-mediated hydrophobic interactions are ubiquitous in biology, yet faithfully describing them has been challenging, due to their inherently non-additive and context-dependent nature. We have recently developed a technique that enables quantitative characterization of these interactions, and will enable the study of a diverse class of problems ranging from amyloids, to protein targeting, and drug discovery. With the tremendous growth in computing power now available and the recent development of enhanced sampling techniques, we on the verge of being able to use computation as a tool for guiding the search for novel materials, with applications in water purification and energy storage. Nano-structured materials, including carbon nanotubes, are promising candidates to tackle the problem of water purification in an energy-efficient manner. Increasing interest in this field has led to a rapid growth in the available chemistries with which these nanotubes can be decorated. However, we need to understand the water-mediated interactions and identify the underlying principles of ion solvation and exclusion in confinement, in order to inform the optimal design of these advanced materials. Another problem in which ion solvation and transport also plays a key role is lithium-ion batteries. Novel polymer electrolytes provide the promise of delivering a lithium battery with high energy-density that could revolutionize the way we consume energy. However, judiciously chosen candidates need to be examined, using state-of-the-art modeling techniques in order to guide the search for the optimal electrolyte material. I propose to employ a combination of statistical mechanics and molecular computation to address these challenging problems that span a wide spectrum of time and length scales.

See more of this Session: Meet the Faculty Candidate Poster Session
See more of this Group/Topical: Education

# (3di) Engineering Materials On Molecular-Level

**Qing Peng**, Argonne National Laboratory, Argonne, IL

Qing Peng, currently a Postdoc fellow in the Energy Systems Division at Argonne National Laboratory. At Argonne, my research is focusing on nanocomposite material synthesis by atomic layer deposition (ALD) for applications in catalysts, photoelectronics, renewable energy, nanoscience and technology. My PhD dissertation is about nanoscale engineering materials with ALD and supercritical fluid process.

My research interests lie in developing step-wise self-limited molecular assembly processes, such as ALD, molecular layer deposition (MLD) and layer-by-layer assembly (LBL) process, to synthesize highly-engineered designer nanostructure materials, including thin films, complex multifunctional composite nanomaterials and hybrid materials. The ultimate goal is to provide advanced materials for addressing the energy crisis and environmental problems.

My research can bring novel molecular-level material and surface engineering approaches into research communities, and can be incorporated as a key element into the energy strategic plans of university. Strong cross-disciplinary collaborations – can be established in the fields of renewable energy, catalyst, surface chemistry, composite, soft matter, biomaterials and material simulation.

See more of this Session: Meet the Faculty Candidate Poster Session
See more of this Group/Topical: Education

# (3dj) Computational Design of a Peptide to Inhibit Macrophage Phagocytosis

**Diego A. Pantano**, Chemical and Biomolecular Engineering, University of Pennsylvania, Philadelphia, PA

Macrophages are the primary policing cells in your body and respond upon contact in a two-step process. In the first step, invariably, the macrophage is activated by contact to anything, even your own cells. However, in the second stage, and just before a macrophage engulfs its target, it checks for identification, that includes the 'marker of self' protein CD47, a widely expressed membrane protein. If CD47 is present, it causes the macrophage to disengage, making the activated phagocytosis process less efficient. Attachment of "self" CD47 on avidin-beads has shown a reduction in the rate of the phagocytosis by the macrophage.

The technical difficulty of expressing CD47's functional domain led us to search for a smaller peptide showing phagocytosis inhibition. In this direction we used molecular dynamics simulations and docking algorithms to understand the binding process and design a useful peptide. We used as a template the crystallographic structure of the complex of CD47 with the N-terminal ligand-binding domain of signal regulatory protein alpha (SIRP-$\alpha$). We tested two different peptides, each of them represent two different strategies aiming to preserve the binding region from diverging from the crystal structure. We found that one of these strategies maintained the structure better than the other. Affinity experiments agreed with our predictions, and suggest the real possibility of more biocompatible materials with a small peptide.

See more of this Session: Meet the Faculty Candidate Poster Session
See more of this Group/Topical: Education

# (3dk) Nanomaterials and Liquid Crystalline Systems: Dispersion and Characterization

**A. Nicholas G. Parra-Vasquez**, Laboratoire Photonique, Numrique et Nanosciences - LP2N, Universit de Bordeaux, Institut d'Optique Graduate School & CNRS, Talence 33405, France

As the discoveries of new nanoparticles, nanorods, nanosheets, and nanotubes continue, all with unique, highly useful properties, there is a push for researchers to develop techniques to characterize and image these particles as well as a push to build materials and applications to exploit these unique properties. My research thus far has focused on the characterization and imaging of carbon nanotubes and gold nanoparticles in liquid crystalline systems. Carbon nanotubes are extremely difficult to disperse in common solvents without functionalization or wrapping the nanotubes with surfactants or DNA; these additives can be detrimental to the properties of the final material. By dispersing carbon nanotubes in superacids no additives are necessary and dispersions result highly, aligned liquid crystalline solutions, which are ideal for creating materials of highly, aligned carbon nanotubes structures. However, in using the highly reactive superacids, characterization of the structures in solution were very challenging, but with the proper care and a bit of ingenuity, already developed techniques were used to characterize these dispersions. Lastly, gold nanoparticles were used to probe the liquid crystalline phase behavior of a common thermotropic liquid crystal at the nanoscale. A photothermal heterodyne imaging technique was implored to create the local phase change. The technique involves heating the nanoparticle with a modulated resonant laser which creates an index of refraction profile that diffracts a nonresonant probing laser to create a measurable signal. I hope to continue research in the area of nanoparticles and nanomaterials with focus in the area of renewable energy.

See more of this Session: Meet the Faculty Candidate Poster Session
See more of this Group/Topical: Education

# (3dl) Structured Nanocomposite Sorbents for CO2 Capture

**Genggeng Qi**, Material Science and Engineering, Cornell University, Ithaca, NY

My PhD research focused on two contiguous subject areas. One was controlled polymerizations and enzymatic polymerizations in heterogeneous media via batch and continuous processes. The other was the design, synthesis and characterization of nanostructured materials, such as polymer latexes and novel aminosilica materials for the enhanced capture of CO2 from anthropogenic sources. My postdoctoral training has expanded my experience on the development of new nanoparticle-ionic materials, mesoporous materials and organic-inorganic hybrids. My recent research is focusing on nanostructured composites for CO2 capture and energy related applications.

See more of this Session: Meet the Faculty Candidate Poster Session
See more of this Group/Topical: Education

# (3dm) Catalytic Conversion, Adsorption and Kinetic Study of Biomass Into Biofuel

**Nafiseh Rajabbeigi** and Michael Tsapatsis, Chemical Engineering & Materials Science, University of Minnesota, Minneapolis, MN

Global population and economic growth is driving a continual increase in the worldwide demand for energy. Closely coupled to this energy demand is the price of petroleum, due largely to the dependence of economic activity on its use. But fossil-fuel use also carries potential environmental consequences, and these concerns motivate us to investigate alternative energy sources. Biomass is one such source of renewable energy and could play a leading role in building future chemical industries. Catalytic conversion of biomass into chemicals and fuels is one key process required to realize this scenario. We recognize that design of sustainable processes involving catalytic conversion of biomass is to be achieved by comprehensive work on efficiency and catalytic performance.

The overall theme of my research is the catalytic conversion of biomass into biofuel and chemicals. We investigate selectivities and activities of certain zeolites as well as ion exchange resins that can efficiently catalyze the reaction. I will discuss a study of the reaction kinetics using zeolite catalysts. One of the important intermediate chemicals from biomass conversion is 5-hydroxymethylfurfural (HMF). I will also present the low-temperature production and separation of HMF from the product mixture of sugars in DMSO for cost effective recovery. I will discuss the types of adsorbents that are appropriate for this separation and the phenomenon of enhanced HMF adsorption in the presence of sugars.

See more of this Session: Meet the Faculty Candidate Poster Session
See more of this Group/Topical: Education

# (3dn) Merging Microfluidics Into Nanoparticle Drug Delivery

**Minsoung Rhee**, Mechanical Engineering, MIT, Cambridge, MA

**Merging Microfluidics into Nanoparticle Drug Delivery**

Minsoung Rhee[1,2]

[1]Massachusetts Institute of Technology, USA.

[2]Brigham and Women's Hospital, Harvard Medical School, USA.

First, I present a novel design for efficient 3D hydrodynamic focusing that has a simple and straightforward structure. Hydrodynamic focusing in a microfluidic channel provides homogeneous reaction conditions that allow for various synthesis applications including polymeric nanoparticle synthesis by nanoprecipitation. Although microfluidic platforms have advantages of rapid mixing and controlled precipitation resulting in homogeneous particles, control of aggregation is a nontrivial issue since polymer particles tend to stick to channel walls and rapidly agglomerate inside, blocking the entire channel. Polymer aggregation can be avoided by 3D hydrodynamic focusing where the polymer stream is constrained both horizontally and vertically. Our system is constructed on a single PDMS layer consisting of three sequential inlets to a conventional 2D hydrodynamic focusing system. The sample flow is first vertically focused by two vertical sheath flows and then horizontally squeezed into a narrow stream flowing in the outlet channel. Computer simulations showed that the size and position of inlet holes play a critical role in determining the shape of hydrodynamic focusing in the channel. When the inlet hole size is comparable to the channel width, the sample stream is successfully constrained in a horizontal band. In case of such complete vertical lamination, two-dimensional analytical models become sufficiently accurate. One can also simply increase the channel height to achieve similar focusing results. Molecules in the vertically focused stream gradually diffuse into both sheath streams, eventually reaching the top and bottom walls. Simulated bottom wall concentrations perfectly matched with the predicted values from two-dimensional mathematical modeling. The performance of 3D focusing was also confirmed experimentally using Poly(lactic-co-glycolic acid) (PLGA) particles known to agglomerate rapidly. Our 3D hydrodynamic focusing can prevent the precipitation at the wall. In addition, confocal micrographs with fluorescent dyes show cross-sectional views with vertical focusing. We have synthesized various polymeric nanoparticles that could neither be synthesized by 2D hydrodynamic focusing nor by bulk mixing in a conventional way. The sample stream containing polymer precursors dissolved in acetonitrile (ACN) becomes vertically squeezed by two vertical ACN sheath streams and then horizontally squeezed by water sheath streams, producing nanoparticles by nanoprecipiation.

Second, a novel single-channel multiple 3D hydrodynamic focusing device for high-throughput systems is presented here. The device consists of two PDMS layers; the bottom layer contains a single channel for multiple focusing and the top layer governs fluidic supply and arrangement, respectively. Our prototypic device can produce 10 multiple focused streams in a parallel manner in a single channel. Each of focused streams can be adjusted to a different flow rate, resulting in a differently squeezed stream. Such multiple parallel focusing is ideal for high-throughput synthesis system where particles are synthesized by diffusive mixing that relies on hydrodynamic focusing. sson one layer and ed PDMS layer consisting of three sequential inlets to a conventional 2D hydrodynamic focusing system. In conclusion, I present here an innovative microfluidic fabrication method, a novel air-operable microfluidic circuit, and new designs for single and multiple 3D hydrodynamic flow focusing systems that have a variety of applications in biological sciences including flow cytometry, drug delivery, drug synthesis, and screening.

Finally, I have designed a microfluidic technology that performs counter-current flow dilution where prepared NPs solution with impurities and washing deionized water are oppositely directed in the microchannel network. This counter-current flow maintains maximum concentration gradients of impurities that in turn facilitate diffusive dilution by washing flow. Since the organic solvent and non-encapsulated drug molecules have higher diffusivities than that of polymeric NPs, the impurities are washed out and NPs can be purified and concentrated with the specially designed microfluidic system where convection between two opposite flows is minimized by imposing high hydrodynamic resistance in connecting bridges. The device operates with input polymer precursors along with drug molecules at programmed flow rates by syringe pumps, and drug-encapsulated NPs assembled in the channel will flow into the purification microchannel network where organic solvent and non-encapsulated drug will be rapidly removed. Such integration of purification functionality eliminates the need for manual collection, delivery, and filtration and thus accelerates the NP preparation and screening process by orders of magnitude. Reduction in processing time also results in faster stabilization of NPs, minimizing the possibility of premature release of encapsulated drug molecules or NP aggregation.

# (3do) Smart Manipulation of Soft Matter for Immunobioengineering

**Kyung-Ho Roh**, Howard Hughes Medical Institute, Stanford University, Stanford, CA

In this poster session, I will display my research interests that can be summarized into three main themes: (1) Technology development for smart manipulation of biomaterials, (2) biological study of the interface between biomaterials and the immune system, and (3) novel immunobioengineering based on the learning from the first two topics. My previous research efforts that lead into my current research interests will be presented accordingly.

First, smart manipulation of biomaterials in micro- and nanometer scale is my research interest extended from my PhD thesis, where the novel applicability of the electrified jetting process was demonstrated for the control of materials distribution in nanometer scale. Particularly, simultaneous jetting of two or three distinct polymer solutions through jetting capillaries in a side-by-side geometry was successfully employed to create biphasic or triphasic nanocolloids. The resulting nanocolloids consist of two or more separate compartments, which generate anisotropy. Selective surface modifications and chemical reactions enhancing this anisotropy enable experiments with mammalian cells to demonstrate that these unique smart nanomaterials can potentially be useful in novel biomedical applications, namely unique drug delivery and bioimaging.

My research interests in biomaterials expanded to pure biological questions through my postdoctoral research experience. I was especially drawn to the immune system, because it is the gateway where the interactions between our body and biomaterials begin. While I studied spatiotemporal dynamics of T-cell receptors and related molecules in the critical T cell activation process at an unprecedented resolution, I became interested in the general principle governing cellular membrane organization and its roles in interfacing with biomaterials, which will be one of my continued efforts. For this, super-resolution fluorescent microscopy, *i.e.* photoactivated localization microscopy (PALM) and stochastic optical reconstruction microscopy (STORM), which I established and applied to my postdoctoral research will continuously be a powerful toolset.

Last but not least, I would like to apply the newly developed biomaterials for various immunotherapies and biological studies of immune systems in general, namely, immunobioengineering. Select topics include creating a smart depot system for the treatment of autoimmune and degenerative diseases, and employing a biohybrid of lymphocytes and multifunctional nanoparticles for directed immune trafficking.

See more of this Session: Meet the Faculty Candidate Poster Session
See more of this Group/Topical: Education

# (3dp) Influence of Confinement and Interfacial Interactions On the Behavior of Membranes and Materials for Energy Technologies

**Brandon W. Rowe**, Polymers Division, National Institute of Standards and Technology, Gaithersburg, MD

The rapid growth of energy-centric research efforts have stemmed from the clear and present need to improve energy efficiency, reduce greenhouse gas emissions, and to define alternative, renewable energy sources. Membrane materials offer potential solutions to many of these issues. Large scale, energy intensive industrial separation processes can benefit from the inherently low energy requirements of membrane-based separations, which do not require phase transitions. Gas separation membranes also offer competitive opportunities for carbon capture. Additionally, membranes are an integral part of energy storage and delivery technologies such as batteries and fuel cells.

Efforts to maximize the productivity of membrane materials typically require developing structures with nano-scale dimensions. In separation membranes, for example, because flux is inversely proportional to the separating layer thickness, productivity increases as film thickness is reduced. Composite thin-film technologies offer the additional benefit of reducing the amount of high-performance material required in production of multi-component membranes. In addition to the significant deviations from bulk behavior that may occur in these nanostructured materials, they are typically comprised of highly non-equilibrium materials whose properties can change dramatically with time.

There is currently a large gap in the fundamental understanding of how confinement and interfacial interactions influence performance and subsequent material stability. Furthermore, the development of realistic accelerated aging tests and long-term predictive models will be required to prove the long term viability of these technologies. If these factors were better understood, strategies to manipulate structure and interfacial interactions could be developed to further improve membrane material performance. In addition to ideas for addressing these current opportunities, this poster will present research aimed at understanding the accelerated aging behavior of ultrathin films used in gas separation membranes and the influence of confinement and interfacial interactions on the performance of polymer electrolyte membrane (PEM) materials.

See more of this Session: Meet the Faculty Candidate Poster Session
See more of this Group/Topical: Education

# (3dq) Nanoparticles for Biomedicine: Development of a Family of Novel Nanocomposites and Fundamental Research Into Bio-Transport Phenomena

**Gang Ruan**, Department of Chemical and Biomolecular Engineering, The Ohio State University, Columbus, OH, Jessica Winter, Chemical and Biomolecular Engineering, The Ohio State University, Columbus, OH, R. Sooryakumar, Department of Physics, The Ohio State University, Columbus, OH, Jeffrey Chalmers, Chemical and Biomolecular Engineering, Ohio State University, Columbus, OH and Shuming Nie, Biomedical Engineering, Emory University and Georgia Institute of Technology

The potential of nanomaterials with special optical, magnetic, electronic and thermal properties for biomedical applications has been extensively demonstrated. My independent research program has two thrusts: (1) developing a new generation of nanomaterials, which could be loosely defined as nanoscale materials with properties beyond those currently in use. These novel nanomaterials would overcome current challenges (e.g. aggregation, loss in performance over time), or/and exhibit enhanced performance (e.g. greater brightness), or/and integrate functions from individual nano-species as well as micro-devices; (2) investigating fundamental behaviors of nanomaterials in biological systems. A focus is given to how nanomaterials are transported in cells ("transport phenomena" in biological systems). The knowledge obtained would be applied for design of new nanomaterials and new strategies of tackling diseases.

In addition to helping to build the groundwork for an emerging field, i.e. bionanotechnology, my work also contributes to the ongoing transformation of a traditional discipline, i.e. chemical engineering. The following are some highlights of my research accomplishments, which also serve as preliminary results for the two research thrusts of my independent research program described above.

**Nanocomposites: from tracking single molecules to delivering drugs**

A major recent development in bio-nanomaterials research is the emergence of composite nanomaterials, which combine the functions of individual nanomaterials. I have developed a family of novel nanocomposites based on micelle-templated self-assembly for a wide spectrum of biomedical applications.

First, a composite nanoparticle that changes fluorescent color continuously was created by co-encapsulating several quantum dots (semiconductor nanocrystals with extraordinary fluorescent brightness and photostability) of different colors into a micelle. This novel nanocomposite, also known as alternating-color quantum dot nanocomposite, simultaneously solved two seemingly irreconcilable problems in quantum dot-based single particle/molecule tracking, namely blinking-caused interruptions in particle trajectory and difficulty to check single particle status in situ without blinking (Ruan et al. *Nano Letters* 11, 941-945 (2011)).

Second, quantum dots (QDs) and superparamagnetic iron oxide nanoparticles (SPIONs) were co-encapsulated into a micelle, resulting in a composite nanoparticle that posses both fluorescent and magnetic properties. Compared with commercial magnetic beads, QD-SPION-micelles have the advantages of small size (less than 50 nm) and bright fluorescence for imaging. Combining QD-SPION-micelles with magnetic microarrays led to a novel nano-conveyor belt technology, which achieved the first simultaneous magnetic manipulation and fluorescent tracking of multiple individual sub-100 nm nanostructures (Ruan et al. *Nano Letters* 10, 2220-2224 (2010)). Preliminary results with biomolecules and cells indicate that the nano-conveyor belt technology could become a platform technology for new methods in molecular diagnostics, cell separation, and nano-fabrication.

Third, poly (lactic-co-glycolic acid) (PLGA), which is the most commonly used biodegradable polymer for drug delivery, was encapsulated into a micelle, resulting in a novel drug delivery vehicle. Compared with previous techniques, which are essentially top-down approaches, this bottom-up approach offers a facile and robust method to fabricate sub-100 nm PLGA-based drug delivery vehicles. Compared with micelles alone (without PLGA), PLGA-micelles offer a way to control drug release without significantly changing particle size (Ruan et al. *patent pending*, patent no. 61/433,794).

Further, a process to produce the micelle-based family of nanocomposites in continuous mode was developed. 300 times scale-up from the original batch production process was achieved (Doug et al. *manuscript in preparation*).

**Probing intracellular transport of nanoparticles**

As numerous types of nanomaterials are being developed for biomedical applications, it is of fundamental importance to understand how nanomaterials interact with biological systems. In particular, transport processes of nanomaterials inside cells have been poorly understood, to a large extent due to lack of fluorescent probes that are sufficiently photostable and bright to track these dynamic processes. I have used quantum dots (QDs), which are fluorescent nanoparticles with extraordinary brightness and photostability, coupled with spinning disk confocal microscopy to image and track intracellular transport.

First, QDs were conjugated with Tat peptide, which is being widely investigated for its potential as a delivery vehicle and is also the peptide sequence responsible for invasion of cells by HIV virus, to probe the mechanism of Tat-mediated intracellular delivery of nanoparticles. It was found that, in stark contrast to what was previously thought, Tat-QDs did not enter cell nucleus, but were trapped in intracellular vesicles. The motions of Tat-QD-vesicles along cytoskeletons were tracked and the velocities were determined. Furthermore, Tat-QD-vesicles were observed just outside cells. This discovery indicates that Tat-QDs can leave the cells via a process known as vesicle shedding (Ruan et al. *Journal of the American Chemical Society* 129, 14759-14766 (2007)).

Second, single QDs were directly delivered into cytoplasm via a bacteria toxin streptolysin O (SLO). The single (or near single) status of the intracellular QDs was confirmed by their blinking behavior. In cytoplasm single QDs gradually formed aggregates, with the aggregation rate dependent on QD surface chemistry and cell type. Surprisingly, it was found that most of the single QDs were bound with dynein, the motor protein responsible for transporting cargos along microtubules towards the microtubule organizing center (just outside cell nucleus), by non-specific binding. Further, the binding of single QDs with dynein was exploited to track the dynein motion in live cells (Agrawal et al. *manuscript in preparation*).

See more of this Session: Meet the Faculty Candidate Poster Session
See more of this Group/Topical: Education

# (3dr) Engineering In the Microvasculature: The Mechanical Microenvironment's Control of Systemic Metabolism

**Joseph M. Rutkowski**, Touchstone Diabetes Center - Department of Internal Medicine, UT Southwestern Medical Center, Dallas, TX

Lying between the endothelial beds of the body's circulation is the interstial microenviroment wherein the complex biohphysical and biochemical interplay of cells, the extracellular matrix, and the molecules of interstital fluid control both immediate tissue homeostasis and systemic functions. The modulators of interstitial flow rates and shear, tissue hydration and composition, and cellular trafficking are the endothelial microvasculature of the blood and venous capillaries and the lymphatic system. As modulators of interstitial flow and molecular transport, the blood endothelial microvasculature and lymphatic system govern tissue homeostasis. I hypothesize that the blood and lymphatic microvasculature, as gatekeepers to the interstitium and governors of molecular and fluid fluxes, modulate metabolism through their endothelial physiology.

In my doctoral studies under Prof. Melody Swartz at the EPFL, we demonstrated that the primary modulator of interstitial drainage, the initial lymphatic vascualture, both controls the interstitium and cells found within and is itself modulated by the tissue mechnical environment. Generation of new functional lymphatic capillaries, lymphangiogenesis, was found to be directional and fully dependent on interstitial flow. Blocked lymphatic drainage, clinically the pathology of lymphedema, resulted in hyperproliferation, yet poor organization and decreased function of lymphatic capillaries. In the pathology of lymphedema, where intersitial flow rates slow towards zero, inflammatory chemokines accumulated, interstitial immune cells failed to traffick, and adipose tissue depots expanded in adipogenesis. Indeed, we quantified lymphatic clearance rates in genetic models of lymphatic dysfunction and found a correlation with fluid transport and remodeling of the extracellular matrix and adipose expansion on tissue hydraulic conductivities and subsequent treatment and potential resolution of the disease. Blood capillary extravasation, as controlled by Starling's Law, was reduced in these instances due to the increased hydraulic and osmotic pressures of the interstitum. Concurrently, adipose expansion or hypercholesterolemia strongly reduced lymphatic drainage capability by inducing vessel degeneration. As lymphatic vessels are the primary route of lipid transport throughout the body, these studies combined to demonstrate the potential that the interplay of microvasculature cells with their surrounding tissues has to influence systemic metabolism.

My current postdoctoral research focuses on adipose tissue physiology and how the microvasculature may control its systemic functions via adipokine and metabolite transport. No longer considered to be a passive reservoir, adipose tissue is an active metabolic player and dynamic endocrine organ. The adipokine adiponectin, intruiging in that it is secreted in multiple oligomer sizes, has exhibited positive effects in regulation of systemic metabolism and insulin sensitivity through tissues such as liver, skeletal muscle, and centrally in the brain. In the laboratory of Prof. Philipp Scherer at UT Southwestern Medical Center we have found that oligomer size, particularly that of the high molecular weight form, presents transport limitations across endothelial barriers with low intercellular permeabilities such as in muscle or in the brain. We have quantified size-dependent exclusion of adiponectin oligomers in model endothelial cells *in vitro*, expressed adiponectin receptors in these cells to determine the role of ligand-receptor binding, and used genetic mouse models of receptor deletion *in vivo* to discern that oligomer function likely depends on microvasculature permeabilities (Podium 65e). We have altered these permeabilities, and subsequently adiponectin transport and potentially function, pharmacologically, pathologically, and with inducible transgenic *in vivo* models. Metabolic stresses modulate the permeability of the vasdculature for increased or decreased lipolytic fluxes and systemic adipokine effects. The critical barrier function of the endothelium thus modualtes multiple aspects of adipose function.

Ongoing studies are aimed at discerning the interplay between adipose tissue and its vaculature. One concomitant pathology with obesity is chronic kidney disease; hanges in renal function alter transvascular fluxes and protein concentrations in the periphery, resulting in excess fluid accumulation. In a recently developed inducible model of renal podocyte apoptosis (Poster 623ai), adiponectin altered renal recovery while ascites led to adipose wasting. Changes in lipid uptake or secretion coincide with altered adipokine profiles: all of which alter or are controlled by endothelial transport. I propose that the biophysical environment of adipose tissue, specifically matrix composition, interstitial flow, and lymphatic drainage, influence adipocyte behaviors and systemic metabolism and that changes in cell infiltration and chemokine levels in the interstitial fluid have a direct effect on adipogenesis, adipokine secretion, and metabolic lipid fluxes. My laboratory will develop and utilize *in vitro* tissue engineered models to study the adipose biophysical environment and relevant, targeted transgenic *in vivo* models to demonstrate the physiologic relevance of our *in vitro* advances.

***Relevant Peer-Reviewed Publications:***

Fischer-Posovszky P, Wang QA, Asterholm IW, **Rutkowski JM**, Scherer PE. *Endocrinology*. Targeted deletion of adipocytes by apoptosis leads to adipose tissue recruitment of alternatively activated m2 macrophages. **2011** Aug;152 (8):3074-81.

Holland WL, Miller RA, Wang ZV, Sun K, Barth BM, Bui HH, Davis KE, Bikman BT, Halberg N, **Rutkowski JM**, Wade MR, Tenorio VM, Kuo MS, Brozinick JT, Zhang BB, Birnbaum MJ, Summers SA, Scherer PE. Receptor-mediated activation of ceramidase activity initiates the pleiotropic actions of adiponectin. *Nat Med* **2011** Jan;17(1):55-63.

Miteva DO, **Rutkowski JM**, Dixon JB, Kilarski W, Shields JD, Swartz MA. Transmural flow modulates cell and fluid transport functions of lymphatic endothelium. *Circ Res*. **2010** Mar 19;106(5):920-31.

**Rutkowski JM**, Markhus CE, Gyenge CC, Alitalo K, Wiig H, Swartz MA. Dermal matrix remodeling and fat accumulation control tissue swelling and hydraulic conductivity during murine primary lymphedema. *Am J Pathol.* **2010** Mar;176(3):1122-9.

**Rutkowski JM**, Davis KE, Scherer PE. Mechanisms of obesity and related pathologies: the macro- and microcirculation of adipose tissue. *FEBS J.* **2009** Oct;276(20):5738-46.

Lim HY, **Rutkowski JM**, Helft J, Reddy ST, Swartz MA, Randolph GJ, Angeli VA. Hypercholesterolemic mice exhibit lymphatic vessel dysfunction and degeneration. *Am J Pathol.* **2009** Sep;175(3):1328-37.

Goldman JG*, **Rutkowski JM***, Shields JD*, Pasquier M, Cui Y, Schmökel HG, Pytowski B, Swartz MA. Co-operative and redundant roles of VEGFR-2 and VEGFR-3 signaling in adult lymphangiogenesis. *FASEB J.* **2007** Apr;21(4):1003-12. (* equal contribution)

**Rutkowski JM** and Swartz MA. A driving force for change: interstitial flow as a morphoregulator. *Trends Cell Biol.* **2007** Jan;17(1):44-50.

**Rutkowski JM,** Moya M, Johannes J, Goldman J, Swartz MA. Secondary lymphedema in the mouse tail: lymphatic hyperplasia, VEGF-C upregulation, and the protective role of MMP-9. *Microvasc Res.* **2006** Nov;72(3):161-71.

**Rutkowski JM,** Boardman KC, Swartz MA. Characterization of lymphangiogenesis in a model of adult skin regeneration. *Am J Physiol Heart Circ Physiol.* **2006** Sep;291(3):H1402-10.

See more of this Session: Meet the Faculty Candidate Poster Session
See more of this Group/Topical: Education

# (3ds) Understanding the Impact of Reaction Parameters On Macromolecular Structure and Binding/ Transport Properties of Recognitive Crosslinked Polymers

**Vishal D. Salian**, Biomimetic & Biohybrid Materials, Biomedical Devices, & Drug Delivery Laboratories, Department of Chemical Engineering, Auburn University, Auburn, AL

In this work, living/controlled polymerization (LRP) is compared to conventional free radical polymerization (FRP) in the creation of highly and weakly crosslinked imprinted poly(MAA-co-EGDMA) networks. It elucidates, for the first time, the effect of living/controlled polymerization on the chain level and begins to explain why the efficiency of the imprinting process is improved using living/controlled polymerization. Imprinted polymers produced via LRP exhibited significantly higher template affinity and capacity compared to polymers prepared using conventional methods. Use of LRP in the creation of highly crosslinked imprinted polymers resulted in a 4-fold increase in binding capacity without a decrease in affinity; while weakly crosslinked gels demonstrated a 3-fold increase in binding capacity at equivalent affinity when LRP was used. In addition, by adjusting the double bond conversion we can choose to increase either the capacity or the affinity in highly crosslinked imprinted polymers, thus, allowing the creation of imprinted polymers with tailorable binding parameters.

With the use of FRP in the creation of polymer chains, as the template/monomer ratio increased, the average molecular weight of the polymer chains decreased despite a slight increase in the double bond conversion. Thus, a larger number of shorter chains were formed. With LRP, the addition of chain transfer agent to the polymerization negated the retardation of kinetic chain length caused by the diffusional limitations presented by the macromer/template complexes. Reaction analysis revealed that propagation time was extended 3-fold in the formation of highly crosslinked polymers when LRP techniques were used. This delayed the transition to the diffusion-controlled stage of the reaction, which in turn led to the observed enhanced binding properties, decreased polydispersity in the chains, and a more homogeneous macromolecular architecture.

# (3du) Drug Delivery with Control of Polymer Architecture and Chemical Composition: Combinatorial Synthesis of Diverse Nanoparticles for Intracellular Delivery

**Daniel J. Siegwart**, Department of Chemical Engineering, Massachusetts Institute of Technology, Cambridge, MA

During my postdoctoral fellowship, we employed a modular design for the high-throughput study of >1,500 structurally distinct nanoparticles with cationic cores and variable shells, enabling elucidation of complexation, internalization, and delivery trends. Using robotic automation, epoxide-functionalized block polymers were combinatorially cross-linked with a diverse library of amines, followed by measurement of molecular weight, diameter, RNA complexation, cellular internalization, and *in vitro* siRNA and pDNA delivery. Analysis revealed structure-function relationships and beneficial design guidelines, including a higher reactive block weight fraction, stoichiometric equivalence between epoxides and amines, and thin hydrophilic shells. Cross-linkers optimally possessed tertiary dimethylamine or piperazine groups, and buffering capacity. Covalent cholesterol attachment allowed for transfection *in vivo* to liver hepatocytes in mice. The ability to tune the chemical nature of the core and shell may afford utility of these materials in additional applications.

Overall, my research involves studies at the Chemistry/Biology/Engineering/ Nanoscience interface and strives to answer fundamental scientific questions and create innovative technologies and therapies that can benefit human health. In particular, I believe that the development of future polymer therapeutics will require the ability to precisely control macromolecular architecture, order, and responsiveness. My PhD research background in advanced synthetic polymer techniques, combined with my postdoctoral training in high-throughput methods, robotics, drug delivery, and animal studies will enable me to probe the effect of polymer architecture and chemical composition on the delivery of small molecule and biomacromolecular drugs.

See more of this Session: Meet the Faculty Candidate Poster Session

# (3dv) Refactoring Biosynthetic Pathways Via Synthetic Biology

**Zengyi Shao**, Chemical & Biomolecular Engineering, University of Illinois at Urbana-Champaign, Urbana, IL and Huimin Zhao, Department of Chemical and Biomolecular Engineering, Department of Chemistry and Biochemistry, Institute for Genomic Biology, University of Illinois at Urbana-Champaign, Urbana, IL

The overall theme of my postdoctoral research is to develop synthetic biology strategies to address the challenges associated with engineering natural product gene clusters for both new drug discovery and product yield optimization. As transiting from a PhD candidate to a more independent researcher, my research focus was correspondingly shifted from engineering and characterization of single enzymes to constructing biological systems composed of multiple components.

***Projects in My Postdoctoral Study***

**DNA Assembler, an *in vivo* Genetic Method for Rapid Construction of Biochemical Pathways.** A highly efficiently approach, named "DNA assembler", was developed, which allows the rapid assembly of an entire biochemical pathway in a single step *via in vivo* homologous recombination in *Saccharomyces cerevisiae*. Three distinct functional biochemical pathways with sizes ranging from 9 to 19 kb were assembled on both a plasmid and a chromosome with a high efficiency. As this method only requires simple DNA preparation and one-step yeast transformation, and circumvents the potential problems associated with the existing cloning methods, it represents a versatile approach in the construction of biochemical pathways for synthetic biology, metabolic engineering, and functional genomics studies.

**Rapid Characterization and Engineering of Natural Product Biosynthetic Pathways *via* DNA Assembler.** Sequenced genomes and metagenomes provide a tremendously rich source for discovery of new genes and pathways. However, the discovery and sustainable production of valuable natural products are often hampered owing to our limited ability to manipulate the corresponding biosynthetic pathways. We redesigned the DNA assembler as a genomics-driven, synthetic biology-based method, for discovery, characterization, and engineering of natural product biosynthetic pathways. This method synthesizes an entire expression vector containing the target biosynthetic pathway and the genetic elements needed for DNA maintenance and replication in *S. cerevisiae*,

*Escherichia coli*, and a target heterologous expression host in a single-step manner. Two important biosynthetic pathways from *Streptomyces* including the aureothin and the spectinabilin biosynthetic pathways were chosen for study. It was demonstrated that since all the fragments are prepared by PCR, various manipulations including site-directed mutagenesis, scar-less gene deletion and artificial gene cluster construction can be easily achieved. Such a method with unprecedented flexibility and versatility has many applications in heterologous pathway expression, pathway functional studies, and combinatorial biosynthesis.

**Refactored Biosynthetic Pathways: Delineating Pathway Expression from Sophisticated Regulation Cascades via Synthetic Biology.** The production of natural product is subjected to sophisticated regulation cascades in which target gene and multiple regulators interact with each other, resulting in a complex regulatory network that responds to a variety of physiological and environmental signals. This highly regulated process complicates the discovery and characterization of new natural products with important biological activities and hampers sustainable production of the high-value ones. We developed a general synthetic biology-based strategy to resolve the complexity and delineate the pathway expression from sophisticated regulation cascades. As proof of concept, three important biosynthetic pathways from *Streptomyces* including the spectinabilin gene cluster, the fosfomycin gene cluster, and the FR900098 gene cluster were chosen for study. We propose to construct refactored pathways by inserting a constitutive promoter in front of each pathway gene. Approximately 40 constitutive promoter candidates in front of the house-keeping genes were cloned from the genomes of *Streptomyces* and other related actinobacteria and fused with reporter genes. Currently, they are being evaluated in different growth conditions. The strong constitutive promoters will be inserted in front of the genes in the spectinabilin gene cluster in order to activate the silent spectinabilin biosynthesis in *S. lividans*. If successful, it will be very useful in activating numerous uncharacterized cryptic pathways for discovering new potential natural products with biological and pharmaceutical importance. In addition, the fosfomycin biosynthetic pathway and the FR900098 biosynthetic pathway, both having important biomedical applications are chosen and will be reconstructed with strong and constitutive promoters for higher production levels.

***Selected Projects in My PhD Study***

**Biosynthesis of Triacetic Acid Lactone (TAL) from D-Glucose.** Rational design was applied to establish the biosynthesis of TAL from a renewable feedstock D-glucose *in vivo*. Based on the catalytic mechanism of 6-methylsalycylic acid synthase (MSAS), we hypothesized that inactivating the ketoreductase (KR) domain would lead to TAL formation *in vivo*. By using bioinformatics tools for BLAST search and sequence alignment, we identified the key catalytic residue of the KR domain. Heterologous expression of a variant of MSAS, Y1572F, in *S. cerevisaie*, enabled the synthesis of TAL at a maximal titer of 0.28 g/L in flask. TAL production was further optimized to 1.7 g/L by fed-batch fermentation. This study demonstrates the power of rational design in engineering of multifunctional proteins.

**Characterization of Alcohol Dehydrogenases in the Biosynthesis of Phosphonic Acid Antibiotics.** Phosphonates represent a potent, yet underexploited, group of bioactive compounds, mainly produced by *Streptomyces.* Despite the significant structural differences among many of phosphonic acids, their biosynthetic routes contain an unexpected common intermediate, 2-hydroxyethyl-phosphonate (HEP), which is synthesized from phosphonoacetaldeyhyde (PnAA) by a distinct family of iron-dependent alcohol dehydrogenases (ADHs). Although the sequence identity of the ADH family members is relatively low (34–37%), *in vitro* biochemical characterization of the homologs involved in biosynthesis of the antibiotics fosfomycin, bialaphos, and dehydrophos unequivocally confirms their enzymatic activities. These unique ADHs have exquisite substrate specificity, unusual metal requirements, and an unprecedented monomeric quaternary structure. Further, sequence analysis shows that these ADHs form a monophyletic group along with additional family members encoded by putative phosphonate biosynthetic gene clusters. Thus, the reduction of PnAA to HEP may represent a common step in the biosynthesis of many phosphonate natural products, a finding that lends insight into the evolution of phosphonate biosynthetic pathways and the chemical structures of new phosphonic acids containing secondary metabolites.

See more of this Session: Meet the Faculty Candidate Poster Session
See more of this Group/Topical: Education

# (3dw) Soft Matter, Interfaces and Complex Fluids: Optics, Dynamics, Elasticity and Self-Assembly (ODES)

**Vivek Sharma**, Mechanical Engineering, Massachusetts Institute of Technology, Cambridge, MA

Polymers, colloids, surfactants, liquid crystals and gels are all termed as complex fluids or soft materials, for they respond to stress like semi-solids or squishy materials. A combination of spring-like elasticity and ability to flow makes their behavior viscoelastic. Their ability to flow is itself distinguished by behavior unlike that of common liquids or gases: the complexity comes from their time dependent, or stress dependent behavior. A large fraction of the industrial and biological materials exist or are processed in form of multicomponent, microstructured fluids, and the ultimate function and use of these materials is dependent on their structure and properties. The relation of the optics, structure and dynamics of complex fluids to order and entropy, thermodynamics and kinetics of phase transitions, elasticity and self-assembly is of both scientific and technological interest.

My goal is to establish a research program to develop multicomponent, multifunctional materials incorporating colloids, polymers and liquid crystals for use in applications ranging from fluids for inkjet printing and spraying applications to the understanding of multicomponent biological fluids. By understanding and using design principles observed in nature (like in iridescent beetles and in butterflies), we will develop materials that combine structural integrity and robustness with the required optical functionality (light reflection, transmission) for harvesting solar energy. While the understanding of dynamics and principles of self-assembly is crucial for developing practically useful processing methods for materials used in photovoltaic applications, the knowledge of interaction of light with matter is essential not only for observing changes in microstructure during flow, but also for use of materials in optical or photonic devices.

See more of this Session: Meet the Faculty Candidate Poster Session
See more of this Group/Topical: Education

# (3dx) Techno-Economic Feasibility Analysis of Sustainable Bioenergy Feedstock Production Using Optimization and Simulation Models

**Yogendra Shastri**, Department of Agricultural and Biological Engineering, University of Illinois at Urbana-Champaign, Energy Biosciences Institute, Urbana, IL

Biomass feedstock production and provision are important components of the biomass based energy sector. Low energy density, seasonal availability and distributed supply create unique challenges that need to be addressed effectively. Multiple inter-dependent operations as well as novel technologies make the determination of the optimal system configuration non-trivial. It must also be ensured that the feedstock production system is sustainable and resilient. Systems theory provides an ideal platform to address these challenges by enabling a systematic analysis using a computational and informational science platform. This work conducts such an analysis of the feedstock production system using systems theory based tools such as optimization and simulation. An optimization based decision making framework named BioFeed has been developed that optimizes various feedstock production operations as a single system. A unique feature of this mixed integer linear programming (MILP) model is the integration of the design as well as operational decisions in a single framework. The model was successfully used to study the production of switchgrass and Miscanthus in Illinois (US). The results indicated that biomass pre-processing significantly affected the total production cost which varied between 45-65$/Mg. Sensitivity analysis quantified the systemic impact of technology improvements and thus identified important technologies within the system. Trade-offs between cost and energy

consumption were also highlighted. A novel computational scheme combining mathematical programming and
agent-based modeling has been developed to solve this highly complex model in a computationally efficient
manner. Results showed an order of magnitude reduction in simulation time. Moreover, the optimal solution was
within ±5\% of the rigorous MILP solution. The presentation will also discuss the development of an
agent-based model using the theory of complex adaptive systems to study the dynamics of the production system.
Results showed that up to 15 years might be required to reach steady regional productivity and the dynamics
are strongly dependent on the stakeholder participation and attributes.

See more of this Session: Meet the Faculty Candidate Poster Session
See more of this Group/Topical: Education

# (3dy) Molecular Modeling of Complex Biological Systems

**Diwakar Shukla**, Chemical Engineering, Massachusetts Institute of Technology, Cambridge, MA

Complex biological systems participate in processes which span a variety of length and time scales. These systems are being investigated by a spectrum of computational techniques ranging from vanilla molecular dynamics to multi-scale coarse graining and bridging between relevant scales. In this poster, I will discuss some recent computatational methodologies used to study long time scale phenomena and their applications to problems like protein folding, aggregation, allosteric transitions in proteins, amyloid fibril formation etc. More details can be found on my website: http://www.stanford.edu/~shukla

See more of this Session: Meet the Faculty Candidate Poster Session
See more of this Group/Topical: Education

# (3dz) Controlled Release Films and Functional Surfaces Targeting Infection, Inflammation, and Bleeding

**Anita Shukla** and Paula Hammond, Chemical Engineering, Massachusetts Institute of Technology, Cambridge, MA

Uncontrolled bleeding and infection are leading causes of patient morbidity and mortality following traumatic injury. Traditional pressure based methods of hemorrhage management are not suitable for incompressible wounds. Existing non-pressure based devices, however, are often ineffective in complex sites and are frequently accompanied by adverse side effects. Additionally, patients are typically administered broad-spectrum antibiotics to prevent and eliminate existing infection. The systemic overuse of antibiotics has led to a worldwide increase in drug-resistant bacteria. As an alternative to these conventional treatments, local therapeutic delivery has the potential to effectively treat cellular dysfunction while avoiding drug toxicity. We have developed degradable layer-by-layer (LbL) assembled films as local delivery coatings to address infection, inflammation, and bleeding. These films were engineered to deliver potent antibiotics such as vancomycin and exploratory drugs such as antimicrobial peptides, which prevent the development of drug resistant bacteria. Active films with large drug loadings and a range of drug release profiles were developed by taking advantage of film architectures, assembly techniques (spray versus dip LbL), and film component interactions. Due to the prevalence of infection and inflammation, degradable coatings for the concurrent release of antibiotics and anti-inflammatory therapeutics were also designed. These films have the potential to address a wide range of infection and inflammation requirements, from short term infection and inflammation eradication for trauma relief to infection prevention and long term inflammation mitigation from biomedical implants. All films were successfully applied to medically relevant substrates, including bandages and sutures, and were shown to be active *in vitro* against *Staphylococcus aureus* and cyclooxygenase. To address current complications with bleeding control, multilayer films were developed based on hydrogen bonding interactions found to occur between a polyphenol and an essential clotting factor, thrombin. These thin films were used to coat a common clinically applied absorbent and porous gelatin sponge without reducing its liquid absorption capabilities. Coated sponges were shown to be highly effective in promoting hemostasis in a porcine spleen injury model. These films have the potential to be applied to any clinical substrate. Additionally, drug loading and release can be tuned based on the desired application. Building upon this work, I will also discuss my future research plans regarding the application of self-assembly and patterned surfaces to develop new biomaterials for translational medicine.

See more of this Session: Meet the Faculty Candidate Poster Session
See more of this Group/Topical: Education

# (3e) Barriers, Hollow Fiber Membranes, and Hybrid Sorbents: A Path to Energy Efficient Technology

**Jong Suk Lee**, Koros Research Group, Georgia Institute of Technology, Atlanta, GA

The understanding of gas/vapor transport in polymeric membranes can be applied for many applications: barrier material development, membrane based gas separation, and hybrid sorbent development. In fact, all of these applications are related to the same fundamental principle, the sorption-diffusion mechanism. The benefits to our society of doing research in these areas are enormous in terms of the environment as well as energy. As for barrier material development, the barrier properties of poly(ethylene terephthalate), the current state of the art, can be significantly improved by the combination of antiplasticization and crystallization for the next generation packaging industry. Furthermore, the effect of flavor molecules on the barrier properties of PET can be evaluated by using a vapor/gas permeation system installed with a highly accurate syringe pump. Membranes have also been utilized for separations involving gaseous feed streams and their industrial growth will be driven by the application of membrane based separations to increasingly aggressive feed streams. The performance of defect-free Matrimid® asymmetric hollow fiber membranes in the presence of high pressure carbon dioxide and trace levels of toluene and n-heptane in the feed was studied. Lastly, a novel class of carbon capture system is being developed via hollow fiber-supported ionic liquid sorbents. Our emergent catch-and-release system has the potential to considerably cut the cost and energy associated with gas uptake and release. All the scientific advances mentioned above will lead our society to a technological leader in developing and deploying advanced energy technologies.

See more of this Session: Meet the Faculty Candidate Poster Session
See more of this Group/Topical: Education

# (3ea) Nanoscale Bio-Recognition: From Proteins Separation to Proteins Aggregation

**Mirco Sorci**, Chemical and Biological Engineering and Center for Biotechnology and Interdisciplinary Studies, Rensselaer Polytechnic Institute, Troy, NY

Bio-recognition, in particular protein-protein interaction, is essential at the nanoscale level in many biological processes (enzyme-subtract interaction, immune response, signaling pathways). Even though our knowledge of the molecular/structural basis of these interactions is limited, scientists have found a way to apply these interactions to many different applications: Affinity chromatography, bio-sensors, drug delivery, etc. My PhD work focused on affinity membranes for bioseparations. The research centered around the development (selection of ligands, modification of membranes) and testing (batch and dynamic adsorption experiments, modeling, estimation of binding capacity, dissociation and kinetic constants, effect of ionic strength and flux) of affinity membranes for the purification of (i) monoclonal antibodies, (ii) lectins and (iii) MBP-fusion proteins. More recently, during my post-doc, I have continued to work on several projects, focused on protein separation as well as protein-protein interactions. The poster will report some recent results on the following topics: Synthesis and characterization of customized membranes for the separation of proteins with similar molecular weight and charge; biophysical characterization of fibrillar precursors, also called oligomers and protofibrils, during amyloidosis process; direct interaction between three different pairs of split-inteins. I have broadened my background as a chemical engineer, integrating many bio-tools. I plan to continue my career keeping these two entities together and studying protein-protein recognition both as basic science (e.g. better understand the mechanism of human diseases) and applied science (e.g. improve bioseparation).

See more of this Session: Meet the Faculty Candidate Poster Session
See more of this Group/Topical: Education

# (3eb) Engineering Nanosilver As Antibacterial, Biosensor and Bioimaging Material

**Georgios A. Sotiriou**, Department of Mechanical and Process Engineering, Particle Technology Laboratory, ETH Zurich, Zurich, Switzerland

This research involves the synthesis of nanostructured materials and the investigation of their properties targeting applications within the bio-medical field. Main focus of the research is to understand the fundamental principles and mechanisms of certain phenomena and to use this knowledge to design and synthesize nanostructured materials that possess the desired properties for their target application and at the same time overcome their shortcomings (e.g. toxicity or flocculation tendency). This would result in sustainable engineered nanomaterials in a rather "eco-friendly" way that offers its sought-out performance and minimizes risks (e.g. toxicity) to human health and environment [1]. Nanosilver particles are attractive for their use in bio-applications because of their plasmonic properties. For example, plasmonic Ag nanoparticles can be used in bio-imaging and/or *in-vivo* therapy (e.g. photothermal treatment of tumors). Even though silver has the lowest plasmonic properties in the UV-visible wavelength range, the more expensive gold is preferred for such bio-applications because of its lower toxicity. Nanosilver exhibits strong toxicity against biological systems because of its direct contact with the biological system and the released toxic $Ag^+$ ions. However it was unclear whether the $Ag^+$ ions or the direct contact with the particle surface plays the dominant role for this toxicity.

This motivated the initial target of the present research. By synthesizing nanocomposite $Ag/SiO_2$ nanoparticles, with closely controlled nanosilver size and content in the $Ag/SiO_2$ particles, it was possible to prevent nanosilver agglomeration or flocculation when dispersed in aqueous suspensions. This, in addition to the identification of the released $Ag^+$ ion concentration by electrochemical measurements, led to decipher the role of the released $Ag^+$ ions in nanosilver toxicity. It was discovered, for the first time, that when small (< 10 nm) nanosilver particles are employed, the $Ag^+$ ion release is significant and thus $Ag^+$ ions dominate the toxicity. In contrast, when nanosilver particles are rather large (>10 nm) the effect of $Ag^+$ ions and the particles is comparable [2]. Furthermore, the origin of the released $Ag^+$ ions was tracked to the leaching of the nanosilver surface layers [3].

The next step capitalized on this understanding. More specifically, synthesis of rather large nanosilver particles (> 30 nm) in order to avoid significant $Ag^+$ ion release was performed. This time, however, these particles were hermetically-coated *in-situ* during their flame-synthesis by a nanothin silica film. The toxicity of these $SiO_2$-coated nanosilver particles was eliminated. This enabled creation of non-toxic plasmonic particles, ready to be employed as agents for bio-imaging. The nanothin $SiO_2$ shell also prevented nanosilver flocculation while it facilitated the bio-functionalization of such nanoparticles with a protein (bovine-serum-albumin). This was the first study to report the utilization of plasmonic nanosilver that is non-toxic as a potential biosensor [4].

When plasmonic nanoparticles are combined with another material, for example a magnetic component, multifunctional nanostructured materials are made. Such nanomaterials can combine more than one functionality, and thus, can be detected by multiple imaging techniques. Magnetic resonance imaging (MRI) is an example of a traditional technique, with which small magnetic particles can be used as contrast agents and for targeted drug delivery, by directing them to organs, tissues or tumors using an external magnetic field or for magnetically assisted cell sorting and separation. Furthermore, a $SiO_2$ coating on the surface of such magnetic nanoparticles facilitates their surface bio-functionalization and prevents their magnetic particle-particle interaction and flocculation or agglomeration. This coating prevents the release of toxic $Ag^+$ ions and minimizes their toxicity, as it was tested against HeLa cells. In addition, their feasibility as superior biomarkers is explored by specifically binding them to live raji and HeLa cells and their detection was possible [5].

Further research involves the investigation of alternative materials suitable for bio-imaging and other bio- applications (e.g. nanophosphors [6]) and the nanotoxicological characterization of engineered nanomaterials [7].

[1] Sotiriou, G. A. & Pratsinis, S. E. Engineering nanosilver as antibacterial, biosensor and bioimaging material. *Curr. Opin. Chem. Eng.* **in press**, doi: 10.1016/j.coche.2011.1007.1001 (2010).

[2] Sotiriou, G. A. & Pratsinis, S. E. Antibacterial activity of nanosilver ions and particles. *Environ. Sci. Technol.* **44**, 5649-5654 (2010).

[3] Sotiriou, G. A., Teleki, A., Camenzind, A., Krumeich, F., Meyer, A., Panke, S.

& Pratsinis, S. E. Nanosilver on nanostructured silica: Antibacterial activity and Ag surface area. *Chem. Eng. J.* **170**, 547-554 (2011).

[4] Sotiriou, G. A., Sannomiya, T., Teleki, A., Krumeich, F., Vörös, J. & Pratsinis, S. E. Non-Toxic Dry-Coated Nanosilver for Plasmonic Biosensors. *Adv. Funct. Mater.* **20**, 4250-4257 (2010).

[5] Sotiriou, G. A., Hirt, A. M., Lozach, P. Y., Teleki, A., Krumeich, F. & Pratsinis, S. E. Hybrid, Silica-Coated, Janus-like Plasmonic-Magnetic Nanoparticles. *Chem. Mater.* **23**, 1985-1992 (2011).

[6] Sotiriou, G. A., Schneider, M. & Pratsinis, S. E. Color-Tunable Nanophosphors by Codoping Flame-Made $Y_2O_3$ with Tb and Eu. *J. Phys. Chem. C* **115**, 1084-1089 (2011).

[7] Sotiriou, G. A., Diaz, E., Long, M. S., Godleski, J., Brain, J., Pratsinis, S. E. & Demokritou, P. A novel platform for pulmonary and cardiovascular toxicological characterization of inhaled engineered nanomaterials. *Nanotoxicology* **in press**, doi:10.3109/17435390.2011.604439 (2011).

See more of this Session: Meet the Faculty Candidate Poster Session
See more of this Group/Topical: Education

# (3ec) Reaction/Metabolite Standardization and Congruency Across Databases and Genome-Scale Metabolic Models

**Patrick Suthers**, Chemical Engineering, The Pennsylvania State University, University Park, PA

The ever-accelerating pace of DNA sequencing and annotation information generation is spearheading the global inventorying of metabolic functions across all kingdoms of life. Increasingly, metabolite and reaction information is organized in the form of community, organism, or even tissue-specific genome-scale metabolic reconstructions. These reconstructions account for reaction stoichiometry and directionality, gene to protein to reaction associations, organelle reaction localization, transporter information, transcriptional regulation and biomass composition. Already over 40 genome-scale models are available for eukaryotic, prokaryotic and archaeal species and are becoming indispensable for computationally driving engineering interventions in microbial strains for targeted overproductions, elucidating the organizing principles of metabolism and even pinpointing drug targets. A key barrier to the pace of extraction of metabolic knowledge from data is our inability to directly make use of metabolite/reaction information from databases (e.g., BRENDA, KEGG, BioCyc, UM-BBD, PubChem, ChEBI, Reactome.org, Rhea, etc.) or other metabolic models due to incompatibilities of representation, duplications and errors. Therefore, the inadvertent inclusion of multiple replicates of the same metabolite, stoichiometrically inconsistent and/or elementally/charge unbalanced reactions can lead to erroneous model predictions and missed opportunities to reveal (synthetic) lethal gene deletions, repair network gaps and quantify metabolic flows. There have already been a number of efforts aimed at addressing some of these limitations such as the Rhea database and Model SEED. Motivated by this challenge we recently developed the MetRxn knowledgebase that integrates, using internally consistent descriptions, metabolite and reaction information from 6 databases and 34 metabolic models Metabolite and reaction data was first downloaded from BRENDA, KEGG and BioCyc using a variety of methods based on protocols such as SOAP, FTP and HTTP. We subsequently pre-processed the data into flat files that were imported into MetRxn. All original information pertaining to metabolite name, abbreviations, metabolite geometry, related reactions, catalyzing enzyme and organism name, gene-protein-reaction associations, and compartmentalization was retained. For all 34 genome-scale models ancillary information culled form the corresponding publications was also imported. The "raw data" from both databases and models was unified using standard SQL scripts on a MySQL server. We used Marvin (Chemaxon) to analyze all 231,085 raw metabolite entries containing structural information (out of a total of 322,936 entries). Metabolite atom bond connectivity was calculated at a fixed pH of 7.2 and converted into standard Isomeric SMILES format. Metabolites were also annotated with Canonical SMILES using the OpenBabel Interface from ChemSpider. Metabolites with missing structural information were re-visited during the reaction reconciliation step. After generating the initial metabolite associations, we identified reaction overlaps using the reaction synonyms and reaction strings along with the metabolite SMILES representations. During this step, reactions were flagged as single-compartment or two-compartment (i.e., transport reactions). Using the corrected metabolite elemental composition and protonation states, reactions are evaluated for charge and elementally balance. We used a linear optimization program to charge and elementally balance all reactions. MetRxn contains over 62,000 unique metabolites and 56,000 unique reactions. In this poster, I present the web-based access to MetRxn, describing updates in its increased scope of standardized metabolic models. Specifically, the impact standardization has on genome-scale models and the resulting flux balance analyses of the models is explored. I also examine some interesting observations with model content overlaps and contrast these to phylogenetic trees, and finally discuss SBML export and a web-based API for integration with external applications.

See more of this Session: Meet the Faculty Candidate Poster Session
See more of this Group/Topical: Education

# (3ed) Micro and Nanosystems for Biotechnology and Health Applications

**Melikhan Tanyeri** and Charles M. Schroeder, Chemical and Biomolecular Engineering, University of Illinois at Urbana-Champaign, Urbana, IL

Micro and nanotechnology hasradically transformed the fundamental studies in many fields of biology andmedicine by catalyzing a paradigm shift in methods toolbox available toresearchers. However, its full potential has yet to be explored and harnessed forapplications in translational medicine, in specific, diagnostics andtherapeutics. To this end, I plan to combine ultra-sensitive optical imagingand detection techniques with microfluidic methods to engineer and implementenabling technologies towards addressing key questions and challenges inclinical diagnostics, systems biology, developmental biology and translationalmedicine. I am specifically interested in developing integrated micro and nanosystems for i) Molecular diagnostics ii) Micro and nanomanipulation of particles, cells and embryos for highthroughput screening iii) Studying microbial communities related to infectiousdisease and immunology. In this presentation, I will summarize my previousresearch efforts on developing microsystems with applications to biosensors andmicromanipulation of micro and nanoparticles; and briefly provide an overviewof my proposed research program as a tenure-track faculty member.

Over the past two decades,development of micro and nanomanipulation methodsbased on optical, electrokinetic, magnetic, acousticfields has revolutionized many fields of science and engineering by enablinginvestigations of a wide array of physical and biological problems includingbiophysical studies of individual molecules, particles and living cells.Recently, we developed an alternative flow-based method to confine, manipulateor isolate single particles and cells in free-solution [1, 4-5]. Based on astagnation point flow generated in a microfluidic device, this novel methodallows for trapping, 2-D manipulation, stretching and sorting of objectsranging from single molecules to individual cells in free-solution.Importantly, hydrodynamic trapping is feasible for any particle with nospecific requirements on the material composition or the chemical/ physicalnature (optical, magnetic, surface charge) of the trapped object. In addition,hydrodynamic trap enables confinement of nanoscale particles because the trappingforce scales linearly with particle radius, whereas the trapping force foroptical or magnetic traps scales with particle volume. Hydrodynamic trappinginherently enables confinement of a single target nanoparticle in aconcentrated sample suspension, provides the ability to change the surroundingmedium of a trapped nanoparticle in real time and enables fine-scale particlemanipulation in free solution for extended time scales. Overall, thehydrodynamic trap offers a new platform for observation of nanoparticleswithout surface immobilization. We anticipate that this microfluidic-basedtechnique will enable new scientific studies in the fields of cellularmechanics, colloidal science and fluid dynamics.

Integrated micro and nanosystems offerseveral key advantages for diagnostic and therapeutic applications, includingreduced amounts of sample and reagents for analysis, and integration of assaysteps such as sample preparation (processing of physiological fluids),post-processing (e.g. reagent mixing, labeling, separation), detection andanalysis into a single device. Specifically, rapid multiplexed assays based onintegrated microdevices could significantly improve the quality of healthcareby reducing turnaround time and analysis cost, and speeding up the decisionmaking process for diagnosis and treatment. Towards this goal, we developed acombinatorial screening chip for high sensitivity, sequence-specific screeningand detection of viral nucleic acid markers (ssDNA) encoding for conservedregions of the genomes of four common viruses [2]. The combinatorial chiprepresents a versatile platform for the development of clinical diagnostictests for simultaneous screening, detection and monitoring of a wide range ofbiological markers of health and disease using minimal sample size. Inaddition, during my graduate studies, I developed an ultrasensitive label-freesensing scheme based on optical resonances in microdroplets [7] which wasemployed in a number of chemical and biological sensing applications includingsingle bacterial cell detection [6].

Postdoctoral Advisors:

Prof. Charles M. Schroeder, Chemical and Biomolecular Engineering, Universityof Illinois at Urbana-Champaign
Prof. Paul R. Selvin, Physics, University of Illinois at Urbana-Champaign

PhD Advisor:

Prof. Ian M. Kennedy, Mechanical and Aeronautical Engineering, University ofCalifornia, Davis

Publications:

1. **M. Tanyeri**, M. Ranka, N. Sittipolkul, C. M.Schroeder, "A Microfluidic-based

Hydrodynamic Trap: Design andImplementation", **Lab Chip**, 11 (10), 1786-1794 (2011).

2. B. R. Schudel, **M. Tanyeri**, A. Mukherjee, C. M. Schroeder,P. J. A. Kenis, "Multiplexed Detection of Nucleic Acids in a CombinatorialScreening Chip", (in press, **Lab on a Chip**).

3. Y. Kim, S. Kim, **M. Tanyeri**, J. A. Katzenellenbogen,C. M. Schroeder, "Dye-conjugated Dendrimers as Bright and PhotostableMolecular Tags for Fluorescence Microscopy and Imaging", (submitted, **Journalof the American Chemical Society**).

4. E. M. Johnson-Chavarria, **M. Tanyeri**, C. M. Schroeder, AMicrofluidic-based Hydrodynamic Trap for Single Particles(http://www.jove.com/details.stp?id=2517), **Journal of Visualized Experiments**,47 (2011).

5. **M. Tanyeri**, E. M. Johnson-Chavarria, C. M. Schroeder,"Hydrodynamic Trap for Single Particles and Cells", **AppliedPhysics Letters**, 96, 224101 (2010).

6. **M. Tanyeri**, I. M. Kennedy, Detecting SingleBacterial Cells through Optical Resonances in Microdroplets (cover article), **SensorLetters**, 6 (2), 326-329 (2008).

7. **M. Tanyeri**, R. Perron, I.M. Kennedy, Lasing Droplets in a Microfabricated Channel, **Optics Letters**,32 (17), 2529-2531 (2007) (Highlighted in photonics. com and Photonics Spectra,October 2007).

See more of this Session: Meet the Faculty Candidate Poster Session
See more of this Group/Topical: Education

# (3ee) Externally-Triggered Nanoparticle Composite Membranes for Drug Delivery

**Brian P. Timko**, Chemical Engineering, MIT / Children's Hospital Boston, Cambridge, MA

Sustained drug release technology has been applied to many medical fields. However, drug release kinetics from devices is usually monotonic over time. To address this limitation, we synthesized externally-triggerable membranes based on the thermoreversible polymer poly(N-isopropylacrylamide)(PNIPAm) and superparamagnetic iron oxide nanoparticles. Drug flux across the membrane can be turned "on" by applying an external magnetic field, with continuous release rate and good reproducibility over multiple cycles. We examine several membrane design parameters, such as thickness and PNIPAm nanoparticle concentration, and evaluate how they affect drug release characteristics.

See more of this Session: Meet the Faculty Candidate Poster Session
See more of this Group/Topical: Education

# (3ef) Visualizing Drugs In Vitro and In Vivo - Application of Novel Chemistry for Drug Development

**Greg M. Thurber**, Harvard Medical School / Mass General Hospital, Boston, MA

Candidate Background

Post-Doctoral Training, Harvard Medical School

PhD in Chemical Engineering, Massachusetts Institute of Technology

B.S. in Chemical Engineering, Pennsylvania State University

Poor drug distribution in tissues has been implicated in treatment resistance for diseases such as cancer, where a large fraction of tumor cells are exposed to sub-therapeutic or even zero drug concentrations in vivo. Due to the inherent difficulties in measuring drug concentrations in biological systems, it has been extremely difficult to determine the impact of this poor distribution on the treatment of cancer. The inability to track small molecule drugs is a major reason why predictions of their distribution have lagged far behind macromolecules and nanoparticles. In this presentation, I will elucidate our current efforts on visualizing drug distribution in live mice and three-dimensional reconstruction of ex vivo tumors providing a more realistic setting for studying cancer treatment. Novel bio-orthogonal chemistry and small animal imaging is used to visualize the distribution of drugs on the microscopic scale. Fundamental chemical engineering transport principles are used to describe the behavior of the drugs in biological systems. These results provide the foundation for developing novel drugs and combination therapies with better tumoral distribution and efficacy.

See more of this Session: Meet the Faculty Candidate Poster Session
See more of this Group/Topical: Education

# (3eg) Combinatorial Design and Evaluation of Multi-Dimensional, Immunomodulatory Scaffolds for Tissue Engineering Applications

**Bret D. Ulery**, Orthopaedic Surgery, University of Connecticut Health Center, Farmington, CT

As the field of tissue engineering continues to move forward and tackle greater challenges like organ regeneration, the need for multi-dimensional, composite degradable polymeric biomaterials that can facilitate complex and integrated tissue development becomes ever more necessary. Current tissue engineering solutions comprised of one material with basic structure releasing a single bioactive factor cultured with one cell type will more than likely be suboptimal for high-order systems. Instead it is necessary to create composite scaffolds with complex micro- and nano-architecture releasing a number of bioactive factors to a mixture of incubated cell types to create artificial organs. But with a significant increase in construct variables comes a nearly untestable number of experimental groups to evaluate by conventional methods. In order to adequately research these new materials the development of combinatorial techniques must be employed. This approach requires that material synthesis, complex construct fabrication and biological evaluation all be conducted in a parallel, combinatorial fashion to expedite research progress. While promising, complex constructs demonstrating remarkable *in vitro* behavior have been reported in the literature, but often these materials fail *in vivo* due to host immune responses. Although some research has focused on creating tissue engineering constructs that evade the immune system, the design of materials that can modulate the immune response in order to facilitate tissue regeneration is an overwhelmingly under-investigated field. By creating tissue engineering solutions that possess immunomodulatory and immunoattractant elements, new constructs can be designed that will yield much more positive *in vivo* outcomes. With graduate research experience in immunology, controlled delivery, polymer synthesis and biomaterials development and postdoctoral training in tissue engineering, cell culture techniques and stem cell biology, I am uniquely qualified to undertake this research. Building on my previous accomplishments, I look forward to conducting novel research in these emerging areas as an independent investigator.

# (3eh) Biopreservation: From Single Cells to Organs

**O. Berk Usta**, Surgery, Massachusetts General Hospital / Harvard Medical School, Boston, MA

I have a background on theoretical and computational modeling of complex and multiphase flows, and elastic solids using mesoscopic approaches such as lattice-Boltzmann and lattice-spring methodologies. I have worked with the pioneers in this field and have investigated questions of both fundamental and biological relevance ranging from colloidal self diffusion, polymer migration in microchannels, separation of microcapsules, viscoelastic droplet deformation to self propulsion and communication of microcapsules. I have recently added an experimental component to my research toolbox working at the Center for Engineering in Medicine with Dr. Martin Yarmush and Dr. Mehmet Toner. We are currently investigating several questions in the broad field of biopreservation, highthroughput single cell analysis, microfluidic separations and drug delivery in efforts to solve problems related to liver and burn injuries.

During this session I will primarily discuss biopreservation schemes such as cryopreservation and supercooling to preserve liver, its cells and tissues where our groups has made great strides recently. I will also introduce the use of microfabricated microwell arrays in the quest to understand the effect of cell heterogeneity on successful preservation; this platform combined with powerful statistical analysis is a potent tool for describing many hidden relations when one looks only at average population responses.

See more of this Session: Meet the Faculty Candidate Poster Session
See more of this Group/Topical: Education

# (3ei) Biomolecular Simulations with Applications In Diabetes, Cancer, and Protein Synthesis

**Harish Vashisth**, Chemistry and Biophysics, University of Michigan, Ann Arbor, MI

Atomic-level understanding of the biomolecular function is an important requirement for progress in development of drugs against a variety of devastating diseases, such as cancer and diabetes. Although many biophysical experimental approaches can provide data on the structure and limited dynamics of biomolecules, conducting an atomistic simulation is an increasingly important route to explore their vast conformational repertoire, which can be exploited for various pharmaceutical applications. Using theory, modeling, and atomistic simulation approaches rooted in statistical mechanics and thermodynamics, I have developed a comprehensive mechanistic understanding of various biomolecules. The first question I address is, how does insulin bind to its cognate receptor, and how the binding of insulin-like growth factors is similar to that of insulin? Next, I address the question of how the receptors for insulin and insulin-like growth factors activated? From my postdoctoral work, I address how a class of sugar/vitamin transporters allow unidirectional transport of substrates against a substantial concentration gradient? Lastly, I will discuss how ribosomes, the protein producing machinery of all cells, assemble into a productive translational apparatus. Understanding the mechanism of protein-synthesis by ribosomes is key to the development of future antibacterial drugs

See more of this Session: Meet the Faculty Candidate Poster Session
See more of this Group/Topical: Education

# (3ek) Statistical Process Inference, Control and Engineering

**Kris Villez**, School of Chemical Engineering, Purdue University, West Lafayette, IN

The field of systems engineering incorporates a broad range of techniques for advanced process inference, control and design. These techniques are commonly deployed to obtain better understanding, tracking and/or manipulation of complex processes in view of economical and energy efficiency, improved safety and reduction of environmental impacts. Historically, first-principle models have been most popular for the tasks of process state estimation, inference and control. Data-driven, statistical techniques have been developed more recently and require less fundamental process knowledge. As such, they have become increasingly popular since the nineties for ill-understood processes. Today, a tremendous amount of techniques is available within the Statistical Process Control (SPC) field.

Despite the boom in SPC research, adoption of these techniques in the industries is limited. This can at least in part be accounted to the lack of oversight in the field. Often, different techniques are aimed to tackle the same problem yet without elaborate comparison or real-life testing. In other cases, the absence of guidelines for practical implementation is a significant barrier for technology transfer. Therefore, I expect to devote a large part of my research career to bringing structure and guidelines to the SPC field.

Even if the amount of SPC techniques is vast, Fault Detection and Identification is largely based on the on-line analysis of data streams without direct inference with the process. As such, they can be labelled as Passive FDI techniques. However, it is technically possible to manipulate process variables in such a way that the produced data stream is more informative about the status of the monitored process. By doing so, the outcomes of on-line inference procedures can be made more accurate. To search for optimal manipulations of process with respect to on-line process inference accuracy, Optimal Experimental Design (OED) techniques will be adopted, refined and integrated into FDI theory and applications. As the active manipulation of a process for improved FDI is a new concept, I expect to take a lead in the emerging field of Active Fault Detection and Identification (Active FDI).

From a theoretical angle, it can also be said that there is no consensus on the theory and application of non-linear data-driven techniques for process monitoring and diagnosis. At this point, the leading research groups in this niche each have their own technique and a status quo is observed on the theoretical level. A better analysis of existing techniques and further theoretical development is essential to make further advances.

Based on the above analysis, a research plan is set out along three tracks. Each of these tracks relates to the field of applied statistics in process systems modelling, inference and control. The three tracks are listed as follows, from the most theoretical track to the most application-oriented track:

Track I will result in the establishment of proper techniques for non-linear process monitoring. It will result in the specification of new model structures and algorithms.

Track II will deliver techniques for improved on-line fault detection and identification (FDI) by means of the development and application of Active Fault Detection and Isolation methods. This will require the integration of Optimal Experimental Design (OED) techniques within FDI theory as well as specific adjustments to enable real-time computation. Extensive testing in both simulation and laboratory will be important.

Track III is oriented at bridging the gap between theory and practice in Statistical Process Control (SPC). This track relies on comparative studies, validation and integration into real-life systems. Industrial partnerships are considered important to ensure impact on the field.

Together, the three tracks provide a balanced plan for research in systems engineering. The first track is largely theoretical. The second and third are more practice-oriented with the third focusing the removal of the gap between statistical theory and industrial process systems engineering.

See more of this Session: Meet the Faculty Candidate Poster Session
See more of this Group/Topical: Education

# (3el) Improving Our Understanding of Atmospheric Particulate Matter Concentrations and Feedbacks

**Kristina M. Wagstrom**, Department of Civil Engineering, University of Minnesota - Twin Cities, Minneapolis, MN

I am currently a postdoctoral associate in the Department of Civil Engineering at the University of Minnesota – Twin Cities working with Dr. Julian Marshall. I am part of a team working on studying the potential air quality impacts associated with using biofuels in place of gasoline to meet future transportation energy demands. We are using an integrated life cycle analysis, regional chemical transport modeling and societal cost estimation approaches. I completed my Ph.D. in Chemical Engineering in May 2009 at Carnegie Mellon University while working with Dr. Spyros Pandis. My doctoral research focused on the development, evaluation and application of an efficient source apportionment algorithm which I used to study different aspects of the atmospheric aerosol population.

My poster will focus on my future interests continuing research in air quality modeling both through model development and also new extensions and applications of these models. Some of the areas that I most hope to expand my research into is the study of health impacts, environmental justice, air quality in the developing world and the interactions between air quality and climate change. For many years, I have viewed chemical transport models as a natural, but underutilized, tool for studying these issues.

My interest in pursuing a position within academia focuses not only on doing research but also the opportunity to teach both in the classroom and through time spent advising both graduate and undergraduate students doing research. I view my time spent working with students doing research as teaching them to be strong researchers.

See more of this Session: Meet the Faculty Candidate Poster Session
See more of this Group/Topical: Education

# (3em) Dynamics and Aggregation of Proteins At Fluid Interfaces

**Robert Walder** and Daniel K. Schwartz, Chemical and Biological Engineering, University of Colorado at Boulder, Boulder, CO

The dynamic behavior of proteins at liquid-liquid interfaces is of fundamental interest to many applications, including complex fluids formulation, pharmaceuticals, and biosensing. However, the methods for understanding proteins at fluid interfaces are typically limited to a small number of traditional methods, such as interfacial rheology. While these methods are useful in characterizing bulk materials properties of protein films, they are of limited utility in understanding the mechanisms that lead to the formation of protein films. In this poster, the use of single molecule TIRF has been used to characterize the dynamic behavior of proteins at the oil-water interface. For isolated single proteins, we find that individual proteins are reversibly bound to the surface. In fact, diffusion and desorption dynamics are characteristic of the aggregate state of an individual protein at the oil-water interface. For protein layer formation to occur at the oil-water interface, we add many orders of magnitude of non-fluorescent protein, so that the effect of protein-protein interactions can be characterized. By studying the dynamics of tracer fluorescent proteins, protein aggregation is found to be the primary mechanism of protein layer formation. Furthermore, the kinetics of protein layer fomation depend on bulk protein concentration and use a kinetic population balance model to explain the non-linear concentration dependence.

See more of this Session: Meet the Faculty Candidate Poster Session
See more of this Group/Topical: Education

Massachusetts Institute of Technology

HARVARD MEDICAL SCHOOL

# Use Intestinal Stem Cells to Treat Colorectal Cancer through Regenerative Medicine

Qun Wang, Jeffrey Karp, Robert Langer

Massachusetts Institute of Technology

HARVARD MEDICAL SCHOOL

## Ulcerative Colitis (UC)

- UC is a form of inflammatory bowel diseases (IBD), it mainly happens in intestine, specifically the large intestine or colon.
- UC will cause confluent superficial ulceration and loss of mucosal epithelium barrier of intestine.
- Over 1,000,000 patients in the Unite States
- About 5 % of patients with ulcerative colitis develop colon cancer.

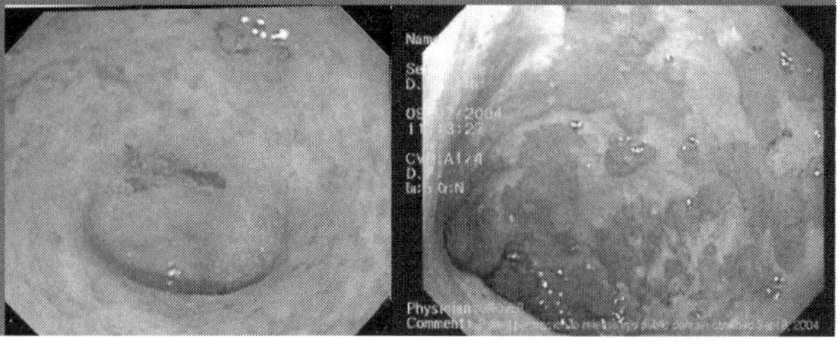

Image showed a bowel section afflicted with ulcerative colitis

Massachusetts Institute of Technology

HARVARD MEDICAL SCHOOL

## Current treatments and their disadvantages

- **Drug Therapy (aminosalicylates, corticosteroids, immuno-modulators, infliximab)**
  - --Weight gain, bone mass loss, diabetes…
  - --Not recommended for long-term use
- **Surgery (total colectomy + Ileostomy or Ileoanal anastomosis)**
  - --peristomal skin retraction, prolapsed rectum.
  - --leakage of stool, stenosis, pouchitis, and etc

Massachusetts Institute of Technology

HARVARD MEDICAL SCHOOL

## Novel approach to the problem: regenerative medicine through delivery of Intestinal Stem Cells

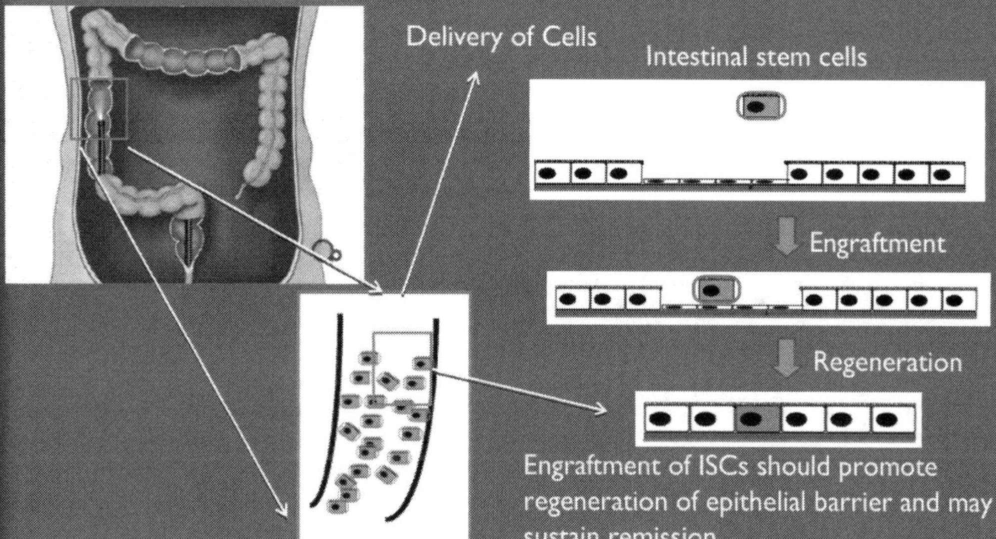


Engraftment of ISCs should promote regeneration of epithelial barrier and may sustain remission

Massachusetts Institute of Technology

HARVARD MEDICAL SCHOOL

## Single intestinal stem cell can build crypt-villus structures of intestine

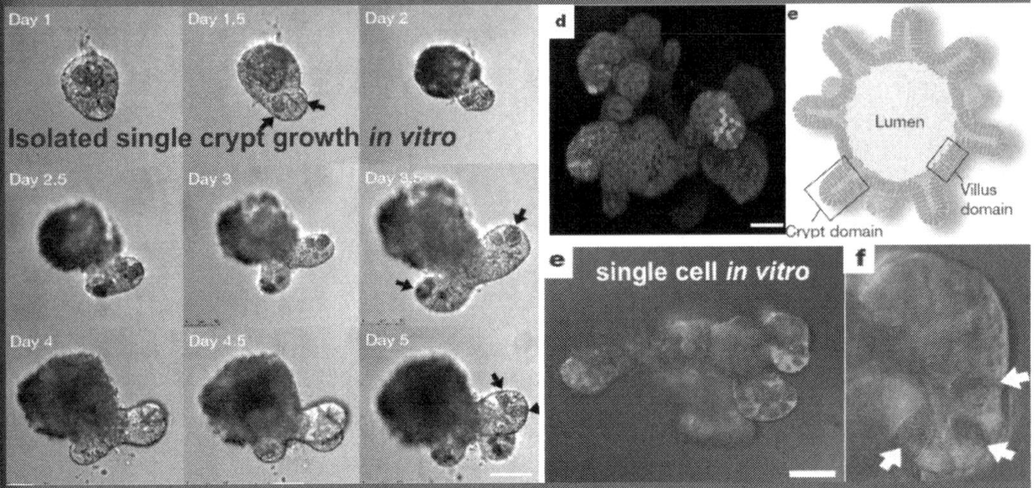


Delivery of intestinal stem cells or organoids could be an novel approach to treat the diseases that affect the intestine.

*Toshiro Sato, et al. Nature, 459, 262*

Massachusetts Institute of Technology

HARVARD MEDICAL SCHOOL

## Isolation and expansion of intestinal stem cells in vitro

Currently, Matrigel was used to culture intestinal stem cells in vitro, however there are several limitations:

1. Undefined compositions
2. Batch-to-batch variations
3. Clinically, can't be used in vivo

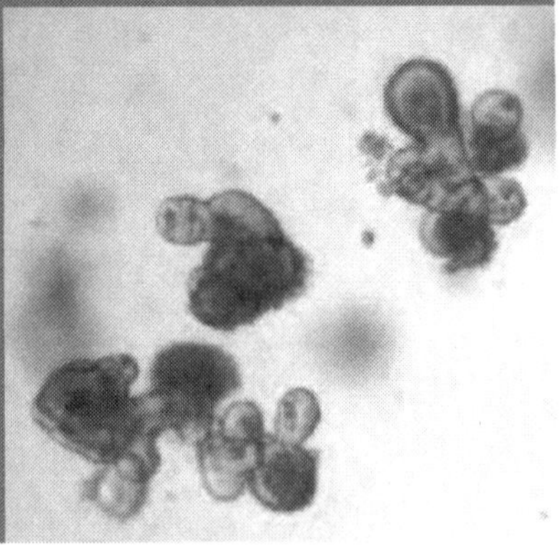

Massachusetts Institute of Technology

HARVARD MEDICAL SCHOOL

# Develop alternative to Matrigel for expanding and implanting intestinal stem cells (ISCs)

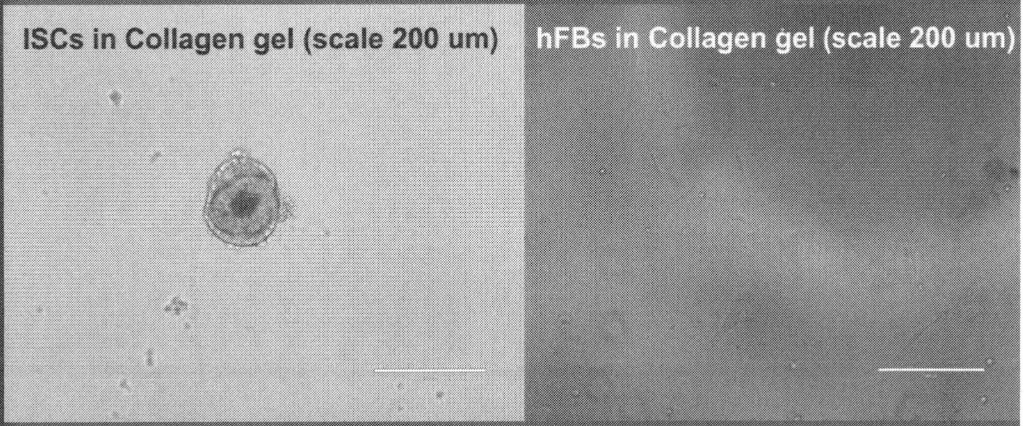


- We screen the suitable gel for ISCs from a materials library
- Human fibroblast has been used as control

Massachusetts Institute of Technology

HARVARD MEDICAL SCHOOL

# Develop vehicle for delivery of intestinal stem cells (ISCs)

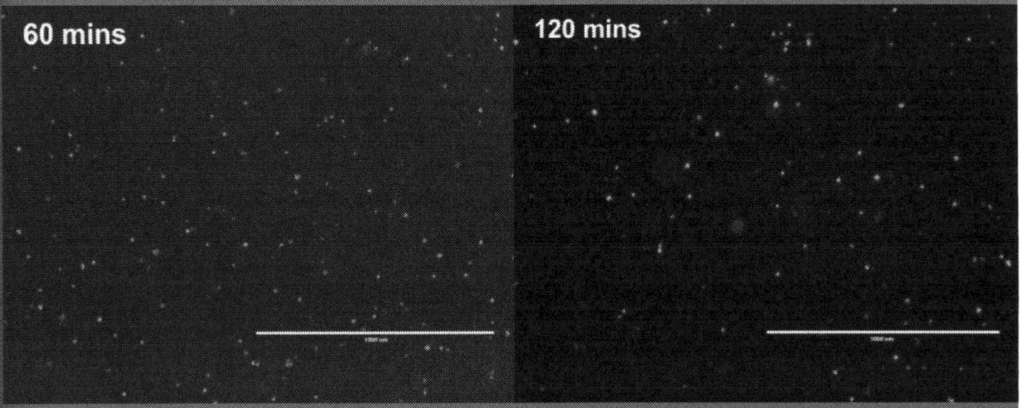


- A new fluid vehicle has been developed to deliver ISCs
- In an in vitro model, most ISCs lives after 2 hours of delivery

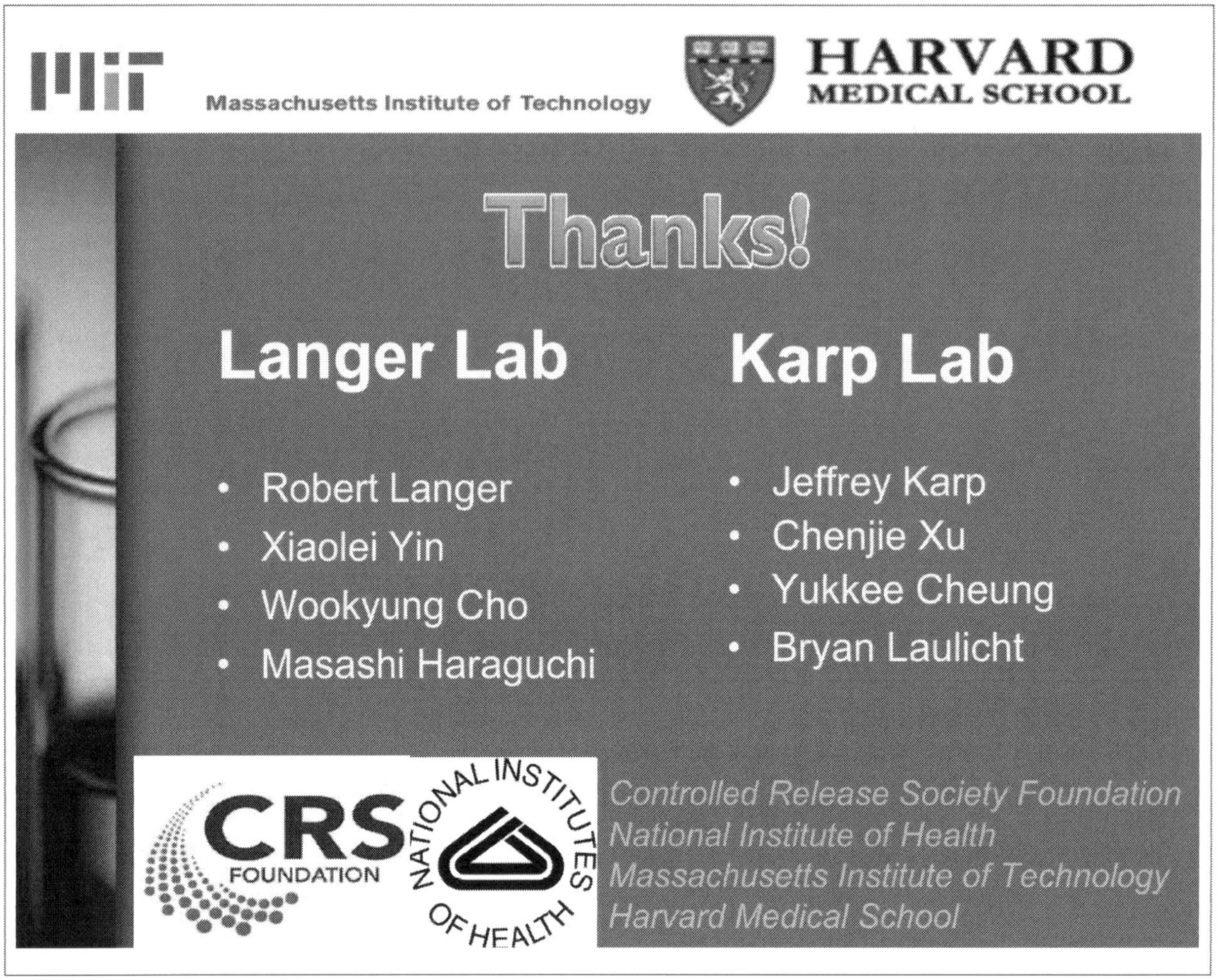
Massachusetts Institute of Technology
HARVARD
MEDICAL SCHOOL
Thanks!
Langer Lab
Robert Langer
Xiaolei Yin
Wookyung Cho
Masashi Haraguchi
Karp Lab
Jeffrey Karp
Chenjie Xu
Yukkee Cheung
Bryan Laulicht
CRS
FOUNDATION
NATIONAL INSTITUTES OF HEALTH
Controlled Release Society Foundation
National Institute of Health
Massachusetts Institute of Technology
Harvard Medical School

# (3eo) Aerosol Processing of Advanced Functional Materials for Energy and Environmental Applications

**Wei-Ning Wang**, Department of Energy, Environmental and Chemical Engineering, Washington University in St. Louis, St. Louis, MO

There are increasing concerns on global energy and environmental issues, such as fossil fuel shortage and global climate change. It is recognized without a doubt that research and innovation in the development of sustainable energy-related materials by environmental-friendly technologies will play a critical role in meeting these challenges. I have a multidisciplinary educational background in chemistry, chemical engineering, and materials science, and professional research expertise in aerosol science and engineering as well as nanoscience and nanotechnology, which are indispensable in achieving the above stated goal. My research interests are focused on aerosol processing of advanced functional materials, starting from gas, liquid, and solid precursors, and their applications to energy and environmental technologies, ranging from lab synthesis to industrial scaled production and from kinetics modeling to formation mechanism studies. Specific examples are nanostructured photocatalyst development for energy conversion from carbon dioxide and biomass, thin film deposition for solar and sensor applications, and functional absorbent material development for $CO_2/NO_x/SO_x$ capture in clean energy-related research. In my future research, I will continue working on developing environmentally benign aerosol processes for sustainable and renewable energy applications. I will also try to advance the technologies I develop in the lab to larger scale applications and build partnerships with industry through technology transfer. I will describe my research plans and give several specific examples during the poster session.

See more of this Session: Meet the Faculty Candidate Poster Session
See more of this Group/Topical: Education

# (3ep) Computational Screening of Metal-Organic Frameworks for Flue Gas Separation

**Taku Watanabe**, School of Chemical and Biomolecular Engineering, Georgia Institute of Technology, Atlanta, GA

Metal-organic frameworks (MOFs) have been extensively studied over the last decade especially for gas adsorption and separation applications. Due to the intense research effort, there are more than 5000 MOFs known today and material selection has become a daunting task. We have applied multi-level computational screening process to identify promising candidate MOFs to be used in a mixed-matrix-membrane. Of all the structures extracted from the database, we examined over 300 structures with detailed molecular simulations. We found some MOFs with exceptionally high diffusion selectivities (>100). Adsorption isotherms were also computed to obtain the sorption selectivity of $CO_2$ and $N_2$. Finally the ideal selectivities were calculated, and materials were ranked based on their performance. The information from these calculations has been used as the basis for complementary experimental efforts to fabricate MOF-based membranes for large scale carbon capture.

See more of this Session: Meet the Faculty Candidate Poster Session
See more of this Group/Topical: Education

# (3eq) Investigating Biofilm-Material Interactions: From Biofouling to Bioremedation to Pathogenesis

**Michael S. Waters**, Polymers Division, Biomaterials Group, National Institute of Standards and Technology, Gaithersburg, MD

Diagnostic measurements of biofilm-material interactions currently struggles to identify changes in biofilms beyond species presence and live/dead. However, most bacteria responsible for disease are intermittently pathogenic, which are controlled by distinct signals. In fact, many bacteria associated with pathogenesis exist in a non-pathogenic form in the natural human microbial flora. Current methods in evaluating bacterial biofilm pathogenicity are focused on evaluating overall biofilm presence and/or viability. This approach implements bulk evaluation that limits the ability discern if a live biofilm is exhibiting pathogenic properties, or is surviving in a non-pathogenic capacity. In an attempt to evolve analytical techniques to distinguish living biofilms in pathogenic and non-pathogenic modes, acidogenic dental pathogen *Streptococcus mutans* was used as a model. In this study, sucrose was used to control the exhibition of *S. mutans* pathogenic properties, which were then evaluated by surface enhanced Raman spectroscopy (SERS) to identify unique chemical signatures and rheology to evaluate the mechanical properties of biofilms grown directly on the instrument in a specially designed bioreactor. These techniques require minimal sample preparation and offer a rapid, low-cost, high throughput system for clinical evaluations, or evaluating of the capacity of various materials to affect bacterial-mediated biofilm pathogenesis.

Evaluating the efficacy of dental materials to protect human teeth requires the capacity to measure tooth accurately and precisely decay. Current practices for determining tooth decay are destructive, qualitative to lowly quantitative, and/or measure bulk changes that have low to no spatial resolution. The combination of the highly variable nature of tooth enamel and the inability to perform serial analyses on the same spatial location limits the capacity to access reproducible information from any experimental set. This study explores the potential of interferometric optical profilometry to make rapid precision spatial measurements in 3-dimensions (3-D) over time (4-D) of human tooth enamel decay as a complementary measurement to existing techniques. In this study, using bioinert 3-D alignment translation stages, human tooth decay was measured with respect to pathogenic dental bacterial biofilms. These investigations demonstrate the ability to quantitatively determine the rate of tooth decay in previously unseen spatial and temporal scales. These new, rapid, low-cost techniques minimize effort for sample preparation and use very few consumables, opening the feasibility for high-throughput investigations of clinical dental materials to a wider international community.

See more of this Session: Meet the Faculty Candidate Poster Session
See more of this Group/Topical: Education

# (3er) Using Cellular Engineering to Understand Cell Behaviors for Improved Health Care

**Erik S. Welf**, Chemical and Biomolecular Engineering, North Carolina State University, Raleigh, NC

As health care costs increase, it becomes increasingly important for us to develop effective disease treatment strategies that improve our quality of life without crippling personal or governmental finances. This aim requires a fundamental understanding of how mammalian cells use information about their surroundings to decide if they will grow, divide, migrate or die: a conceptual cellular process model. In traditional engineering applications, use of a poor process model to design control strategies will almost always result in poor system performance; likewise, in health care applications an inaccurate conceptual model of how cells respond to different environmental conditions can lead to ineffective or harmful therapeutic interventions.

Billions of dollars are spent each year on medical research, yet often the results of this work are inconclusive because they do not contribute to development of a useful cellular process model. I propose coordinated experimental and computational approaches that make the most of each experiment by designing and analyzing cell perturbations and responses with a goal of identifying how cells react to changing environmental and intracellular conditions. Using cutting edge experimental technologies, we can manipulate the cellular microenvironment by tuning physical and chemical variables of the cell culture system, and we can measure biochemical and phenomenological cell responses on various timescales while accounting for variables such as cell-cell heterogeneity and subcellular spatial distributions of signaling molecules. To this end, I will discuss how the techniques I developed during my doctoral and postdoctoral work lend themselves to this approach, and I will describe my plans for developing a world class laboratory in the area of cellular engineering.

See more of this Session: Meet the Faculty Candidate Poster Session
See more of this Group/Topical: Education

# (3et) Converting Lignocellulosic Biomass Into Sustainable Chemicals and Fuels

**Stephanie G. Wettstein**, Chemical and Biological Engineering, University of Wisconsin-Madison, Madison, WI

Converting lignocellulosic biomass into sustainable chemicals and fuels is of major interest due to environmental and economic issues associated with petroleum and the desire to use a non-food source to produce fuels. The conversion of lignocellulosic biomass to produce fuels and chemicals involves a balance between production of high-value chemicals with the production of lower-value, but higher-volume transportation fuels. My research interest is developing novel pathways to convert biomass to fuels and chemicals from lignocellulosic biomass, which combines my knowledge of separations and catalysis. My undergraduate degree in paper science and work experience at Kimberly-Clark provided me experience with pulp/paper feedstocks and manufacturing processes. As a graduate student under Professors Falconer and Noble (University of Colorado – Boulder), I investigated the influence of adsorbate-induced crystal changes on zeolite membranes using XRD, gas separations, and pervaporation. My post-doc research, under the guidance of Prof. James Dumesic, involves using catalysis and novel platform chemicals to produce sustainable biofuels and chemicals. I hope to build on my materials science and catalysis knowledge to create a strong research program to study novel combinations of separation and catalytic processes.

See more of this Session: Meet the Faculty Candidate Poster Session
See more of this Group/Topical: Education

# (3eu) Lipid-Like Materials for siRNA Delivery: Giving Genes the Silent Treatment

**Kathryn A. Whitehead**, Robert Langer and Daniel G. Anderson, Department of Chemical Engineering, Massachusetts Institute of Technology, Cambridge, MA

Despite the promise of RNA interference therapeutics, progress towards the clinic has been slowed by the difficulty of delivering short interfering RNA (siRNA) into cellular targets within the body. Over the past four years, a class of lipid-like materials termed 'lipidoids' has been studied for applications in siRNA-mediated gene silencing. It has been demonstrated that lipidoids have the potential to achieve therapeutic effect at low siRNA doses in a variety of biological systems in vivo, including hepatocytes, endothelial cells, macrophages, and leukemia and lymphoma cells. Here, I will report on the biological effect of lipidoids as well as the varied mechanisms of action through which these materials can act.

See more of this Session: Meet the Faculty Candidate Poster Session
See more of this Group/Topical: Education

# (3ev) Electrokinetic Motion of Particles near An Electrode: Theory, Experiments, and Applications

**Christopher L. Wirth**, Department of Chemical Engineering, Carnegie Mellon University, Pittsburgh, PA

The motion of particles proximate to an electrode stimulated by an electric field has been investigated for nearly two decades because of its potential impact on the assembly of colloidal crystals and on the high throughput evaluation of electrocatalysts. A rich mixture of electrokinetic phenomena has motivated an abundance of theoretical and experimental work in this area.[1] Despite the wealth of knowledge built over the last twenty years, a number of questions remained that required answers to leverage this technology. At the start of my PhD, I was charged with answering these questions; this poster will display snapshots of this undertaking.

1. **What causes symmetry breaking in a single particle's longitudinal motion?** The equilibrium spacing of a colloidal crystal depends on a broken symmetry between a single particle's longitudinal motion and the electric field. Symmetry breaking arises from the dynamic forces acting on the particle during AC electrophoresis and not on an additional out of phase element as previously expected. New modeling suggests that inertial effects and dynamic drag on the particle play a crucial role in determining the phase angle (i.e. evidence of the broken symmetry) between the electric field and the resulting particle motion.

2. **Are electrode reactions responsible for electrolyte dependent particle motion?** No. Using a new tool called the Electrochemical Total Internal Reflection Microscope, I measured single particle longitudinal and pairwise lateral motion engendered by an electric field while the electrode was ideally polarizable (IPE).[2] Despite removing any possibility of electrode reactions on the IPE, particle motion depended on the choice of dispersing electrolyte. Isolated particle pairs aggregated in 0.15 mM $NaHCO_3$ and separated in 0.15 mM KOH with every other experimental condition kept identical.

3. **Can single particle motion be an effective tool for probing local current density on an electrode?** Yes. Instead of considering the particles to be objects of

manipulation, I employed the particles as probes of local current density.[3,4] Local current density as predicted by the particle's motion was within a factor of two of the nominal current density even with a simple model using no adjustable parameters. In principle, this allows for the local measurement of current density on spatially heterogeneous electrocatalysts, such as composition spread alloy films, paving the way for high throughput electrochemistry.

In the future, I will build on my experience in electrokinetics and complex fluids to conduct research in responsive fluids, namely electrorheological, magnetorheological, and electrowetting fluids. I'm also interested in extending Total Internal Reflection Microscopy to a broader experimental space, particularly to anisotropic particles.


1. DC Prieve, PJ Sides, CL Wirth; 2D particle assembly of colloidal particles on a planar electrode, Current Opinion in Colloid & Interface Science (2010) *15* (3).

2. CL Wirth, RM Rock, PJ Sides and DC Prieve; Single and pairwise motion of particles near an ideally polarizable electrode, in preparation.

3. CL Wirth, PJ Sides, DC Prieve; The imaging ammeter, Journal of Colloid and Interface Science (2011) *357* (1).

4. PJ Sides, CL Wirth, DC Prieve; An imaging ammeter for electrochemical measurements, Electrochemical and Solid-State Letters (2010) *13* (8).


See more of this Session: Meet the Faculty Candidate Poster Session
See more of this Group/Topical: Education

# (3ew) Discrete Particle Modeling of Dense Multiphase Flows

**Chunliang Wu**, Cain Department of Chemical Engineering, Louisiana State University, Baton Rouge, LA

Dense dispersed flow has been identified as a complexsystem, which consists of a large number of finitesolid particles, gaseous bubbles or liquid droplets that are convected by a laminar or turbulent flow of a continuous fluid.Of significance are the ongoing research efforts to develop numericalmodels and to understand the flow behavior of such complexsystem. Several models at different resolutionscale levels have been formulated, among which thediscrete particle model (DPM) shows a great potential to unearth the systemcomplexity for its merit of modeling the dispersed phase in a natural way.

During my PhD study I developed a 2D hard-spherediscrete particle/bubble model and coupled it to computational fluid dynamics. LaterI solved several difficulties in 3D code development, including the accurate voidfraction calculation under unstructured grid and the implicit two-phasecoupling algorithm. I also proposed an efficient particle collision handlingalgorithm which allows a full 3D simulation of gas-particle systems havingseveral millions of particles without parallel computing. The developed 3D DPMcode has been integrated into the commercial CFD software FLUENT byuser-defined functions. The code has been validated through a comprehensivecomparison to the experiments and can capture many important characteristics ofgas-solid fluidization systems, such as bubbling, spouting, particle clusteringand core-annulus flow structures under specific operating conditions.

I have a great interest in developing numerical models and have gained alot experiences in scientific computation, especially in high performancecomputing since I worked as a research assistant on the project of meso-scale air pollution modeling at the Hong KongPolytechnic University. My research goal is to develop an efficient, versatileand robust 3D-DPM code that can be used to study the dispersed multiphasesystems. This relies not only on programming skills but also on a thoroughunderstanding of the physics inside, such as all kinds of interactions takingplace in a dispersed system. The latter actually helps a lot on designingnumerical algorithms. My future research mainly focuses on the following topics

1) 3D-DPM code development for simulation of bubblyflow with free surface

2)Three-phase flow and particle-scale heat transfer in fluidized bed byDPM simulations

3) Particle mixing and segregation in granularflows

In this poster session, I will present myrecent work on the 3D parallel DPM code development under TeraGridand LONI HPC environment. By designing a specific interface, the developed parallelcode can run separately for granular flow simulation or corporate with the CFDsolver for dispersed gas-solid flow simulation based on the user's demand. BothMPI and multithreading techniques are adopted in the code development. Theparallel code is scalable up to 128 CPUs.

**Post-doctoral advisor**

K. Nandakumar,Gordon A. and Mary Cain endowed chair professor, Cain Department of ChemicalEngineering, Louisiana State University, Baton Rouge, LA 70803

**Selected publications**

[1] Wu C.L., Berrouk A. S., Nandakumar K.,Three-dimensional discrete particle model for gas-solid fluidized beds onunstructured mesh. *Chemical Engineering Journal* 152: 514-529, 2009

[2] Wu C. L.,Zhan J.M., Li Y.S., Lam K.S., Berrouk A.S., Accuratevoid fraction calculation for three-dimensional discrete particle model forgas-solid fluidized bed on unstructured mesh, *Chemical Engineering Science* 64:1260-1266,2009.

[3] Wu C. L., Berrouk A. S., Nandakumar K., Anefficient chained-hash-table strategy for collision handling in hard-spherediscrete particle modeling, *Powder Technology* 197: 58-67, 2010

[4] Wu C. L.,Yang Y., Wong S.L., Alvin Lai C. K., A new mathematical model for irradiancefield prediction of Upper-Room Ultraviolet Germicidal Systems, *Journal ofHazardous Materials* 189: 173-185, 2011

[5] Wu C.L.,Zhan J.M., Li Y.S., Lam K.S., Denseparticulate flow model on unstructured mesh, *Chemical Engineering Science* 61:5726-5741,2006

# (3ex) Engineering Microorganisms for Biosynthesis

**Kang Wu**, Energy Biosciences Institute, University of Illinois, Urbana, IL and Christopher V. Rao, Chemical and Biomolecular Engineering, University of Illinois, Urbana Champaign, Urbana, IL

Synthetic biology is an emerging field that aims to design and construct new biological parts or systems and redesign existing ones to carry out novel tasks. Its potential for biosynthesis of valuable products is based on our understanding and ability to engineer biological systems, which are often complex functional networks of genes and biomolecules that mediate life processes. My research interests are focused on deciphering and engineering these regulatory networks to create recombinant microorganisms for production of industrially important chemicals and fuels. During the poster session, I will present examples from my graduate research that investigated problems involving genetic and signaling networks in bacteria and a project from my postdoctoral research that engineer a thermophile for biodiesel production. I will also discuss my plans to develop an independent research program.

See more of this Session: Meet the Faculty Candidate Poster Session
See more of this Group/Topical: Education

# (3ey) Theranostic Nanomedicine for Cancer Imaging and Therapy

**Yun Wu**, Nanotechnology Science and Engineering Center, The Ohio State University, Columbus, OH

My research interests are focused on developing multifunctional nanoparticle based theranostic nanomedicine for cancer imaging and therapy. The composition and structure uniformity of the multifunctional nanoparticles play an extremely important role in the success and efficacy of their applications *in vitro* and *in vivo*. To maximize theranostic efficacy, my research has focused on (1) designing the formulation of non-toxic and targeted nanoparticle carrier systems that carry and stabilize various therapeutic agents (microRNAs, oligonucleotides, siRNA and small molecules) and imaging agents (quantum dots, gold nanoparticles, magnetic nanoparticles); (2) developing novel microfluidic-based production method to better control structure uniformity and quality of multifunctional nanoparticles.

During this poster session, I will highlight specific projects of my PhD and postdoctoral research including formulating nanoparticles for efficienct microRNAs delivery in lung cancer treatment, utilizing quantum dots mediated fluorescence resonance energy transfer (FRET) to study the intracellular fate of nanoparticles and using microfluidic coaxial electrospray to produce theranostic nanoparticles in a well controlled manner. Future challenges and my research plans will be also discussed in the poster session.

See more of this Session: Meet the Faculty Candidate Poster Session
See more of this Group/Topical: Education

# (3ez) Dynamics of Polymer Materials and Complex Fluids: Multiscale Simulations

**Li Xi**, Department of Chemical Engineering, Massachusetts Institute of Technology, Cambridge, MA

My general research interest lies in the study of soft matter, polymer materials and complex fluid flows, using computer simulations ranging from the atomistic scale to the continuum level. Behaviors in these systems are determined by complex dynamical systems with numerous degrees of freedom, the understanding of which usually requires advanced and innovative simulation techniques and data analysis strategies. My graduate work at University of Wisconsin-Madison focused on the dynamics of non-Newtonian fluid flows. One particular project I worked on is the study of polymer-induced friction drag reduction in turbulent flows. A physical framework was proposed which consistently explains all major transitions from turbulence in Newtonian flow to that in highly viscoelastic ones. Understanding of this system could lead to the development of novel flow control strategies that greatly reduce energy consumption in fluid transportation. After moving to Massachusetts Institute of Technology as a postdoctoral associate, I started to investigate the molecular mechanism of the diffusion of small molecules in amorphous polymer matrices. Methods are being developed to systematically study and quantitatively describe the effects of polymer configurations and polymer-penetrant interactions on the movement of penetrant molecules. This insight could provide theoretical guidelines in various applications, such as solvent selection in the coating of polymer films and design of polymer membranes for liquid or gas separations. It also paves the way for the development of simulation techniques that provide better prediction of penetrant diffusivity.

See more of this Session: Meet the Faculty Candidate Poster Session
See more of this Group/Topical: Education

# (3f) Simulations of the Shear Mixing of Bidisperse Cohesive Particles with a Large Size Range

**Lee R. Aarons**, Mechanical and Aerospace Engineering, University of Florida, Shalimar, FL

It is well-known that mixing granular materials consisting of particles with different properties (e.g. size, density, etc.) is generally a difficult task as the different particles will tend to segregate. An additional difficulty in granular mixing arises when particle cohesion is significant, as particles form agglomerates. Of course, homogeneous granular materials feature an even distribution of their components and physical and chemical properties that are constant throughout the materials, and so they are generally more desirable than inhomogeneous materials. In addition, when the range of grain size is sufficiently large so the smaller particles can fit in the gaps between packed larger particles, improving homogeneity can potentially result in increasing solids loading. As such, it is of great interest to explore how the homogeneity of cohesive granular materials featuring particles with a large size range can be controlled. The ultimate goal of my research is to develop models that can be used to predict the homogeneity of such materials given the properties of the particles and interstitial fluid and mixing conditions.

Through the use of discrete element method (DEM) simulations, I have modeled the shear mixing of bidisperse collections of spherical particles with a 7:1 diameter ratio. In these simulations, cohesion arose by having the particles attract each other via the van der Waals force, and I have examined the influence of particle cohesion on mixture homogeneity by performing simulations with various particle Hamaker constants. I have also investigated the influences of gravity, shear rate, and the viscosity of an interstitial fluid on mixture homogeneity. Through the use of several order statistics as well as microstructure images, I will show how the homogeneity of these mixtures can be controlled.

See more of this Session: Meet the Faculty Candidate Poster Session
See more of this Group/Topical: Education

# (3fa) Atomic-Scale Resolution of Interfacial Structure and Dynamics In Nano- and Biomaterials

**Peter N. Yaron**, Department of Chemical Engineering, Carnegie Mellon University, Pittsburgh, PA

My research interests are in the atomic-scale resolution of interfacial structure and dynamics in nano- and biomaterials and encompass the fields of membrane biophysics, protein and materials synthesis and colloidal soft-matter. At the forefront of these research efforts is the use of x-ray and neutron scattering and complementary bench top techniques coupled with computationally aided analysis developed especially to characterize interfacial behavior. An atomic understanding of these systems will be utilized to identify the underlying mechanisms of behavior and function. My research will focus in three directions:

- Biophysical characterization to elucidate function and behavior of cellular membrane-bound proteins
- Interfacial behavior of nanoparticle-protein complexes
- Rheological properties of emulsions stabilized by nano- and biomaterials

My past research in the above areas have been done in collaboration with researchers whom specialize in pushing the spatial and temporal experimental resolution limits to study such systems. These collaborations have involved researchers at universities and at national facilities including National Institutes of Science and Technology, Oak Ridge and Los Alamos National Laboratories as well as international collaborations with facilities in U.K., France and Australia. I am seeking a tenure-track position to continue this multi-disciplinary research and to push our understanding of these systems at the limit of our current understanding.

See more of this Session: Meet the Faculty Candidate Poster Session
See more of this Group/Topical: Education

# (3fb) Finding Our Way by Following Electrons: Energy Chemistries Through Electronic Structure Theory

**Paul M. Zimmerman**, Dept. of Chemistry, University of California at Berkeley, Berkeley, CA

-As global energy consumption continues to rapidly expand, clean energy technologies become increasingly necessary. In this context, my research employs electronic structure simulations in two promising areas: chemical catalysis and solar energy. In addition to gaining fundamental descriptions of new chemistries, these studies have specific implications for the advancement of energy technologies, including hydrogen storage, carbon dioxide reduction, hydrocarbon conversion, and organic solar cells. Breakthroughs in these technologies could lead to zero emission vehicles, carbon neutral fuels, better utilization of fossil fuel feedstock, and inexpensive solar electricity.

In recent years, electronic structure simulations have become standard tools for providing molecular level information. The widespread use of simulations results from their ability to provide otherwise unattainable chemical mechanisms, as well as to supply guidance and interpretation to experiment. Although many simulation techniques have proven qualitatively useful, further development is needed to attain quantitative predictions. For instance, detailed investigation of quantum mechanical/molecular mechanical simulations of zeolites (Zimmerman et al., JCTC, in press) demonstrates that accurate simulations are achievable, but only with careful (and transferable) parameterization. Our new, fast and accurate model is actively being utilized to simulate catalysis in zeolites, including natural gas upgrading and alkane cracking. For solar research, simulations have typically relied on time-dependent density functional theory (TD-DFT) to investigate responses of molecules to light. This approach, however, is qualitatively incorrect for describing multiple exciton generation in organic materials. More sophisticated methods, such as restricted active space double spin flip configuration interaction (RAS-2SF), are required to gain fundamental insight into these processes. RAS-2SF breaks through limitations of TD-DFT and is able to explain the intermolecular mechanism that leads to efficient multiple exciton generation (Zimmerman et al., Nature Chem. 2010). The mechanisms elucidated in these studies are able to suggest design principles for exploiting the underlying chemical processes. Continued development and application of novel molecular simulation methods will be essential for the design of practical clean energy solutions.

See more of this Session: Meet the Faculty Candidate Poster Session
See more of this Group/Topical: Education

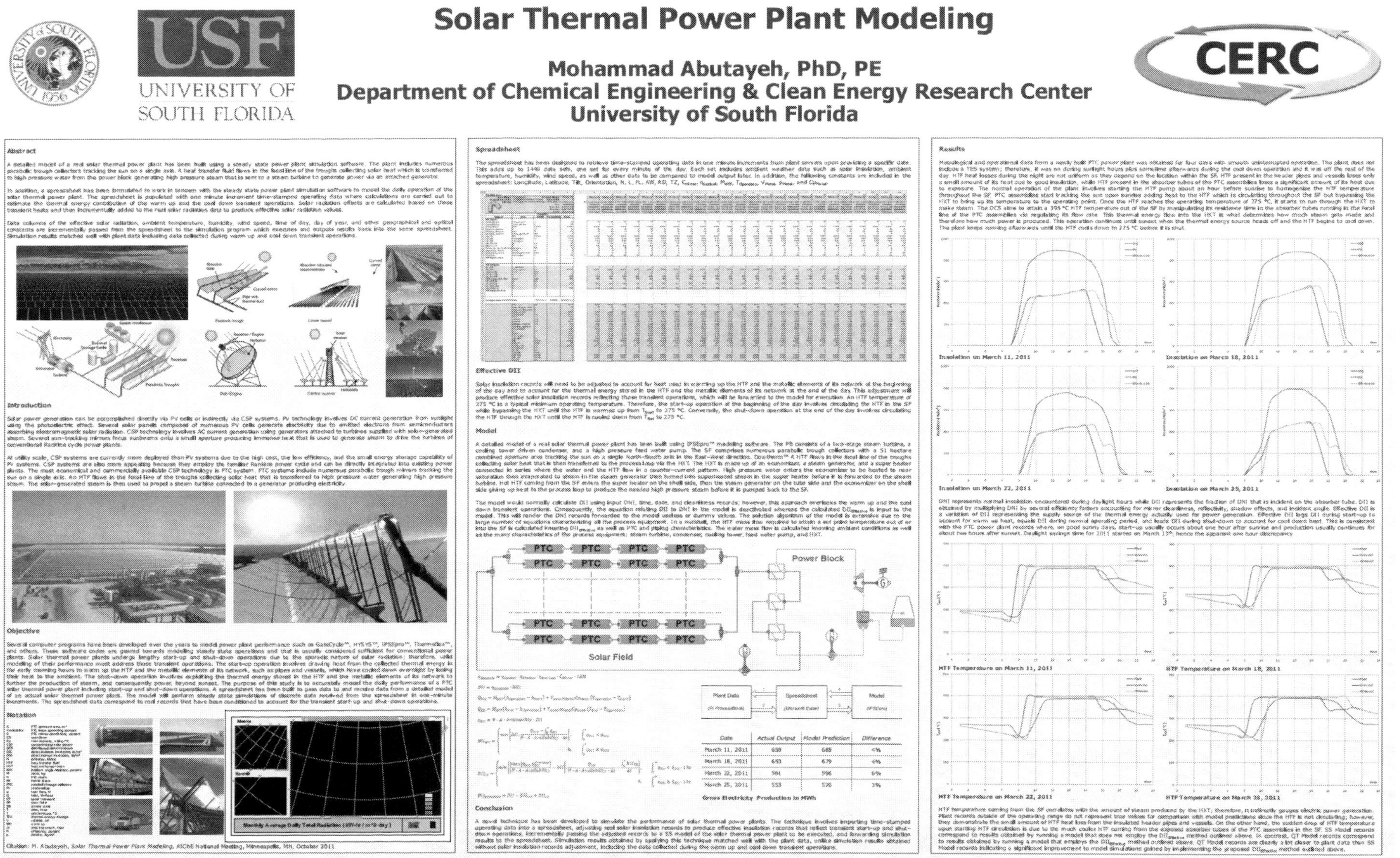
Solar Thermal Power Plant Modeling
Mohammad Abutayeh, PhD, PE
Department of Chemical Engineering & Clean Energy Research Center
University of South Florida
UNIVERSITY OF SOUTH FLORIDA
CERC
Abstract
Introduction
Objective
Notation
Spreadsheet
Effective DII
Model
Power Block
Solar Field
PTC
Plant Data
Spreadsheet
Model
Date
Actual Output
Model Prediction
Difference
March 11, 2011
March 18, 2011
March 22, 2011
March 25, 2011
Gross Electricity Production in MWh
Conclusion
Results
Insolation on March 11, 2011
Insolation on March 18, 2011
Insolation on March 22, 2011
Insolation on March 25, 2011
HTF Temperature on March 11, 2011
HTF Temperature on March 18, 2011
HTF Temperature on March 22, 2011
HTF Temperature on March 25, 2011

# (3h) Cycloaddition Polymerizations: New Uses for Old Reactions

**Brian Adzima**, Dept. of Chemical and Biological Engineering, University of Colorado, Boulder, CO

Whether the system of interest is a protein or microchip, working on the nanometer scale requires highly efficient and well controlled reactions that can be externally manipulated. The development of such reactions presents a key opportunity for realizing improved materials and devices. Cycloaddition reactions such as the Diels–Alder reaction and the copper-catalyzed azide–alkyne cycloaddition (CuAAC reaction) are highly efficient for bond formation and allow selective reactions in the presence of other functional groups. My work focuses on understanding the relationships between the underlying chemical mechanisms and the resulting material properties. Moreover, I am interested in novel strategies for remotely triggering these reactions that allow for the creation of self-healing, lithographic, and other functional mateirals. In one example, I have utilized magnetically susceptible particles to trigger reversible material healing in a Diels-Alder network. This novel process enables the material to retain its native properties over ten cycles of fracture and repair, demonstrating the Diels–Alder reactions utility for creating a selectively healable material. In a different, but related example, masked or focused laser light can be used to selectively eliminate the reversibility of the Diels–Alder adducts. Upon heating depolymerization occurs and the unexposed material can be removed. Unlike conventional approaches this technique allows for the fabrication of arbitrary three dimensional objects.

See more of this Session: Meet the Faculty Candidate Poster Session
See more of this Group/Topical: Education

# (3i) Nanoscale Ionic Materials

**Praveen Agarwal**, Chemical and Biomolecular Engineering, Cornell University, Ithaca, NY and Lynden A. Archer, Chemical & Biomolecular Engineering, Cornell University, Ithaca, NY

Nanoscale ionic Materials (NIMs) are a new class of organic inorganic hybrid materials which are able to display fluid like behavior in absence of any solvent. Created by densely grafting polymer chains on the nanoparticle surface, these materials are termed as self suspended fluids where suspending medium and suspended phase are essentially the same. Originating from their hybrid chemical structure, NIMs provide a unique platform for studying colloids and polymers in confinement. Additional advantage comes from the fact these are bulk systems and their fluid like nature which can provide insight into their rheology. We have investigated the dynamics of colloidal glasses and polymers in confinement using NIMs as model systems. Moreover we have extended this material platform to create hybrid polymer networks displaying shape memory properties.

See more of this Session: Meet the Faculty Candidate Poster Session
See more of this Group/Topical: Education

# (3j) Angiogenic and Immuno-Suppressive Scaffold for Cell Transplantation with Magnetic Resonance and X-Ray Imaging of Graft Viability

**Dian R. Arifin**, Institute for Cell Engineering/Department of Radiology, Johns Hopkins University School of Medicine, Baltimore, MD

**Background.** Currently, I am an instructor with a joint appointment in the Institute for Cell Engineering and the Department of Radiology at the Johns Hopkins University School of Medicine with Dr. Jeff W.M. Bulte as my supervisor. My postdoctoral work focuses on pancreatic beta islet cell engraftment. Here, beta islet cells are encapsulated inside biocompatible hydrogel microcapsules that prevent the opsonization of host antibodies and the subsequent destruction of the graft by the host immune system. The microcapsules are trackable by MRI, X-ray/CT and ultrasonography for image-guided cell transplantation in real-time, and non-invasive imaging of the graft distribution *in vivo*. I obtained my doctoral degree from the Department of Chemical and Biomolecular Engineering at the University of Notre Dame under the guidance of Dr. Andre F. Palmer. In my graduate research, I synthesized artificial red blood cells using surface-engineered liposomes and polymeric vesicles for blood substitute applications and treatment of anemia. Applying a combination of my backgrounds, I would like to pursue a research in tissue engineering and therapeutic agent delivery with medical imaging capabilities to guide, study and monitor the efficacy of treatment *in vivo*.

**Angiogenic and Immuno-suppressive Scaffold to Support Cell Transplantation.** Transplantation of organ or cells holds great potential to treat and cure various diseases, such as diabetes, liver failure, stroke and heart disorders. Cell transplantation in humans, however, has been hampered by severe immune rejection, potential toxicity of immunosuppressive drugs used to suppress the immune responses, insufficient cell graft revascularization, and the lack of ability to monitor the fate of graft post-transplantation. Furthermore, grafts should survive the initial hypoxic condition following engraftment and eventually connect with the host vasculature system in order to have a therapeutic benefit.

My target disease is type I diabetes and engraftment of pancreatic beta islet cells is currently the most promising therapy to treat this disorder. Here, I propose to design and develop a scaffold system to promote vascularization of the graft, and provide anti- inflammatory and immuno-suppressive properties without the use of drugs. My biomaterial of choice for the scaffold is a biodegradable and anti-inflammatory hydrogel that has tunable physical properties and encourages integration of the graft into host tissues. Supportive stem cells will be co-encapsulated inside the scaffold to promote intra- and extra-islet cell vascularization. Unlike previous cell encapsulating scaffolds that rely on co-encapsulated growth factors and immuno-suppressive agents, these supportive cells 1) produce a steady supply of growth factors *in vivo* to support the formation and the subsequent maturation of blood vessels, 2) support the graft during the initial hypoxia following transplantation, and 3) modulate the host immune responses by secreting immuno-suppressive factors.

**Non-Invasive Monitoring and Quantification of Graft Viability Using Magnetic Resonance Imaging and X-ray Computed Tomography.** Prior to encapsulation into scaffold, cell grafts will be labeled with bioinert perfluorocarbon (PFC) nano-emulsions for monitoring of the grafts using magnetic resonance imaging (MRI) and/or X-ray computed tomography (CT). Non-viable cells release their PFC agents, which are subsequently removed by the body. The mass of viable cell graft correlates with the signals of MRI and CT. Therefore, this strategy offers a non-invasive method to monitor and quantify the graft survival over time, and is particularly beneficial in clinical cases.

**Drug Delivery Vehicles with Imaging Capabilities.** I am also interested to explore the applications of PFC nano-emulsions as drug delivery vehicles that can be tracked by MR-, CT- and/or ultrasound imaging. Drugs and other therapeutic compounds can be loaded inside PFC emulsions. The surface of the emulsions can be engineered to target specific *in vivo* sites or cells. Using PFC, the delivery and *in vivo* distribution of therapeutic agents can be guided, serially tracked and/or quantified by clinical medical imagers.

See more of this Session: Meet the Faculty Candidate Poster Session
See more of this Group/Topical: Education

# (3k) Characterization and Control of Functional Nucleic Acid System

**Victor A. Beck**, Bioengineering, California Institute of Technology, Pasadena, CA

The programmable chemistry of nucleic acid base pairing has led to the development of entire new fields of scientific inquiry collectively known as DNA nanotechnology. From little more than knowledge of the canonical Watson-Crick base pairs and an appreciation for the topology of the double-helix, researchers have been able to create nearly arbitrary nanometer scale two- and three-dimensional shapes, biological circuits capable of logical computation, and autonomous machines composed of several hundred molecules and operating at the nanometer length scale. More recently, these technologies have been playing an expanded role as both sensing and regulatory elements capable of directly interacting with biological substrates.

As a specific example, in our lab we developed the hybridization chain reaction (HCR). Employed as programmable in situ amplifiers, HCR enables simultaneous mapping of multiple target mRNAs within intact vertebrate embryos. Employed as programmable mechanical transducers, HCR enables selective killing of cultured human cancer cells containing targeted cancer mutations. At the core of these efforts has been the use of a combined theoretical and experimental approach to elucidate and improve the properties of HCR systems. By successive iteration between theoretical postulation, experimental observation, and model refinement we have been able to develop a theoretical framework sophisticated enough to re-engineer equilibrium and kinetic properties of the assembly mechanisms.

Future progress will require that we continue to develop accurate models, but the staggering complexity of phenomena in biology necessarily makes a full theoretical description impossible. Our efforts to develop accurate models, abstractions, and simulation methods will not succeed if we look only for experimental validation. Experimental observation must accompany the theoretical efforts at every step, allowing us to inform and refine these models. It is this iteration between approaches that will offer us the most opportunity for success and which is at the core of my proposed research program.

See more of this Session: Meet the Faculty Candidate Poster Session

# (3I) Soft Intermolecular Interactions for Engineering Molecular Order In Organic Photonic Materials

**Stephanie J. Benight**, Chemisty, University of Washington, Seattle, WA and Larry R. Dalton, Chemistry, University of Washington, Seattle, WA

Molecular order is known to play a major role in defining bulk material processes specific to photovoltaics, electronics, and electro-optics (i.e., carrier mobility, nonlinear optical processes, exciton dissociation and charge collection, etc.). A grand challenge that unites these fields is to develop a fundamental understanding of how spatially anisotropic intermolecular interactions may be used to influence long-range molecular order. Understanding the behavior of these anisotropic interactions would enable the engineering of ordered materials through rational design of supramolecular interactions.

Organic EO materials for photonic applications are currently of great interest due to superior properties and processing advantages (large area processability, low loss) that promise reductions in the size, weight, and power requirements of economically vital technologies, such as in telecommunications, computing, and sensing applications. By utilizing an effective rational design approach encompassing synthesis, characterization, and theoretical simulations, great steps toward understanding these "soft" intermolecular interactions and producing further enhanced materials can be achieved. A detailed approach toward understanding these spatially anisotropic intermolecular interactions and proposed implementation into next generation photonics, plasmonics, and metamaterials-based device structures will be discussed. These devices represent the future in chip scale integration of electrical and photonic devices when incorporated with organic molecules.

See more of this Session: Meet the Faculty Candidate Poster Session
See more of this Group/Topical: Education

# (3m) Sustainable Life: Solutions to Renewable Energy, Protections From Chemical and Biological Threats, and Improved Biomedical Devices

**Dhiman Bhattacharyya**, Chemical Engineering, Massachusetts Institute of Technology, Cambridge, MA

**Sustainable Life: Solutions to Renewable Energy, Protections from Chemical and Biological Threats, and Improved Biomedical Devices**

Dhiman Bhattacharyya

Department of Chemical Engineering, Massachusetts Institute of Technology, Cambridge, MA 02139

In order to live a green, safe and healthy life; in one word *to sustain life*, we need to develop innovative technologies for renewable supplies of energies, detection of threats from chemical and biological agents and improvement of currently available state-of-the-art biomedical devices.

The following proposed research studies on these three fields are predicated on recent advances involving development of improved vapor phase based coating technology. These advances now permit precise surface modification of virtually any solid substrate material. Specifically, chemical vapor deposition processing involving pulsed plasma and/or oxidative polymerizations will be applied to research topics focused on developing new materials and improving existing technologies for renewable energy, sensors and biomaterial applications. I have extensive experience in development and application of these new surface modification approaches during my Ph.D. and Postdoctoral Studies. The research areas that I will pursue in my near-future academic career are mentioned below.

***1. A Nano/Micro-structure Based Thermoelectric Materials to Harvest Renewable Energy.***

***2. The Development of Chemiresistive Sensor for Detection of Biological and Chemical Threats.***

### *3. The Development of Total Artificial Lung and Kidney.*

My goal as a faculty member is to build a high-energy, interdisciplinary research program at the interface of surface engineering and materials for above mentioned applications. I envision that hybrid nanocomposites, whether it is organic-inorganic, bio-organic or bio-inorganic, are the solutions for next generation advanced materials. In addition, my research will also focus on basic researches for better understanding of the mechanisms of the CVD techniques. I am interested in working in a fast paced, highly collaborative environment where I can make a significant impact through translating ideas into reality and through mentoring students to become the next generation leaders in this field.

***Patents:***

- **K. K. Gleason, D. Bhattacharyya, M. C. Barr. *Reactor Design of A Single Unit CVD Chamber for Various Applications*. Patent application in process.**
- **R. B. Timmons, D. Bhattacharyya, K. T. Nguyen, H. Xu. *Surface Tailored Materials to Enhance the Growth Rates of Cells under Both in vitro and in vivo Conditions*. U.S. Patent Filed.**
- **R. B. Timmons, D. S. Wavhal, D. Bhattacharyya, N. Mukherjee. *Covalently Functionalized Particles for Synthesis of New Composite Materials*. U.S. Patent Filed.**
- **R. B. Timmons, W. Chyan, D. Bhattacharyya. *Improved Controlled Drug Release from Molecularly Tailored Hydrogel Films*. U.S. Patent Filed.**
- **T. Reid, R. B. Timmons, R. R. Deshmukh, D. Bhattacharyya, T. Mosley. *Plasma Surface Depositions of Thin Film Coatings Containing Selenium*. U.S. Patent Filed.**

***Selected publications:***

- **D. Bhattacharyya, K. Senecal, K. K. Gleason. "High Surface Area Flexible Chemiresistive Biosensor by Oxidative Chemical Vapor Deposition." *Submitted.***
- **D. Bhattacharyya, K. K. Gleason. "Single-Step Oxidative Chemical Vapor Deposition of –COOH Functional Conducting Copolymer and Immobilization of Biomolecule for Sensor Application." *Chemistry of Materials.* (Published online on April 26, 2011 as Article ASAP), DOI: 10.1021/cm2002397.**

- **W.-J. Yoon, D. Bhattacharyya, R. B. Timmons, P. R. Berger. "Plasma-Polymerized Multistacked Bipolar Gate Dielectric for Organic Thin-Film Transistors."** ***Organic Electronics*****, 2010, 11, 1767-1771.**
- **D. Bhattacharyya, H. Xu, R. B. Timmons, K. T. Nguyen. "Surface Chemistry and Polymer Film Thickness Effects on Endothelial Cell Adhesion and Proliferation."** ***Journal of Biomedical Materials Research Part-A,*** **2010, 94A, 640-648.**
- **D. Bhattacharyya, W.-J. Yoon, P. R. Berger, R. B. Timmons. "Plasma Polymerized Multistacked Organic Bipolar Films: A New Approach to Flexible High – κ Dielectrics."** ***Advanced Materials,*** **2008, 20, 2383-2388.**
- **C. L. Chapman, D. Bhattacharyya, R. C. Eberhart, R. B. Timmons, C.-J. Chuong. "Plasma Polymer Thin Film Depositions to Regulate Gas Permeability through Nanoporous Track Etched Membranes."** ***Journal of Membrane Science*****, 2008, 318, 137-144.**
- **D. Bhattacharyya, K. Pillai, O. M.R. Chyan, L. Tang, R. B. Timmons. "A New Class of Thin Film Hydrogels Produced by Plasma Polymerization."** ***Chemistry of Materials*****, 2007, 19, 2222-2228.**

**See more of this Session:** Meet the Faculty Candidate Poster Session
**See more of this Group/Topical:** Education

# (3n) Integrating Alternative Solvent Systems with Electrocatalysis for Energy and Environmental Applications

**Elizabeth J. Biddinger**[1], Umit S. Ozkan[2], Charles L. Liotta[3] and Charles A. Eckert[1], (1)School of Chemical & Biomolecular Engineering, Georgia Institute of Technology, Atlanta, GA, (2)William G. Lowrie Department of Chemical & Biomolecular Engineering, The Ohio State University, Columbus, OH, (3)School of Chemistry & Biochemistry, Georgia Institute of Technology, Atlanta, GA

Energy and environmentalapplications will remain main thrusts of interest in chemical engineering forthe foreseeable future and are more-and-more becoming tied hand-and-hand witheach other. Resource type and uses, utilization efficiencies, and emissionsand other wastes all need to be addressed. All of these issues have to beaddressed in a cost-effective and technically feasible manner. I proposeadding tools from my PhD and post-doctoral experiences to my intellectual toolboxto address these problems. I have a background in electrocatalysis,heterogeneous catalysis and alternative solvent systems and will show how theycan be integrated together to give a stronger tool set to tackle the difficultproblems that lay ahead.

In my PhD work at Ohio State, Iinvestigated nitrogen-containing carbon nanostructures ($CN_x$) asalternatives to platinum for oxygen reduction reaction (ORR) electrocatalystsin PEM and direct methanol fuel cells. In this work, $CN_x$electrocatalysts were grown over removable oxide supports or unsupported metalcatalysts. The nanostructure growth catalyst and growth reactant gases wereselected to control the nanostructure and nitrogen content of the resulting $CN_x$.The nature of the active site for this class of ORR electrocatalysts has beenheavily debated in the literature. $CN_x$ nanostructure, nitrogen-typeand ° Content were tuned to examine the nature of the active site. Thecatalysts were extensively characterized using traditional heterogeneouscatalysis experimental techniques including X-ray photoelectron spectroscopy(XPS), transmission electron microscopy (TEM), surface area analysis,thermogravimetric analyses and other temperature-programmed experiments. Thecatalysts were electrochemically tested for activity and selectivity tocorrelate the catalyst properties with desired activity and selectivity toORR. Results of this work suggested that an active site containing nitrogenand not a metal ion existed on the graphitic edge plane of the activeelectrocatalysts.

During my post-doctoral experience atGeorgia Tech, I have worked on using alternative solvent systems to reduceemissions, decrease process energy consumption, increase product separationsand improve desired reaction pathways and kinetics. In this poster, I willfocus on the use of reversible ionic liquids (RevILs) as a tunable medium forseparations. The main application that will be highlighted with these RevILsis $CO_2$ capture from coal-fired power plants. Here we useone-component silylamines as the RevIL. In this system, the excess watertypically used in amine-based $CO_2$ capture absorption systems is notrequired. This results in the possibility of significantly reducing the energyof regeneration of the absorbant system while maintaining high $CO_2$capture capacities. The silylamines selected are a molecular liquid before $CO_2$capture. Upon reaction with $CO_2$, they form an ionic liquid with anammonium cation and a carbamate anion. With a moderate increase in temperature,the RevIL will reverse back to the molecular liquid, resulting in the releaseof a purified $CO_2$ stream. The chemical structure of thesesilylamines can be altered to tune the desired properties of the absorptionsystem including $CO_2$ capture capacity, energy of regeneration, viscosity and $CO_2$ release purity. The modified silylamines havebeen studied to develop structure-property relationships. With theseestablished structure-property relationships, viable RevIL candidates for $CO_2$capture absorption systems can be identified and further studied in thelaboratory.

The two research areas presented inthe poster can be combined by using the alternate solvent systems in thepre-processing of reactants, post-processing of products or as the electrolyteitself in the electrochemical system. Electrocatalysts will be developed thatare both active and selective to the desired reaction. The synergetic effectsof use of the electrocatalysts with the solvent systems will be examined in thesystems studied.

See more of this Session: Meet the Faculty Candidate Poster Session
See more of this Group/Topical: Education

# (3o) Biological Decision Making In Normal and Cancer Cells

**Marc R. Birtwistle**, Systems Biology, University College Dublin, Belfield, Ireland

Cancer is a disease of deregulated cellular decision making, and in the past few decades we have uncovered a myriad of biochemical molecules involved with such decisions. But how do observable biochemical events encode biological decisions such as proliferation and migration, and is such coding different between normal and cancer cells? How can we manipulate biochemical networks to rewire cell fate decisions? These questions lie at the heart of future research in my lab and will be the glue that holds together three main research areas: (1) developing new and improving existing methodologies for how to construct, simulate, discriminate, and validate mathematical models of mammalian signaling and cell-fate decision processes such as those involved in transformation and progression to metastasis, (2) understanding the sources and controllability of biological noise, and the consequences of this noise for practical manipulation of signaling and phenotypes on single cell and population levels, and (3) generating predictive mathematical models of cellular signaling and cell-fate decision processes to enable rational design and control of natural and engineered biological systems. Our research will be applicable to many fields in medicine, where understanding how to manipulate mammalian cell fates is crucial for applications ranging from cancer to diabetes to stem cell therapy, and also biotechnology, where production of biologic pharmaceuticals by mammalian cell lines is becoming increasingly important. We will be one of the few labs in the world that combine cutting edge experimental biology with deep theoretical and computational research. This will be enabled by my unique expertise in both computational modeling and wet biology, which is a result of my PhD guidance from theorists Boris Kholodenko and Tunde Ogunnaike, and postdoctoral guidance from experimentalist Walter Kolch. This poster will first briefly describe my previous and current work, and then how this previous experience will enable me to lead the proposed future projects intended to germinate my independent research.

See more of this Session: Meet the Faculty Candidate Poster Session
See more of this Group/Topical: Education

# (3p) Engineering Responsive Proteins for Synthetic Biology

**Mark A. Blenner**, Department of Pathology & Immune Disease Institute, Harvard Medical School & Children's Hospital Boston, Boston, MA

Chemical engineers have made great advances in heterologous expression of proteins and synthetic pathways used to create complex biomolecules and reprogram cell behavior. However, understanding and engineering dynamic control of these proteins and pathways has been less explored. Nature provides many examples in which environmental signals (temperature, pH, and shear force) or specific stimuli (metal ions, small molecules, and other proteins) are used to control protein conformation and spatial organization, resulting in regulation of protein and pathway functions. New tools for conformational and spatial control of proteins and pathways will have broad impact on synthetic biology, leading to modular components that allow microbes to sense and respond to environmental changes, construction of pathways that integrate multiple signals, more selective therapeutic targeting by spatial regulation of protein function, and spatial control of enzymatic reactions to increase product formation. My research group will focus on using protein engineering and directed evolution to design responsive behavior into proteins and pathways in order to address critical problems in bioenergy, heath, and the environment. We will apply techniques and concepts from my doctoral and postdoctoral research when possible and forge new concepts and technology towards engineering responsive proteins.

Here I present my work towards engineering stimulus responsive proteins and peptides as well as developing tools for examining their mechanisms of conformational change. My doctoral research (with Scott Banta, Columbia University – Chemical Engineering) examined stimulus responsive peptides, including engineered beta roll scaffolds that fold and unfold by binding and releasing calcium, and utilized single-chain antibodies as conformational change sensors. My postdoctoral research (with Timothy Springer, Harvard Medical School) focuses on using directed evolution in combination with a novel selection method to engineer mutants of von Willebrand Factor (VWF) binding to platelet glycoprotein Ib (GPIb) stabilized in their alternative force-activated conformation. VWF initiates primary hemostasis at the site of vascular injury by binding to platelets through GPIb only in the presence of elevated shear stress. Consequently, this interaction is spatially controlled to where clotting is needed.

# (3q) Unlocking the Functional Potential of Microbial Communities

**James Boedicker**, Applied Physics, Caltech, Pasadena, CA

Natural microbial ecosystems are often complex mixtures of hundreds or thousands of species, making it difficult to reduce such systems into its functional components and processes. Understanding how these communities function is further complicated by how the activity of each individual species interacts with its local environment, the space it inhabits, and its neighbors. Yet it is these interactions within the community that are essential for many of the important functions of microbial systems, including the degradation of environmental wastes, nitrogen cycling in the soil, and the digestion of complex metabolites in the gut. Interactions within communities also regulate community composition and help to regulate the flux of nutrients through networks of biochemical pathways. The ability to identify the sets of microorganisms essential to function and to quantify how their interactions regulate activity could enable both control over microbial community function and the ability to design communities to perform specific functions.

In order to discover and study microbial communities, we first must develop new methods of studying microbes at the community level. This poster will highlight my work thus far in this direction, including how the activity of microbial communities is dependent upon spatial structure and my work in modeling and experimentally validating quantitative predictions of microbial activity. The poster will also introduce future research directions that will answer key questions in this area including: How does community composition influences activity? Can the community composition and function be controlled by adjusting chemical and biological parameters of the environment? What are the key mechanisms that regulate interactions within the community?

See more of this Session: Meet the Faculty Candidate Poster Session
See more of this Group/Topical: Education

# (3r) Membrane-Mimetics and Protein Assemblies for Drug Discovery, Delivery and Energy Applications

**Mohan B. Boggara**, Center for Biotechnology and Interdisciplinary Studies, Rensselaer Polytechnic Institute, Troy, NY

Both in my doctoral and postdoctoral work I have been using a combination of computational and experimental biophysical tools (*including neutron scattering techniques, molecular simulations, docking and NMR*) with particular focus on understanding drug-biomembrane interactions and protein-protein interactions. I will demonstrate this by presenting specific projects I have been working on during the poster session and intend to develop an independent research program focusing on three core areas.

(1) *Membrane-mimetics*: Here I plan to design, develop and characterize lipid and polymer-based self-assembled nano-structures in the form of bilayers, thin films, vesicles or polymerosomes as mimics of biological membranes to address drug/ nanoparticle/protein-membrane interactions. Such systems will also serve as platforms for bio-energy and bio-sensor applications.

(2) *Protein-Protein Interactions*: Here I intend to identify and study protein-protein systems that are relevant to diseases with particular focus on neuro-degenerative diseases (e.g. tau-tubulin) and cancer (e.g. EGF/EGFR or DNA-repair proteins). Understanding the molecular details of such protein-protein or protein-ligand interactions is critical to estabilishing functional relations between proteins in biological systems as well as the structure-based rational drug-discovery.

(3) *Large Protein Assemblies*: Here I intend to study large protein assemblies of both functional and pathological relevance. Most biological processes are accompanied by assembly of proteins into oligomers and higher order clusters of such oligomers either at membrane surface (e.g. EGFR in cancer cells) or in cytoplasm (e.g. amyloid fibrils). Understanding the molecular details of both structure (symmetry relations) and dynamics (kinetics of formation of intermediate oligomers) of such large complexes will be critical to the development of drugs and might serve as means to creating novel materials.

# (3s) Pattern Formation In Active Fluids

**Justin S. Bois**, Chemistry and Biochemistry, UCLA, Los Angeles, CA

Motivated by the dynamics of the actomyosin cell cortex in which stress generation is under biochemical regulation, we develop a hydrodynamic theory for pattern formation in active fluids. Nonhomogeneous active stress profiles drive fluid motion which transports diffusing stress regulators by advection. Based on this principle, we present a mechanism for pattern formation in which a single diffusing species up-regulates active stress, resulting in steady nonhomogeneous flow and concentration profiles. We also investigate general pattern-forming behaviors of reaction diffusion systems embedded in active fluids. In particular, we find that the presence of active stress-driven flow greatly expands the region of parameter space in which patterns may form in classic Turing-type systems. The theory is applied to the polarization of the one-cell *C. elegans* embryo.

See more of this Session: Meet the Faculty Candidate Poster Session
See more of this Group/Topical: Education

# (3t) Theory and Modeling of Confined, Templated Colloidal Systems

**Lorenzo Botto**, Chemical & Biomolecular Engineering, University of Pennsylvania, Philadelphia, PA

Confinement and templating allow experimentalists unprecedented control over colloidal systems, and will dominate colloidal and nanoparticle self-assembly and synthesis of responsive materials in the near future. I will develop theory and modeling of these systems, including three-dimensional, unsteady behavior of soft matter in complex geometries, to guide experimentalists, and shed light into new physics. A particular area of interest is anisotropic particle assembly. Tremendous advancements in the synthesis of complex-shaped colloids have created a driving need for numerical and theoretical modeling. The finite size of these colloidal "building blocks" allows continuum simulation techniques to be used, giving the opportunity to explore length and time scales well out of the reach of atomistic simulations and to provide predictions amenable to experimental confirmation. My initial research will focus on simulations of highly coupled problems involving anisotropic colloids and the physico-chemistry of interfaces, solid substrates and membranes.

See more of this Session: Meet the Faculty Candidate Poster Session
See more of this Group/Topical: Education

# (3u) Polymer Nanoengineering for Biomedical and Fluidic Applications

**Pouyan E. Boukany**, Nanoscale Science and Engineering Center, The Ohio State University, Columbus, OH

Nanodevices and Nanoprobes have established a new paradigm for diagnosis and treatment of disease and study of cellular functions
and pathways by combining nanotechnology and biotechnology.My research objective is comprised of two main thrusts. The first one is devoted to use nanofluidics-based devices for providing quantitative insights into the fundamental mechanism of drug delivery, disease treatment, gene therapy and response of individual cells to therapeutic/biomolecular reagents.The second one is to understand the molecular dynamics of complex fluids using DNA as a model and advanced visualization techniques.

My past research at the University of Akron, advised by Prof. Shi-Qing Wang in polymer science, focuses on investigating fundamental questions in polymer/DNA dynamics such as how entangled polymeric liquids response fast external deformation.By integrating confocal fluorescent microscope with rotational rheometer, the moleualr picture behind flow instabilities such wall slip and shear banding was captured, with simultaneous velocimteric and rheometric measurement.This unique set-up provides new insights into the phenomenology of entangled fluids in presence of flow [1-9].

In one of my unique nanostructure, arrays of DNA nanostrands were generated on micro-patterned PDMS stamp and they were transferred to confined nanoscopic elements by imprinting method, to fabricate sealed arrays of nanochannels/ nanowires [10]. This fabricated nanochip was used for localized gene, drug and nanoparticle delivery to individual cells with precise control of dose, composition, and location. The delivery is achieved by the focused electric field through a nano-channel.By adjusting the voltage level, pulse duration and pulse number, dose and location control of the delivery can be precisely controlled.This approach is a benign process that can achieve perfect cell viability and transfection with much less gene/drug/nanoparticles than conventional delivery methods [11].

References:

1. Wang, S. Q.; Rvindranath, S.; Boukany, P.; Olechnowics, M.; Quirk, R. P.; Halasa, A.; Mays, J. Phys. Rev. Lett. 2006, 97, 187801.

2.Wang, Y.; Boukany, P.; Wang, S. Q.; Wang, X. Phys. Rev. Lett. 2007, 99, 237801.
3. Boukany, P. E.; Wang, S. Q. Macromolecules, 2008, 41(4), 1455.
4. Boukany, P. E.; Hu, Y. T.; Wang, S. Q. Macromolecules, 2008, 41(7), 2644.
5. Boukany, P. E.; Wang, S. Q. Soft Matter, 2009, 5 (4), 780.
6. Boukany, P. E.; Wang, S. Q. J. Rheol. 2009, 53 (1), 73.
7. Boukany, P. E.; Wang, S. Q.; Wang, X. Macromolecules 2009, 42(16), 6261.
8. Boukany, P. E.; Wang, S. Q. Macromolecules 2009, 42 (6), 2222.
9. Boukany, P. E.; Hemminger, O.; Wang, S. Q.; Lee, J. L. Phys. Rev. Lett. 2010, 105, 027802.
10. Guan, J.; Boukany, P. E.; Chiou, N.; Hemminger, O.; Lee, J. L. Adv. Mater. 2010, 22, 3997.
11. Boukany, P. E.; Mross, A.; Zhang, X.; Hu, X.; Bo, Y.; Lafyatis, G.; Lee, J. L. Nature Nanotechnology 2010, (under review).

See more of this Session: Meet the Faculty Candidate Poster Session
See more of this Group/Topical: Education

# (3v) Biochar As Part of Thermochemical Platform for Processing Biomass

**Catherine E. Brewer**, Center for Sustainable Environmental Technologies, Iowa State University, Ames, IA

Slow pyrolysis, fast pyrolysis and gasification are thermochemical processes to produce fuel and chemicals from biomass. The char co-products from these processes have much potential as biochars: sustainably produced biomass charcoals used for amending soils and sequestering carbon. As biochar properties vary significantly with feedstock and reaction conditions, biochar characterization is critical for understanding these variations, for obtaining meaningful data from biochar agronomic studies and for determining the most beneficial and economical application of a given char co-product (fuel, sorbent or soil amendment). This presentation describes the biochar characterization work I have done for my PhD research in chemical engineering and Biorenewable Resources & Technology (BRT) as part of Iowa State University's thermochemical processing research group in collaboration with the Departments of Chemistry and Agronomy. The methods used in this research include proximate analysis, CHNS elemental analysis, BET surface area gas sorption analysis, X-ray fluorescence spectroscopy (XRF), FT infrared spectroscopy with photoacoustic detection (FTIR-PAS), solid-state $^{13}C$ nuclear magnetic resonance spectroscopy (NMR), cation exchange capacity (CEC), and short-term soil incubations.

See more of this Session: Meet the Faculty Candidate Poster Session
See more of this Group/Topical: Education

# (3w) Interdisciplinary Approach to the Design, Synthesis, and Evaluation of Inhalable Therapeutics

**Timothy Brenza**, Chemical and Biological Engineering, Iowa State University, Ames, IA

Infections and diseases of the respiratory system contributeto morbidity and mortality in both the developing and the developed world. The therapeutic advantages of inhalabletherapeutic treatment of respiratory diseases include the application of the therapeuticto the site of infection which enables a high localized concentration oftherapeutic with low systemic exposure. However, the successful development ofinhalable therapeutics requires the ability to successfully formulate thetherapeutic into an inhalable form, the successful deposition of the materialinto specific respiratory regions, and controlled interactions of the materialwith extracellular and cellular environments. The solution to these challengesrequires the expertise and interaction of multiple disciplines including:material science, engineering, molecular biology, immunology, pharmaceutics,and physiology.

My graduate and post-doctoral experiences in chemicalengineering and pharmaceutics have uniquely prepared me to tackle these challenges.For drug delivery vehicle design, I have synthesized biodegradable polymers andused these to encapsulate antibiotics and proteins into particles on both themicron and nano scales. I have also usedoligosaccharides from the surface of non-typeable*Haemophilus influenzae*and polyethylene glycol to modify the surfaces of polystyrene nanospheres to provide cell targeting and mucosal avoidance, respectively. The extracellular and cellular interactions of these particleswere evaluated utilizing *in vitro* modelsof the upper respiratory tract, mucosal secretions, and alveolar macrophages.

To supplement the *invitro* experience, translational studies into particle aerosolizationand aggregation of dry powder formulations have provided an understanding of depositionand distribution of inhalable therapeutic formulations. Combining thisknowledge with my experience with human respiratory physiological measurements provides me with the tools required to address the challengeof designing inhalable therapeutic delivery systems for treatments ofrespiratory diseases.

# (3x) Studying Host-Pathogen Interactions In Model Membrane Systems: Lessons Learned From a Bacterial Toxin

**Angela C. Brown**, Department of Pathology, University of Pennsylvania, School of Dental Medicine, Philadelphia, PA

During my Ph.D. work at Drexel University, I worked in the Biological Colloids Laboratory, where I applied a colloidal science/engineering approach to the study of a biochemical problem, that of characterizing lipid rafts. (These cholesterol-rich regions of the cell membrane are proposed to be involved in cellular processes, such as signal transduction and protein trafficking.) This work not only introduced me to the fascinating world of lipids and membrane biophysics but also revealed a division that exists between researchers studying the same problem with different approaches: biologists employing "real" biological systems (live cells) and engineers/colloid scientists studying "model" (liposome) systems. This division was particularly acute in the lipid raft field, where there remains a lack of consensus about the size of lipid rafts, 18 years after their first discovery. I feel strongly that each side of this debate has its merits and flaws and that to truly solve this problem, the two fields, biology and engineering, must work together. My research goal is to bridge the gap between the biological and engineering worlds, employing engineering philosophies, techniques, and tools to solve biological problems.

To gain the skills necessary to achieve this goal, I pursued a postdoctoral research project at the University of Pennsylvania, School of Dental Medicine. In this project, I have applied my background in lipid phase behavior to study the mechanism of action of a bacterial protein toxin, the leukotoxin (LtxA) produced by *Aggregatibacter actinomycetemcomitans*. LtxA kills human white blood cells in a manner that has long been assumed to involve pore formation. However, by focusing on the lipid-specific interactions of the toxin, we were able to determine that the cytolytic mechanism is more complex than simple pore formation and instead involves several types of lipid phase changes. Using the expertise of the laboratory, we then related these model results to our results obtained in human immune cells.

In this poster, I will describe our two-part approach (using both model systems and natural cells) and demonstrate what it has allowed us to determine about the mechanism of action of LtxA to illustrate the power of this approach to study host-pathogen interactions. I will also describe my future plans to use these techniques to study the interesting behavior of LtxA and other toxins and extend this work to future studies involving other types of pathogens, such as the herpes simplex virus type 1 (HSV-1). I am particularly interested in applying this method to study mechanisms by which pathogens "hijack" elements of the host defense system (particularly the cell membrane) for their own purposes.

This cross-disciplinary nature of this approach to study disease pathology has been met with great interest in the oral microbiology and immunology field, resulting in an F32 postdoctoral fellowship from the NIH NIDCR and a number of colleagues expressing interest in pursuing collaborations. The types of studies I intend to pursue in a future faculty position will allow me to take advantage of some of the new funding opportunities that have been created recently as both governmental funding agencies and private foundations have recognized the importance of these types of cross-disciplinary studies.

See more of this Session: Meet the Faculty Candidate Poster Session
See more of this Group/Topical: Education

# (3y) Integrating Nanoscale Phenomena to Catalytic Applications Through Material Design

**Nicholas Brunelli**, School of Chemical & Biomolecular Engineeering, Georgia Institute of Technology, Atlanta, GA

Advances in the synthesis of nanoscale materials have permitted more direct assignment of property changes that occur due to quantum and nanoscale size effects. In particular, the size of nanoscale materials changes physical, chemical, optical, and catalytic properties. Exploiting the novel properties of nanoscale materials for catalytic applications will reveal interesting properties that can be tuned in order to enhance catalytic performance.

My PhD research experience involved a plasma-based aerosol synthesis method for nanoparticles and characterization of nanomaterials. The limited ability of current technology to determine nanoparticle size *in situ* led me to design, build, simulate and validate the nano-RDMA, a new instrument capable of classifying nanoparticle size in the one to ten nanometer size range. The instrument performance permitted *in situ* investigation of size effects on surface dehydrogenation of silicon nanoparticles, demonstrating a lower activation barrier for dehydrogenation from small clusters. Additionally, the capabilities of the instrument were tested through demonstrating the sputter production of atomic clusters (<1 nm in size) from an atmospheric plasma source.

In addition to synthesizing and characterizing nanoparticles, my research involved designing, building, and characterizing an electrospray source that was used to incorporate platinum nanoparticles in membrane electrode assemblies (MEAs) for low catalyst loading, solid acid fuel cell applications. The electrospray source produced crystallites of the fuel cell membrane material on the scale of 100 nanometers – smaller than ever produced before - while simultaneously mixing platinum nanoparticles to create the MEA. The reduced size discrepancy between the catalyst particle (i.e. 10 nm) and the membrane material permitted low catalyst loading for the fuel cell.

My post-doctoral research examines the cooperativity between organic acid and base sites in bio-inspired cooperative catalysis. Through designed immobilization of acid and base sites on a rigid support, the interaction of organic functional groups (that would otherwise quench each other) can be controlled, permitting the ability to design-in catalytic cooperativity at the nanoscale. Alternatively, a more flexible support can be used, allowing an additional way whereby the acid-base interaction strength can be tuned.

Integrating my background of instrumentation design, nanomaterials, and catalysis, my future research will exploit nanoscale design to control material functionality for applications in the fields of energy, catalysis, and advanced functional materials. Characterizing surface and bulk properties will elucidate properties that develop due to the size of the materials, requiring new techniques and associated instrumentation to probe subtle yet significant changes that occur. The integrated and multidisciplinary probing of materials properties will lead to fundamental discoveries of nanoscale material properties.

**Post-doctoral Advisor**: Christopher W. Jones, Georgia Institute of Technology, School of Chemical & Biomolecular Engineering

**Doctoral Thesis Committee**: Professors Konstantinos P. Giapis, Richard C. Flagan, Sossina M. Haile, J. L. Beauchamp, California Institute of Technology, Division of Chemistry and Chemical Engineering

Publications:

[1] N.A. Brunelli, R.C. Flagan, K.P. Giapis, *Aero. Sci. Tech.* **2009**, *43*, 53-59.

[2] A. Varga, N.A. Brunelli, M. Louie, K.P. Giapis, S.M. Haile, *J. Mat. Chem.* **2010**, *20*, 6309-6315.

[3] J. Jiang, M. Attoui, M. Heim, N.A. Brunelli, P. McMurry, G. Kasper, R.C. Flagan, K.P. Giapis, and G. Mouret, *Aero. Sci. Tech.* **2011**, *45*, 48-492.

See more of this Session: Meet the Faculty Candidate Poster Session
See more of this Group/Topical: Education

# (15a) An Introductory Course Planning In Chemical Engineering

**Yousef Jalali**, Chemical Engineering, Islamic Azad University, Shahrood Branch, Shahrood, Iran

The content of chemical engineering major has been changed in response to the growth of industry, emerging new technologies and demands on personal skills and competencies. Today the difference between offering programs is not significant due to standardized accreditation and international expectations. However development of freshman course is still unsolved problem in the many countries. In this study several strategies for planning the first year courses in eighteen institutions across the world is analyzed. The study then suggests an introductory course based on ethics, sustainability, rich history and simple principles of chemical engineering. The goal of such as this course is to pique the interest in students looking current trend across the world with an emphasis on ethics.

See more of this Session: Chemical Engineering Education Poster Session
See more of this Group/Topical: Education

# (15b) Using Real and Imaginary Pedagogical Parts to Demonstrate Fluid Dynamics and Control Concepts

**Anthony Butterfield**, Chemical Engineering, University of Utah, Salt Lake City, UT

Both hands-on experience with physical systems and the use of web-based simulations have been shown to have significant pedagogical benefits. In this work we created both a physical apparatus and its simulation in order to illustrate several important engineering concepts to our students.

The physical apparatus consists of a computer fan positioned beneath a horizontal acrylic tube, which has an ultrasonic range finder aimed up its center. By controlling the power sent to the fan, objects in the tube may be made to levitate, and their position may be determined by the rangefinder. Using a custom Matlab GUI, we have created a system where students may manually control the power to the fan, or create automatic controllers that measure the object's position and use that measurement as feedback to adjust the flow rate accordingly.

To illustrate concepts of drag, students may place objects of various shapes or weights in the tube and determine the flow rate needed to lift the object; typically weighted and perforated ping-pong balls or paper cones have been used. Though this system is complicated by the tube wall, an approximately square root relationship was found between a spherical object's weight and the flow rate needed to lift it, as one might anticipate from theory regarding falling spheres.

To illustrate the importance of and basic concepts behind automatic control, students may use this system to take on the role of a controller. They may alter the power to the fan using the GUI to achieve a desired object height. However, the most effective pedagogical use of this apparatus was found by setting the power to the fan based on the the position of the student's hand, as detected by a webcam. Students may achieve a very memorable and tactile experience of what it takes to properly control an integrating system by feeling a direct connection to the manipulated variable. Once the difficulties of control have been demonstrated firsthand, then a simple PID controller may be introduced, explained, and implemented to achieve superior control. Furthermore, the effects of poor control can be easily visualized by altering PID parameters to achieve unstable oscillations, slow response times, and so on.

A virtual counterpart to this demonstration was also created in the form of a JavaScript applet which may run on any major browser. This applet simulates the physical system and allows the student to alter the associated constants and witness their effect. With this virtual system the student may also implement automatic control or act as the controller themselves, using their mouse position to adjust the power to the fan. The simulation may be used to prepare students to work on the actual apparatus, or allow a large class to all experiment with a system similar to that used in an instructor's demonstration. Furthermore, use of simulations along with the physical system allows students to gain a clearer picture of the problems engineers may have in reconciling theory with practice.

These twin pedagogical tools have been successfully used as an introduction to basic concepts of control and drag forces, and they have been well received by students. Furthermore, the system has both control and fluid dynamics components that are addressable with a range of theoretical sophistication, and thereby we have created related projects that have been challenging for both undergraduate and graduate students. Finally, this system has been a very useful tool for high school outreach purposes, due to its interactive aspects and the relatable way in which it illustrates otherwise abstract issues of automatic control.

See more of this Session: Chemical Engineering Education Poster Session
See more of this Group/Topical: Education

# (15c) Spicing up An Engineer's Education: Towards Chemical Common Sense

Hanna Praefke, Kamila A. Kreusch and **Marcel A. Liauw**, Itmc, RWTH Aachen University, Aachen, Germany

Chemistry classes for (non-chemical) engineers tend to be theoretical, doing little to teach the "look and feel" of chemistry. A new edition of a "Chemistry for engineers" class has started to expose the students to cinnamic aldehyde, smelling of cinnamon, or vanilline, smelling of vanilla. From their molecular structure, the properties of these substances, like color or melting point, are deduced. In every one of twelve lectures, a host of such qualitative (or even quantitative) structure-property relationships is explored. Selected chemical substances are passed around to make the engineering students acquainted with the chemists' approach to chemicals: visual inspection, smelling, testing reactivities. The attitude of the students with respect to this new, strongly chemicals-based chemistry class is assessed. It appears that the typical "molecular approach" of chemists is better understood and acknowledged.

See more of this Session: Chemical Engineering Education Poster Session
See more of this Group/Topical: Education

# (15d) A MEMS Education Project

**Maru Colbert**, BMS/MIT/BU, Cambridge, MA

In this applied project, which originated from MIT research, a Micro-Electro-Mechanical Systems or MEMS device was designed to detect NOx emissions in vehicles. This micro-fabricated device is a micro-sensor which converts a measured mechanical signal into an electrical signal to detect this greenhouse gas. In Phase I of the project the initial design and function parameters of the MEMS device were determined and a preliminary prototype was developed. Determination of the generator resonator die, the NOx sensitive coatings, packaging solutions and the electronics was done in Phase I. Improving the design and fabrication of the micro-sensor array, developing packaging for operation of the sensor at temperatures above 800°C and compaction of the sensor components to enable sub-second reaction time are critical components of the current Phase II. The chemical component of this fast response, high sensitivity, MEMS based NOx sensor is the coating/sensing material adsorbed onto the resonator. In order to increase selectivity to NOx emissions and temperature sensitivity, there has to be optimal integration of a functional coating on the micro-resonator. Two specific perovskite and one zeolite compound were selected. These coatings were selected due to their ability to catalyze a NO oxidation process as well as for their absorption properties relative to the inorganic matrix of the absorbent. The catalysis and absorption properties of the selected coatings are temperature dependent.

This complex and multi-disciplinary entrepreneurial and research project provides rich and relevant material for a case study in chemistry careers, principles and practice. An extension of the "instructor-as-researcher" component of an NSF grant is the development of a set of learning modules that can be adapted for high school through frosh chemistry. These work sets mimic the research to production questions present in this and many projects. The modules include a novel "chemstory" assessment and a literature search-to-experimental design project-which incorporates critical thinking and writing. It also has a career study which outlines the eight science/engineering disciplines drawn upon for this business endeavor. Other, more standard course segments incorporate the laboratory procedures, data collection and key science concepts involved in this work. There is a value added section as well for the most advanced, possibly AP students. This is an innovative bench-to-pilot to production scale curriculum project. Problem solving, process design, scale-up and instrumentation are incorporated. The chemistry/pre-(chemical) engineering, mathematics, writing, economics and autoCAD components are keyed and identified on a concept map. This can be taught as a complete unit, used as an extracurricular chemical project, or components can be used to augment a math, chemistry or related class. The design and use of this curriculum for outreach will be featured.

This is an example of outreach to instructors, teachers and the recipients, students. When this material is taught in schools through a chemistry club or in conjunction with NOBCChe as an extracurricular mentor project, it reaches students who might not otherwise be exposed to chemistry or chemical engineering in a meaningful way. NSF is providing outreach by even funding this type of project which provides innovative research opportunities for teachers and instructors who may not otherwise gain experience in industry.

See more of this Session: Chemical Engineering Education Poster Session
See more of this Group/Topical: Education

# (15e) Engaging High School Students In Advanced Chemistry and Chemical Engineering Careers

**Christopher J. Barr**[1], Ann HajIbrahim[2], Constance Schall[1] and Carol Stepien[3], (1)Chemical and Environmental Engineering, University of Toledo, Toledo, OH, (2)Central Catholic High School, Toledo, OH, (3)Lake Erie Center, University of Toledo, Oregon, OH

With the impending retirement of many engineers and scientists in the near future and the low matriculation of newly gradated domestic high school students into advanced STEM education, there is a critical need to educate and motivate high school students to pursue entry into these fields. As part of this endeavor, a National Science Foundation GK-12 grant titled "Graduate Teaching Fellows in STEM High School Education: An Environmental Science Learning Community at the Land-Lake Ecosystem Interface" at the University of Toledo's Lake Erie Center was formed. This program pairs graduate students in STEM disciplines with teachers from local high schools. Graduate student fellows exchange their STEM knowledge and research experience through high school classroom and field exercises to exchange teaching strategies with paired teachers, mentor high school students in the development of STEM skills and potential for STEM careers, and to enable high school teachers to gain professional development and experience in inquiry-based state-of-the-art science research.

Educational opportunities in chemistry and chemical engineering have been improved through development and execution of a variety of outreach activities. These educational opportunities are based around newly developed activities and demonstrations as well as advanced research experiences. These activities and demonstrations incorporate environmental, analytical, and biological chemistry in general as well as in specific research-based examples currently being examined at the University of Toledo. A summary of activities and outcomes is presented. Surveys presented to the students before and after student fellow interaction with the high school students provided a metric for examination of these outreach activities.

See more of this Session: Chemical Engineering Education Poster Session
See more of this Group/Topical: Education

# (15f) Introducing High School Freshman Students to Typical Chemical Engineering Research

**Sami Alouani**[1], **Steffano Oyanader**[1], Mario Oyanader[2] and ALI Alouani[3], (1) Cookeville High School, Cookeville, TN, (2)Chemical Engineering, Tennessee Technological University, Cookeville, TN, (3)eLECTRICAL eNGINEERING, TENNESSEE TECHNOLOGICAL UNIVERSITY, Cookeville, TN

Getting high school students interest in a Chemical Engineering carrier represent a great challenge as they may not really relate the profession to something they do in a daily basis. Even more challenging is to capture student interest in Chemical Engineering research without presenting the student an appealing fun experience. In this pilot project, high school freshman students has been challenge to investigate the possibility of using saliva as a potential reactant in the making of ethanol. Although, in the actual experiment the students used $\alpha$-Amylase instead of saliva, the relationship between the two is something they found out by themselves as part of their critical thinking process. Typical practices of Process Design and Design of Experiment were introduced as the students' research journey evolved calling for more robust ideas and explanation of phenomena. Students were closely monitored in their execution of the scientific method without emphasizing the learning of new technical words but in the understanding of new concepts using hand on approaches. A detail description of experiences by the very own actors of this paper is intended to be presented as a case study for future consideration when working with STEM students. Among the many results, the students registered and won a third place at the local University STEM project competition this year. This particular example validates the high quality of the final product and therefore the introduction to research approach used in this case may be of great benefit to recruit potential CHEME students.

See more of this Session: Chemical Engineering Education Poster Session
See more of this Group/Topical: Education

# (15g) On Centrifugal High Volume Separation of Oil and Water

**Kal Renganathan Sharma**, Department of Chemical Engineering, Prairie View A & M University, Prairie View, TX

*Centrifugal separation of oil and water from streams that is concentrated with oil such as the ones from the oil spill disasters is investigated in this study. The existing theories are largely for Stokes settling of oil drops. In this study two layers, one rich in oil and another rich in water are allowed to spin in a centrifuge. The tangential velocity profile is derived from the equations of continuity and motion. The power drawn at the inner rotor is calculated for a set of parameters fo the system and a angular speed w RPM. Each simulation required the solution of four simultaneous equations and simultaneous unknowns. The power draw was found to be linear with angular rotor speed on a log-log plot. The viscosity of the oil was increased five times to study the effect on the power draw. An expression of the interlayer thickness ratio, a was obtained by use of a component mass balance on oil streams that flow in and out of the continuous centrifuge.*

See more of this Session: Chemical Engineering Education Poster Session
See more of this Group/Topical: Education

# (15h) On Instruction of McCabe and Thiele Method for Binary Distillation

**Kal Renganathan Sharma**, Department of Chemical Engineering, Prairie View A & M University, Prairie View, TX

I instruct CHEG 3043, Equilibrium Staged Separations at Prairie View A & M Univeristy, Prairie View, TX. Since Treybal (1980) a number of developments in the field have come about. Desktop computers with spreadsheets and software such as Aspenplus and RADFRAC simulation package has changed the emphasis from a set of procedures to methods that are more versatile. The design specifications in Treybal (1980) is not accurate. For example when product split, $P_{split,}$ W/D ratio and distillate compositions are given teh residue composition can be calculated. W/D ratio can be shown from mass balance to be given in terms of the feed, distillate and residue compositions. So once the residue, distillate compositions and W/D are known the feed composition can be calcualted. It cannot be specified as indicated in Treybal (1980). I go over ${}^6C_3 = 20$ different problem types. The emphasis on calculated of number of ideal stages, $N_p$ needed to achieve a desired level of separation is in excess. Detailied methods are provided for the 20 different problem types. For example;

***Given Product Split, $P_{split}$ ($x_D D/x_w W$), Reboil Ratio, B and Reflux Ratio R calculate Residue composition, $x_W$, $N_p$, W/D***

The 4 unknowns are the coordinates of the intersection point, $x_{int}$, $y_{int}$ and the product compositions, i.e., the distillate composition, $x_D$ and residue composition, $x_w$. 4 equations can be generated from the operating line for rectification, operating line for stripping, feed line and the given product split, $P_{split}$ information. Thus,

(2.77)

Eq. (2.77) along with Eqs. (2.70-2.72) form the 4 simultaneous equations that can be used to solve for the 4 unknowns. Once the unknowns are solved for the operating lines of rectification, stripping can be constructed in addition to the feed line that is specified. The W/D can also be calculated from the now available product compositions

The 4 unknowns are the coordinates of the intersection point, $x_{int}$, $y_{int}$ and the product compositions, i.e., the distillate composition, $x_D$ and residue composition, $x_w$. 4 equations can be generated from the operating line for rectification, operating line for stripping, feed line and the given product split, $P_{split}$ information. Thus,

(2.77)

Eq. (2.77) along with Eqs. (2.70-2.72) form the 4 simultaneous equations that can be used to solve for the 4 unknowns. Once the unknowns are solved for the operating lines of rectification, stripping can be constructed in addition to the feed line that is specified. The W/D can also be calculated from the now available product compositions.

During steam distillation, the number of stages $N_p$ can be more than the minimum number of stages, $N_m$. This is because there in no vapor return. Useof $N_m$ is not recommended when steam is used.

See more of this Session: Chemical Engineering Education Poster Session
See more of this Group/Topical: Education

Engaging Faculty In Assessing and Improving Students' Critical Thinking

Barry S. Stein, Tennessee Tech University, Cookeville, TN

## Introduction

Our society is being transformed by the way information is disseminated and accessed. Wireless connectivity with the internet is providing individuals with immediate and continuous access to a vast amount of information in all disciplines. These information resources are increasing at a phenomenal rate but with the influx of new information there is also misleading and false information. It is therefore essential that people understand how to evaluate this information and effectively use it to solve problems. The critical thinking skills needed to accomplish these tasks are now a cornerstone to both workplace development and effective educational programs. It is not surprising that national polls indicate over 90% of the faculty in this county think critical thinking is the most important part of undergraduate education (1).

Given the importance of these critical thinking skills, it is essential that we find a way to assess these skills to better understand whether our educational programs are effectively preparing our students with the skills they need to be successful. Valid and reliable assessments will also provide a means to assess our improvement efforts in higher education.

Our institution has been exploring existing methods and developing new tools to accomplish these goals since 2000. This work engaged faculty across a broad range of disciplines to identify a core set of critical thinking skills that transcend disciplines. The assessment tool we developed (the *Critical thinking Assessment test* or CAT) was designed to assess these core critical thinking skills and to directly engage faculty in the evaluation of their own students' thinking because this faculty involvement in the assessment process is essential if the outcomes of assessment are going to influence how faculty prepare students.

The CAT instrument and scoring guide have been tested and refined through repeated administrations to students and scoring sessions since 2000. In 2004, NSF provided funding to work with other institutions across the country (the University of Hawaii, the University of Texas, the University of Southern Maine, the University of Colorado, the University of Washington, and Howard University) to use, evaluate, and help refine the CAT instrument. The success of this work led to additional funding in 2007 to train representatives from other institutions to lead CAT scoring workshops on their own campuses and to evaluate the effectiveness of these regional training workshops. The dissemination model has been very successful and in 2010, NSF provided additional funding to expand national dissemination and to expand the use of the CAT instrument for the assessment of project outcomes in other NSF funded projects [1].

1. Partial support for the development and dissemination of the CAT instrument was provided by the National Science Foundation's TUES program (formerly CCLI) under grants 0404911, 0717654, and 1022789. Additional information and project reports can be found at www.CriticalThinkingTest.org . Any opinions, findings, and conclusions or recommendations expressed in this article are those of the author and do not necessarily reflect the views of the National Science Foundation.

Skills Assessed by the CAT

**Evaluating Information**
- Separate factual information from inferences.
- Interpret numerical relationships in graphs.
- Understand the limitations of correlational data.
- Evaluate evidence and identify inappropriate conclusions.

**Creative Thinking**
- Identify alternative interpretations for data or observations.
- Identify new information that might support or contradict a hypothesis.
- Explain how new information can change a problem.

**Learning and Problem Solving**
- Separate relevant from irrelevant information.
- Integrate information to solve problems.
- Learn and apply new information.
- Use mathematical skills to solve real-world problems.

**Communication**
- Communicate ideas effectively.

Face Validity

The CAT instrument was designed to measure the components of critical thinking and problem solving that faculty across disciplines think are most important. The graph below shows the percent of faculty that think each question is a valid measure of critical thinking. These evaluations include a wide variety of disciplines from six diverse institutions involved in the first NSF project to evaluate and refine the CAT instrument (2).

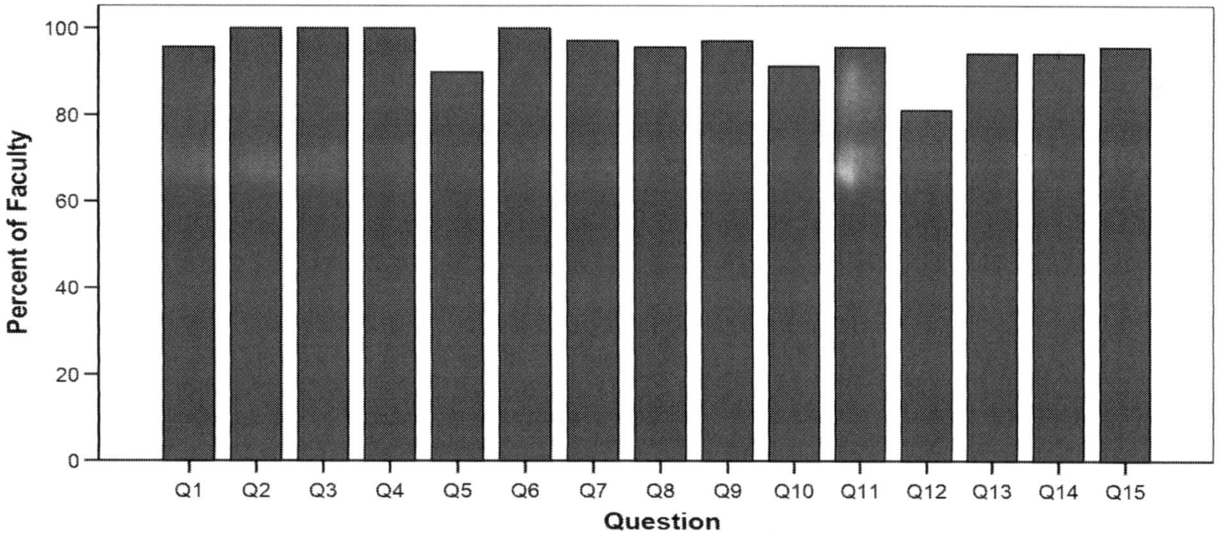


**Figure 1**. Percent of Faculty Rating Questions as Valid Measures of Critical Thinking

## Criterion Validity

Criterion validity for this type of test is difficult to establish, since there are no clearly accepted measures that could be used as a standard for comparison. Since the CAT Instrument is designed to assess a broad range of skills associated with critical thinking, we looked for reasonable but moderate correlations with other measures of academic performance (see Table 1) and slightly higher correlations with other (more narrow) measures of critical thinking (see Table 2).

**Table 1.** Correlation with General Measures of Academic Performance

| | ACT | SAT | Academic Profile | Grade Point Average |
|---|---|---|---|---|
| CAT | 0.501* | 0.516* | 0.562* | 0.295* |

* correlations significant, $p < .01$

**Table 2.** Correlation with Other Measures of Critical Thinking

| | CCTST (California Critical Thinking Skills Tests) | CAAP Critical Thinking Module |
|---|---|---|
| CAT | 0.645* | 0.691* |

* correlations significant, $p < .01$

The relationship between student responses on the National Survey of Student Engagement (NSSE) and performance on the CAT instrument has also been examined (see Table 3). Five items on the NSSE were significant predictors of performance on the CAT instrument (multiple $R = .49$, $p < .01$). The negative relationship between CAT performance and the extent to which students felt that their college courses emphasized rote retention is particularly important and supports both the criterion validity and the construct validity of the CAT instrument (3).

**Table 3.** Correlations with the NSSE

| NSSE Question | Coefficient |
|---|---|
| (2a) Memorizing facts, ideas, or methods from your courses and readings so you can repeat them in pretty much the same form. *(negative relationship)* | **-.341** ** |
| (3b) Number of books read on your own (not assigned) for personal enjoyment or academic enrichment. | .277 ** |
| (11e) Thinking critically and analytically & (11m) Solving complex real-world problems | .244 * |
| (7h) Culminating Senior Experience (thesis, capstone course, project, comprehensive exam, etc.) | .231 * |

* Significant at .01 level; ** Significant at .001 level

## Cultural Fairness

The cultural fairness of the test has been evaluated in two ways. A multiple regression analysis of CAT performance revealed that once the effects of entering SAT score and GPA and whether

English was the primary language were taken into account, neither gender, race, nor ethnic background were significant predictors of overall CAT performance. A cultural differential item functioning (DIF) analysis was also performed to examine question bias. The review of DIF results did not reveal any items with prevalent cultural bias.

### Test Sensitivity

Performance on the CAT instrument reveals neither floor effects nor ceiling effects for any of the participant groups tested thus far. Test takers have included community college students and a very diverse collection of 4-year undergraduate institutions (including regional universities, R-1 institutions, and Ivy League institutions). The sensitivity of the test is also sufficient to reveal differences between freshman and seniors or to reveal the effects of single course that develops critical thinking skills.

### Institutional Use

Over 80 institutions across the country are collaborating on national dissemination of the CAT instrument. The participating institutions range from community colleges to large research institutions and include private and publicly funded institutions. Institutions are using the CAT for a variety of assessment purposes including: evaluating campus-wide outcomes, evaluating general education outcomes, evaluating specific program outcomes, and evaluating specific course outcomes.

### Faculty Development and Emerging Success Stories

Regional training workshops prepare representatives from other institutions to lead scoring sessions on their own campuses and to stimulate discussions of how faculty might address student weaknesses through modifications to their teaching and assessment methods. Faculty are encouraged to develop their own discipline specific analog activities/assessments that engage students in critical thinking. This type of faculty development along with appropriate institutional support can transform instruction and student learning.

Institutional findings are beginning to emerge that provide a better understanding of the types of activities leading to significant gains on the CAT instrument. These findings indicate that a single course or project can impact critical thinking and significantly improve students' performance on the CAT instrument. Faculty members at Clemson University, Sam Houston State University, and the University of Wisconsin at Madison (among others) have implemented courses that significantly improved students' performance on the CAT instrument (4).

### References

1. Bok, D. (2006). *Our Underachieving Colleges: A candid look at how much students learn and why they should be learning more*. Princeton: Princeton University Press.
2. Stein, B., Haynes, A., Redding, M. (2007) Project CAT: Assessing Critical Thinking Skills. In D. Deeds & B. Callen (eds.) *Proceedings of the National STEM Assessment Conference*. National Science Foundation.

3. Stein, B., Haynes, A., Redding, M. (2011) National Dissemination of the Critical thinking Assessment Test (CAT) in Conference Proceedings, Transforming Undergraduate Education in STEM: Making and Measuring Impacts, AAAS/NSF.
4. Stein, B., and Haynes A. (2011) Engaging Faculty in the Assessment and Improvement of Students' Critical Thinking Using the Critical Thinking Assessment Test. *Change: The Magazine of Higher Learning*, 43:2, 44-49, 2011).

# (19b) Some Thoughts and Activities of Critical Thinking for Chemical Engineering Education

**John P. O'Connell**, Chemical Engineering, University of Virginia, Charlottesville, VA

Critical thinking focuses on improving the outcomes of human thought by rational and disciplined analysis. The elements might be framed as fundamental and precise questioning; careful information gathering and processing; validation of conclusions; creative exploration of alternatives; and effective communication of the process and outcomes. These are traits which are not limited to problem solving, because they are habits of thought to be used for any goal of thinking. We assert that engineering graduates should have established such capability, along with the many other skills, knowledge, and values our profession demands, but often are not explicitly included in our curriculum.

These skills might best be inculcated by encounters with, and guidance through, situations for which results depend on the quality of thought exercised in their examination. The educational challenges are selection of the most important elements of critical thinking, and how to incorporate these effectively and efficiently into the teaching/learning experience. There are literature and organizations, for example by the pamplets of the Foundation for Critical Thinking [1] and books by Edward De Bono [2], which describe general concepts and goals, along with some techniques and activities that might be implemented as a part of classes and projects.

The presentation will describe some activities stimulated by these resources that have been incorporated into a foundational engineering course, as wells as chemical engineering thermodynamics and laboratory courses over several years. These efforts have not been systematic in approach, regularity, or assessment, and therefore do not form a program for incorporation into a curriculum. Here, the intention is to contribute to discussion about how the discipline and articulation of critical thinking might be engendered in ChE students to give them the best chance of overcoming the enormous challenges they face in the 21st century and beyond.

[1] http://www.criticalthinking.org/

[2] http://www.edwarddebono.com

See more of this Session: Critical Thinking In the Chemical Engineering Curriculum
See more of this Group/Topical: Education

# (19c) What Is Critical Thinking? A Constructivist Approach

**Elliot P. Douglas**, Department of Materials Science and Engineering, Univresity of Florida, Gainesville, FL

Measurement of critical thinking requires a clear understanding of what critical thinking is. Most definitions of critical thinking are based on philosophical or educational beliefs about what critical thinking "should be". These definitions are often based on a consensus opinion of experts, rather than on empirical evidence of how critical thinking is enacted in practice. This presentation focuses on two studies that shed light on how critical thinking actually occurs. The first was a pilot mixed methods study, which compares the critical thinking skills of undergraduate and graduate materials engineering students using the California Critical Thinking Skills Test. Surprisingly, undergraduates scored higher that graduate students, suggesting they have better critical thinking skills. Examination of the scores and comments from the students revealed that the difference is due to the way each group approached the test. Thus, the results appear to reflect differences in test-taking skills rather than critical thinking skills, and call into question the use of multiple choice instruments to measure critical thinking. The second study was a constructivist qualitative study in which undergraduate civil engineering students were shown problems from the CCTST, both with and without the multiple choice answers, and asked to discuss their approaches to the problems. This then led into a broader discussion of how they viewed critical thinking. The findings illustrate that critical thinking as practiced by students takes a wide variety of forms, and thus how experts view critical thinking may not reflect what actually occurs in the classroom.

See more of this Session: Critical Thinking In the Chemical Engineering Curriculum
See more of this Group/Topical: Education

# (19d) Encouraging Students to Critically Think about the Origins and Assumptions Behind Heat and Mass Transfer Convection Coefficient Correlations Through a Simple Demonstration

**Bradley C. Bundy**, Chemical Engineering, Brigham Young University, Provo, UT

A significant portion (~25-35%) of most undergraduate heat and mass transfer texts and courses are dedicated to instructing students in the use of convection coefficient correlations. While most students become proficient at using the correlations, they have often commented that they do not really understand how the correlations were derived or why so much error (~25%) is associated. To overcome this limitation in student education, a simple classroom demonstration and follow up assignment are presented. The demonstration encourages students to not just think about what correlation to use, but to critically think about 1) how were the correlations in the text experimentally determined, 2) why is there so much error associated with each correlation, 3) where can I go to find the original experiments that led to the development of a given correlation, and 4) after I have finished this course and I need a convection coefficient correlation what resources will be available. Specifically, a simple apparatus was constructed to collect heat transfer data while varying independent variables of convection coefficient correlations such as the Reynolds number. Using the raw data students, generated their own convection correlation and calculated the error associated with their correlation as part of their homework assignment. Students were also assigned to find and describe the use of a heat or mass transfer convection coefficient correlation from a common resource that they may have access to later in their career such as Perry's Chemical Engineer's Handbook. Finally, students were given the opportunity to search archival journals for correlations and explain their use. Following IRB approval, the effectiveness of this approach was assessed by a student survey at the end of the course.

See more of this Session: Critical Thinking In the Chemical Engineering

# (19e) Not Your "Normal Chemistry Lab": Enhancing Critical Thinking Through New Unit Operations Laboratory Approach

**Katherine Ziemer**, Department of Chemical Engineering, Northeastern University, Boston, MA

Throughout this new apporach to UO Lab, the students are led through thinking critically about data, realizing the concept of no single right answer, and learning that data can have multiple levels of possible meanings. Three types of experiential learning (**D**iscovery, **D**evelopment, and **D**esign – **3D**) that integrate the students' maturing level of understanding of fundamental phenomena through the classroom are used to deepen the knowledge of concepts and build engineering problem solving skills through the practice of critical thinking. For the **Discovery** experiments, students role play "discoverer" making observations and postulating general theory and relationships between governing parameters based on their current knowledge and observations, *prior* to a detailed explanation in the classroom. Thus the exercise their observation and analysis skills without expectations – or at least not being entirely sure of the "correct" answer. The **Development** experiments use traditional unit operations experiments to first teach students the key elements of a successful experimental method, and then allow them to create their own experiment to be completed by a different group of students; thus identifying interesting scientific questions based on critical analysis of their own data, and then designing an experiment to help answer those scientific questions. For the **Design** experiment, the students are given an engineering challenge which they research (critical analysis of literature), perform bench scale experiments (critical analysis of their own data), design a solution, test a solution (more critical analysis), and then redesign and test a solution. For example, students build a river, the instructor pollutes the river, and the students have to engineer a way to clean up the river.

See more of this Session: Critical Thinking In the Chemical Engineering Curriculum
See more of this Group/Topical: Education

# (19f) Hands-On, Remote, and Simulated Labs: Is There a Productive Synergy?

**David DiBiasio**[1], Jim Henry[2], Murat Ozkaya Ozkaya[2], Jarrod Henderson[2], Marina Miletic[3] and William Clark[4], (1)Chemical Engineering Department, Worcester Polytechnic Institute, Worcester, MA, (2)Chemical Engineering, University of Tennessee at Chattanooga, Chattanooga, TN, (3)Department of Chemical and Biomolecular Engineering, University of Illinois Urbana-Champaign, Urbana, IL, (4)Chemical Engineering, Worcester Polytechnic Institute, Worcester, MA

**Hands-on, Remote, and Simulated Labs: Is There a Productive Synergy?**

David DiBiasio[1], Jim Henry[2], Murat Ozkaya[2], Jerrod A. Henderson[2], William M. Clark[1], and Marina Miletic[3], [1]Chemical Engineering Department, Worcester Polytechnic Institute, 100 Institute Road, Worcester, MA 01609, [2]Chemical Engineering, University of Tennessee at Chattanooga, 615 McCallie Avenue, Chattanooga, TN 37403, [3]Department of Chemical and Biomolecular Engineering, University of Illinois Urbana-Champaign, 206 Roger Adams Lab, Urbana, IL 61801

**Background**

Student learning outside the classroom may be complemented by a variety of structures. Laboratory and project work are typical modes. However, lab work can be hands-on, remote, or based upon simulations. Ideally, remote experiments can be conducted at any time from any place. They are particularly useful for students at universities where resources are severely limited and there is no access to any significant experimental equipment. Simulations are also accessible 24/7, do not require substantive resources (unless the software is expensive) and can provide efficient and easy ways to study multiple variables. Multiple runs are frequently impossible or too costly with real lab equipment whether remotely operated or not.

Although this work does not directly address the Unit Operations course, it represents an attempt to integrate UO equipment appropriately in the sophomore year. In addition it engages students in issues related to critical thinking (ambiguous problem statements, multiple solutions, teamwork, and difficulties understanding differences between theory and reality.

In all contexts, understanding student learning is critical. Each mode has unique advantages and disadvantages that contribute to learning. However, each format is different and it is not always clear what is optimal for student learning. For example, do the challenges in simulating a batch distillation column enhance or confuse basic concept acquisition?

At WPI distillation is taught in the sophomore year throughout a project-based spiral curriculum. All projects are team-based. We introduce basic concepts early in the sophomore year then revisit distillation throughout the year with successively more complex assignments and projects. This includes at least two lab experiences. The first experiment uses a batch column operated at total reflux to introduce students to multistage distillation including efficiency and energy balances at total reflux. A follow-on course explores pressure swing distillation without a lab component. Near the end of the year, the batch column operated at a constant external reflux ratio is used again in a lab project. This project engages students in process dynamics and challenges them to compare differences between analysis and reality using the Rayleigh analysis and to compare simulated dynamics (using Aspen BatchSep®) to observations.

Large enrollments forced us to "outsource" lab work by taking advantage of the available but remotely-operated column at UTC. The column is the same as WPI's but is accessible and controllable through a web-based interface.

## Methodology

At recent AIChE meetings we presented results that showed student learning in remote and hands-on environments was essentially equivalent and that students who did simulation-only (no lab work) "experiments" might have demonstrated a slight edge in learning concepts related to process dynamics. This past year all student teams conducted hands-on experiments (remote or local) and all were supplemented with simulation runs. We compared learning among these cohorts.

Teams were four students each (teams were self-selected), and they were randomly assigned to cohorts. Six teams ran a local bubble-cap column (hands-on), six teams ran a local sieve tray column (hands-on), six teams ran the UTC column (remote) and all teams used the simulation. All teams conducted identical experiments with identical assignments. Simulations required a base case assignment but made additional runs were made varying reflux ratio, heat input, and initial compositions. Analysis and project reporting requirements were the same for all cohorts.

Evaluation had three components with direct and indirect assessments, giving us some degree of triangulation. All cohorts completed surveys managed by a third partner: UIUC. The online survey included closed-end, Likert-scaled responses and open-ended questions. The open-ended questions included student attitudes but also probed team processing since we wanted to learn more about how each team handled roles, logistics, scheduling, data collection, and analysis.

Course instructors compared final reports from all teams. A pre/post in-class quiz compared individual learning for students in all cohorts in a quantitative manner. Qualitative measures include evaluating student attitudes about the experience and assessing students' improvement in ABET outcomes such as: (b) – design, analysis and interpretation of data; (d) – functioning on multidisciplinary teams; (g) – effective remote communication; (h) - have broad education necessary to understand the impact of engineering solutions in a global, economic, and environmental societal context; and (k) - using techniques, skills, and modern engineering tools (such as computers and web interfaces) necessary for engineering practice.

## Results

### *Logistics*

As previously reported, having completed two cycles of this pilot we have a fairly good understanding of what it takes to put 20 teams through this lab project. It would have been essentially impossible to run all teams through the local WPI lab or the UTC lab, during the time allotted for the course. However, splitting effort between 2 columns at WPI and UTC makes logistics workable. We learned that running either the remote or the simulated column presents different but no more burdensome scheduling issues than running the local column.

### *Student Attitudes*

Survey results will not be available until sometime after May, 2011. Previous results indicated that there are few, if any, serious issues. Students don't mind being assigned to cohorts nor was there any indication of teams feeling they lost out on something because they weren't in a different cohort. New results examining how student perceptions of simulations helped (or didn't) will be available over the summer and will be analyzed in time for the conference.

*Student Learning—Analysis and Critical Thinking*

All teams produced a significant final report that was graded by the WPI instructor using the same rubrics. We evaluated presentation quality, demonstration of appropriate concepts and ability to critically analyze and explain results, particularly system dynamics and differences between theory and observations. We examined learning issues as they specifically related to two important concepts: distillation column dynamics and column analysis using the Rayleigh method. We also gave a pre/post quiz regarding simple concepts in these areas to compare differences among cohorts. The WPI semester ends on May 3 so results from this analysis are not available as of this writing but will be presented at the conference.

Our previous content analysis of the reports showed there were common deficiencies that were independent of cohort. These included mistakes with energy balance calculations; and another typical confusion over quantitative comparison of experiment to Rayleigh analysis. Students expect exact comparisons yet reality is never that predictable. In too many cases, the report presentation quality was marginal. This result is not unusual given the end-of-semester time demands.

Our report analysis is ongoing and details will not be available as of the deadline abstract. However, a preliminary observation is that difficulties with Aspen BatchSep® may have caused unnecessary frustration with some teams. The impact on learning is not known at this time.

The authors note that this is very preliminary and will only be confirmed (or not) when the content analysis is done. Previously we've used the psychology of "presence" (Ma and Nickerson, ACM Computing Surveys, v. 68 no. 3, article 7, 2006) as a framework for interpretation. Student learning in lab environments is connected to their perception of presence and the process of collaboration. Presence can be physical, telepresence, or virtual (for example, a simulation). In our case the remote teams experienced telepresence, local teams experienced physical presence, and simulation adds a virtual presence to all teams. Hence a hands-on lab that has minimal physical interaction may not be significantly different than a remote one where the interaction is mediated via a computer. In fact, the UTC column has three cameras that can be manipulated by students bringing an element of physical presence into the mix. However, the Aspen simulation is viewable only through the standard Aspen interface. It is not clear whether students made synergistic connections between simulation and lab, or if they segmented the two experiences resulting in less than optimal learning. By meeting time we will understand these issues and also know more about students' collaboration process, and how that dynamic affected interactions, and hence learning, within each cohort.

See more of this Session: Critical Thinking In the Chemical Engineering Curriculum
See more of this Group/Topical: Education

# (19g) Promoting Students' Critical Thinking Skills Using the Unit Operations Laboratory Experience

**Ronald L. Miller**, Colorado School of Mines, Golden, CO and John M. Persichetti, Chemical Engineering 1600 Illinois Street, Colorado School of Mines, Golden, CO

The unit operations laboratory represents an idea context for helping students improve their critical thinking skills if experiments, data analysis, and assessments are designed in ways that require students to go beyond routine data collection and reporting. At the Colorado School of Mines, the unit operations laboratory is taught as an intensive 6-week summer session in which critical thinking (with a focus on higher-order thinking) is a key objective that students must demonstrate. An analysis of student oral and written reports shows that students can progress from naïve “black-white” thinking through stages of multiplicity (“all explanations for the data are equally good”) to developing a reasonably sophisticated plausible explanation for explaining data trends but that the development takes weeks of intensive practice and feedback. Techniques for promoting and monitoring this development will be discussed using examples of student work to illustrate the emergence of higher-order thinking skills as the course progresses.

See more of this Session: Critical Thinking In the Chemical Engineering Curriculum
See more of this Group/Topical: Education

# (19h) Developing Critical Thinking Skills for Undergraduates, Professionals and for Graduate Students

**Donald R. Woods**, McMaster University, Hamilton, ON, Canada

In the McMaster Problem Solving program we have developed a workshop to develop critical thinking skills. The elements include three basic components, a 10-step process, separating facts from opinions and the use of diagramming. A brief description with interactive illustrations of the workshop will be given. This workshop has been given to undergraduates, professionals in nursing and engineering and to process trouble shooters. For graduate students my approach is an individualized mentoring approach whereby the students and I critique refereed papers in our research area. The guidelines used in this approach will be given.

See more of this Session: Critical Thinking In the Chemical Engineering Curriculum
See more of this Group/Topical: Education

# (19i) Using Concept Mapping to Teach Students Critical Thinking for Fragmenting Senior Design Projects Into Coherent and Manageable Tasks

**Paul Blowers**, Chemical and Environmental Engineering, The University of Arizona, Tucson, AZ and Armin Sorooshian, Chemical and Environmental Engineering, University of Arizona, Tucson, AZ

The senior capstone design sequence is challenging to students for several reasons, many of them concerning the complexity of topics and the need to balance the large workload around their other schedules. Students seem to be vaguely aware that the capstone experience is to integrate prior topics they have passed into a coherent whole, but have often forgotten details they had previously mastered. This work introduces a systematic approach for teaching students how to quickly make progress on a complex and comprehensive project. A series of scaffolded assignments leads students through the new process. First, students read a process description and draw a block flow diagram representing the major pieces. They then eliminate blocks that would be handled off-site or that represent processed materials that are outside the scope of their own study. They then consider the specific engineering courses they have taken and then identify, in the broadest sense, what courses could be needed to investigate pieces inside the block. The layer of detail increases as they then make a list of topical areas from each class they suspect may be necessary in a block; if they are unsure and do not know enough about a block to apply chemical engineering topics confidently, they leave question marks in that block to indicate they need to quickly assimilate more information in that area. The final cut of detailing at this level is to then list the raw physical properties like density, molecular weight, etc. that would be needed in order to begin calculations. A sample document leads the student through these developmental levels of fragmenting a complex design into components they can begin to analyze prior to the students applying the techniques to another case and then beginning their own design projects. Specifically, discussions are held on how to use the knowledge, or lack thereof, for each block in order to delegate work effort on the team in order to reach their goals. Feedback indicates that students are more focused on their projects and are able to jumpstart quickly to levels of performance that normally took 5-6 weeks of frustrated effort as experienced by students in years prior to implementation of the critical thinking approach for design.

See more of this Session: Critical Thinking In the Chemical Engineering Curriculum
See more of this Group/Topical: Education

# (21a) Thoughts On Unit Ops Lab: What Was, Is, and Ought

**Matthew J. Cline** and B. Erik Ydstie, Chemical Engineering, Carnegie Mellon University, Pittsburgh, PA

The responsibilities of engineering educators continuallyevolve with, adapt to, and even influence industrial and societal conditions. Facultiescontinuously struggle over what, and how, to best prepare students. Prof.Ogden at Bucknell asserted in 1934 that a certain minimum amount ofsemi-plant equipment must be used in order to present more effectively theapplications of the theory of chemical engineering as actually practiced inindustrial operations. The Society for the Promotion of EngineeringEducation, now the American Society for Engineering Education, studied chemicalengineering laboratory instruction and equipment for the next year. The natureof that study (and of subsequent studies), along with the attendant findingsand recommendations, illustrate the status quo to a large extent. Thehistorical development of chemical engineering laboratories, in particular theteaching of methods for developing and scaling industrial production processes, will be traced. Juxtaposed with the underlying motivations for more recentcalls for change, one will see why undergraduate chemical engineeringlaboratory equipment and teaching practices continue to shift focus away frompilot-plant scale operation and process design. The ongoing changes to theUnit Operations lab at Carnegie Mellon University demonstrate increased focus,in particular, to engineering practice skill development and to student-led,research or development projects of a wide variety of topics. Furthermore,each project may have different desired educational outcomes. This *is* and,we assert, *ought* to be. The likely consequences designers of chemicalprocesses and producers of chemical products may see will be discussed.

See more of this Session: Design, Construction and Operation of Unit Operations Labs and Pilot Plants
See more of this Group/Topical: Process Development Division

# (21b) Scale-up of Membranes for Separation of Hydrogen From Syngas for Carbon Capture

**David H. Anderson**, Carl R. Evenson, John D. Faull, Darcie D. Bailey and Doug S. Jack, Eltron Research & Development, Inc., Boulder, CO

In partnership with the Department of Energy (DOE) and Eastman Chemical Company, Eltron Research & Development Inc. is developing a process technology that separates hydrogen from carbon dioxide in syngas, enabling clean power generation and carbon capture in coal-fired Integrated Gasification Combined Cycle (IGCC) power plants. The process is based on membranes that permeate pure hydrogen and retain the $CO_2$ at high pressure to aid its eventual sequestration. In addition, membrane separation can achieve nearly complete carbon capture depending on process configuration. Scale-up is in progress for the membrane technology including pilot plant testing on coal-derived syngas. The status of the scale-up steps and recent results of the testing will be presented.

See more of this Session: Design, Construction and Operation of Unit Operations Labs and Pilot Plants
See more of this Group/Topical: Process Development Division

# (21c) Integrating Membrane Separations Into the Chemical Engineering Laboratory

**Daniel Anastasio**, Chemical, Materials, and Biomolecular Engineering, University of Connecticut, Storrs, CT and Jeffrey R. McCutcheon, Department of Chemical, Materials and Biomolecular Engineering, University of Connecticut, Storrs, CT

Crossflow reverse osmosis (RO) and forward osmosis (FO) systems were constructed to introduce senior-level chemical engineering students to key facets of membrane separations. During the RO experiments, students examined several commercial RO membranes, measuring water flux and salt rejections at various pressures, flow rates, and feed water salinities. During FO experiments, students examined a single commercial FO membrane, measuring water flux and feed conductivity at various draw concentrations, flow rates, and temperatures. Ultimately, students used the data they collected to calculate each membrane's hydraulic permeability constant and a variable indicative of membrane selectivity (salt rejection for RO and salt permeability for FO). The data presented is taken directly from the laboratory reports of undergraduate seniors at the University of Connecticut. These data were compared to the manufacturer's specifications. Boundary layer theory was utilized in both experiments to predict theoretical permeability and/or flux values after accounting for concentration polarization. Together, these versatile systems have granted students a thorough understanding of membrane separations and mass transport phenomena in the context of a capstone undergraduate laboratory course.

See more of this Session: Design, Construction and Operation of Unit Operations Labs and Pilot Plants
See more of this Group/Topical: Process Development Division

# (21d) Project Based Learning In Senior Unit Operation Laboratory At the University of Michigan

**Henry Y. Wang**, Pablo LaValle and Matthew A. Robinson, Department of Chemical Engineering, University of Michigan, Ann Arbor, MI

The University of Michigan has one of the oldest unit operations laboratories in this nation having been part of our core curriculum in Chemical Engineering since the founding of the department. All students must work on several types of process equipment such as a reactor, distillation column and evaporator to enhance their understanding of the fundamental technical knowledge from various lecture courses. During the past few years, we have modified the existing equipment in the laboratory and integrated them to form a simulated biodiesel production facility. In addition to the advantages of providing the students a clear understanding of the interconnectedness of these unit operations in a real production process, this approach also allows the laboratory to become an invaluable teaching tool for various soft skills fundamental to our profession. Skills such as team work, time management, the importance of producing practical information in a timely manner for other groups or teams are emphasized as the students work in an interdependent group environment to develop the specification and design of a large industrial scale facility based on the information gathered from an actual pilot plant or laboratory environment. Additionally, the students are encouraged to produce by the end of the semester a single group report on the viability of the process integration and improvement for the entire plant. We shall share our experience in this transformation of the senior unit operation laboratory in this talk.

See more of this Session: Design, Construction and Operation of Unit Operations Labs and Pilot Plants

See more of this Group/Topical: Process Development Division

# (21e) Using Pilot Plants In a Capstone Unit Operations Laboratory Course to Create a Chemical Manufacturing Experience

**John F. Sandell**, David W. Caspary and Jason M. Keith, Chemical Engineering, Michigan Technological University, Houghton, MI

Unit Operations courses are typically considered "capstone" courses in many undergraduate chemical engineering programs. At Michigan Technological University, each semester of the senior year has a required 3 credit unit operations focused course. This paper details the content for the second of these two courses which is a unique, manufacturing focused Chemical Plant Operations course utilizing two fully automated pilot plants. The pilot plants include a two-story, 30 gallon jacketed, batch polymerization reactor (PDMS reactor) and a three-story, distillation column (Solvent Recovery Unit or SRU). Both pilot plants are part of the Process Simulation and Control Center (PSCC) at Michigan Technological University. The course incorporates participation by numerous industrial speakers as well as utilization of an Emerson Process Management Delta V™ control system, Foundation™ fieldbus and Hart-enabled process instrumentation, an OSIsoft PI™ data acquisition system, Six Sigma quality control tools, and ISA S-88 methods. Lecture material includes quality management, the application of statistical process control, and current trends in quality manufacturing. Students complete two "project cycles" requiring the completion of proposals prior to each run followed by the submission of comprehensive written and oral reports.

See more of this Session: Design, Construction and Operation of Unit Operations Labs and Pilot Plants
See more of this Group/Topical: Process Development Division

# (21f) Distributed Control System Implementation for Undergraduate Labs

**Peyton C. Richmond**, Chemical Engineering Department, Lamar University, Beaumont, TX

A Honeywell Experion DCS system was recently acquired and configured as part of a multi-year plan to expand and remodel the Lamar University Chemical Engineering Department's Unit Operation and Process Control Labs. The first phase is complete: a new "Control Center" with 10 Flex Stations was established and the Department's two Process Control Experiments were converted from independent single loop control to DCS based control. Those Control Experiments were used by our undergraduates for the first time during the spring 2011 semester. The next phase will involve extending the DCS system to the Unit Operations laboratory which is located in a separate building. This presentation will include some of configuration choices made and the lessons learned during the acquisition and configuration of this industrial DCS system for academic use.

See more of this Session: Design, Construction and Operation of Unit Operations Labs and Pilot Plants
See more of this Group/Topical: Process Development Division

# (21g) "I Thought PTFE Tape Was Optional": Teaching Practical Engineering Skills As a Component of Unit Operations

**Eric P. Codner**, Chemical and Biological Engineering, University of Wisconsin - Madison, Madison, WI

Undergraduates are increasingly skilled at computer simulation, seemingly at the expense of mechanical aptitude and knowledge of process fabrication and operation techniques. While this may not be a serious problem in large organizations, many beginning engineers enter small companies in which they are the sole technical resource, and a lack of practical skills may substantially hinder their professional progress.

The unit operations course at the University of Wisconsin – Madison has traditionally required students to design and construct several experiments during the term. Over the past four years our approach has undergone considerable development, resulting in the following initiatives:

- Students in the unit operations and fluid mechanics laboratories are now required to assemble portions of the equipment prior to performing experiments, in order to familiarize them with tools, materials, and equipment commonly found in process environments.
- Many experiments are structured to include a troubleshooting or process improvement component, in which students analyze the operation of the experiment and make determinations concerning maintenance, backup operation, or repair.
- There is increased emphasis on measurement uncertainty and statistics, and most measurements are obtained using data acquisition systems. A number of the experiments employ PLCs for measurement and control purposes.

This paper discusses the current state of the senior laboratory curriculum, student response to the changes, and ongoing challenges such as minimizing equipment damage and successfully training teaching assistants.

See more of this Session: Design, Construction and Operation of Unit Operations

# (82a) Himmelblau Award Presentation

**David A. Kofke**, Chemical and Biological Engineering, University at Buffalo, The State University of New York, Buffalo, NY

The recipient of the David Himmelblau Award for Innovations in Computer-Based Chemical Engineering Education will deliver a lecture based on the work recognized by the award.

See more of this Session: Computing and Technology In Chemical Engineering Education I
See more of this Group/Topical: Computing and Systems Technology Division

# (82b) Expertise In Chemical Process Modeling

**Paul M. Mathias**, Fluor Corporation, Aliso Viejo, CA

Chemical process modeling is based upon a small set of very general equations, and then particular problems are treated as a special case of these fundamental equations. This quantitative approach has resulted in engineers with strong mathematical skills, and the development of powerful software that provides rigorous solutions to highly complex problems. But an unintended consequence is that chemical engineers have lost a tacit understanding of physical properties and process behavior, and, as a result, process designs have suffered and the trouble-shooting skills of chemical engineers have diminished. In other words, chemical engineers are not attaining expertise.

So, how do we improve the expertise of chemical engineers? This presentation will build upon previous work of the author [1, 2, 3], and demonstrate that internalizing the patterns of physical properties and process behavior is a vital component to attain process-modeling expertise. Interestingly, key solutions to the problem are enabled by software advances, namely the easy availability of comprehensive property databases and advances in graphics and visualization.

1 Paul M. Mathias, "Interactive Visualization of the McCabe-Thiele Diagram," Chemical Engineering Progress, December, 105(12), 36-44 (2009).

2 Paul M. Mathias and Suphat Watanasiri, "Some Patterns of Fluid Phase Behavior," J. Chem. Eng. Data, Article ASAP, 15 March 2011.

3 Paul M. Mathias and Suphat Watanasiri, "Patterns of Chemical Process Modeling," presented at AspenTech Global Conference, Washington, D.C., 23-25 May 2011.

See more of this Session: Computing and Technology In Chemical Engineering Education I
See more of this Group/Topical: Computing and Systems Technology Division

# (82c) Students Learn Fundamentals of Process Operations and Control Using Dynamic Simulator of An Integrated Gasification Combined Cycle (IGCC) Plant with CO2 Capture

**Debangsu Bhattacharyya**[1], Richard Turton[1] and Stephen E. Zitney[2], (1) Department of Chemical Engg., West Virginia University, Morgantown, WV, (2) Collaboratory for Process and Dynamic Systems Research, National Energy Technology Laboratory, U.S. Department of Energy, Morgantown, WV

While today's undergraduate chemical engineering students often learn about the importance of dynamic simulation and process control, they typically do not spend considerable time gaining practical experience in plant-wide operations and control. At West Virginia University's (WVU) National Research Center for Coal and Energy, students have access to a real-time dynamic simulator for an integrated gasification combined cycle (IGCC) plant with carbon capture and compression. Fueled with coal, petroleum coke, and/or biomass, the gasification island of the IGCC plant consists of two oxygen-blown, downward-fired, entrained-flow, slagging gasifiers with radiant syngas cooler, two-stage sour shift reactors followed by a dual-stage acid gas removal process. The combined cycle island consists of two F-class gas turbines, steam turbine, and a heat recovery steam generator with three-pressure levels. The IGCC dynamic simulator is operated under the auspices of the National Energy Technology Laboratory's (NETL) Advanced Virtual Energy Simulation Training And Research (AVESTAR™) Center.

Using the IGCC operator training system (OTS), students gain hands-on, interactive, simulator-based experience with normal base load operations, cold, warm, and hot startups, plant shutdown, load shedding and following, and fuel switching. The students also investigate abnormal situations in real-time, including equipment malfunctions and failures, together with changes initiated through actions from the instructor station or from field operators. In addition, students learn how the plant responds dynamically to changes in the manipulated inputs, as well as how the control system impacts plant performance, stability, robustness and disturbance rejection characteristics. The students study the step and ramp response of the simple linear, lower order processes along with more complicated higher-order and nonlinear processes. The students observe the "real-life" transient response of the processes with non-minimum phase behavior and face the challenge of controlling these systems with PID only control. Various open-loop unstable processes are studied and the role of feedback mechanism in making the system closed-loop stable is explained by demonstrating a number of familiar systems in this simulator environment. Other than the conventional control structures, the simulator is utilized for studying more complex control structures such as cascade control, feedforward control, ratio control, override control, split-range control, etc. The system is also capable to monitor, evaluate, and report the performance of the students through a trainee performance monitoring (TPM) module. The students are challenged by asking them to control the plant within a set of operating limits in the face of the disturbances and simulated process faults.

See more of this Session: Computing and Technology In Chemical Engineering Education I
See more of this Group/Topical: Computing and Systems Technology Division

# (82d) Teaching Design Using the CACHE Learning Resource Center

**Warren D. Seider**, Chemical and Biomolecular Engineering, The University of Pennsylvania, Philadelphia, PA

The initial version of a "Guide to Teaching Design with Internet Links" will be introduced. It describes a framework of materials for teaching product and process design with links to recommended materials from many sources including textbooks, monographs, completed design case studies, design problem statements, design software, POWER POINT lectures, and related materials prepared by design instructors. Several area editors have assembled materials in 10 design areas including product design, process integration, process simulation, the environment, safety, design optimization, ... This presentation is intended to introduce some of the features in the guide and show how to use them effectively.

See more of this Session: Computing and Technology In Chemical Engineering Education I
See more of this Group/Topical: Computing and Systems Technology Division

# (82e) Peer Evaluation In Chemical Engineering Design Through Wikis

**Caryn L. Heldt**, Chemical Engineering, Michigan Technological University, Houghton, MI

Practicing engineers are put into situations where their work will be critically evaluated by others and they are also expected to critically evaluate the work of others. To help practice these skills, we used peer evaluation through wikis as a method for the chemical engineering design teams to both evaluate other's work and to be evaluated. Students, in groups of four, were asked to design a process that will manufacture products ranging from acetone and $CO_2$ scrubbing solvents to ibuprofen and glucose. After four weeks, a progress report was due. This progress report was submitted to the instructor for grading, as well as posted on a wiki site. The students were assigned two groups to assess and they were given the suggestion to look at both the technical and communication aspects of the report. The groups were assigned to allow the students to read a variety of projects from different sections of the class and also for the strong groups to be paired with weaker groups, and vice versa. The learning objectives of the exercise were to (1) enhance written communication skills due to (i) peer review of their work and (ii) requirement to express opinions in a written format, (2) develop critical thinking skills when evaluating other projects, and (3) obtain an in-depth understanding of their own projects from peer-review comments. Surveys were also done to assess the students' attitudes toward the wiki assignment. At the conclusion of the project, the individual section instructors were all impressed with the evaluations done by the students. The students were interested and engaging in their comments and often came up with ideas that were not thought of by the instructor. This paper will present some of the positive and negative aspects of using wiki's in engineering education and explore student reaction to this educational model.

See more of this Session: Computing and Technology In Chemical Engineering Education I
See more of this Group/Topical: Computing and Systems Technology Division

# (86a) REU Site: Materials and Systems Biology Research In Biotechnology and Biomedicine

**Juergen Hahn**, Department of Chemical Engineering, Texas A&M University, College Station, TX and Dan Shantz, Chemical Engineering, Texas A&M University, College Station, TX

**PROJECT ELEMENTS**

New REU Site

REU Site: **Materials and Systems Biology Research in Biotechnology and Biomedicine**

Principal Investigator: Juergen Hahn, Ph.D., Associate Professor, Chemical Engineering

Submitting Organization: Texas A&M University

Location: Texas A&M University, College Station, Texas

Main field and subfields: chemical engineering, chemistry, physics, biology, or mathematics

Number of Undergraduates: 10 students annually for a total of 30 students

Participation: Summer—10 weeks

No related international, ethics, or RET component

Contact: Juergen Hahn, (979) 845-3568 begin_of_the_skype_highlighting (979) 845-3568 end_of_the_skype_highlighting; Hahn@tamu.edu

http://www.che.tamu.edu/undergraduate-program/undergrad-research-reu/

## PROJECT SUMMARY

The overarching goal is to engage annually 10 undergraduate students from chemical engineering or closely allied disciplines in materials or systems biology research with faculty serving as mentors to prepare and encourage the students to pursue graduate education.

Objectives include:

1. Annually recruit 10 undergraduates from off-campus with an emphasis on recruitment of students from underrepresented groups;
2. Open new research opportunities to students who have had little or no previous research experience, providing insights into research methodologies, self-discipline, team integration, reporting and exchanging new knowledge;
3. Provide information on benefits of attending graduate school, the application process, and encourage students to attend graduate school in chemical engineering;
4. Enable interactions of the REU students with visiting undergraduate researchers from Tsinghua University and Universidad de los Andes who participate in existing departmental programs;
5. Provide community building opportunities among REU students, TAMU graduate students, visiting international undergraduate researchers, and faculty mentors; and
6. Provide opportunities for REU students to present and publish with faculty mentors.

Targeted student participants include rising junior and senior students in chemical engineering and from chemistry, physics, biology, or mathematics who express an interest in engineering graduate school. Students will be required to have a cumulative GPA of at least 3.3, since that is the minimal requirement for acceptance to graduate school for TAMU Chemical Engineering. Existing relationships with schools in the region including Prairie View A&M University (an HBCU) and two Hispanic Serving Institutions: Texas A&M University Kingsville and the University of Houston-Downtown will be leveraged to maximize participation of students from underrepresented groups. Local recruitment of students supported by non-NSF funds will target underrepresented students at TAMU by recruiting through student engineering organizations such as the Society of Women Engineers, National Society of Black Engineers, Mexican American Engineers and Scientists, and the Society of Hispanic Professional Engineers. TAMU Engineering is one of the leading institutions in graduating

Hispanic engineers at the baccalaureate level, which will lend itself to participation by this increasing minority population in Texas and the nation.

This REU site will have a focus of chemical engineering approaches to materials science and systems biology research in biotechnology and biomedicine. A combination of computational and experimental research will be conducted in these areas. The intellectual merit of the proposed research is to bring the rigorous analytical approach characteristic of chemical engineering to problems of interest in biotechnology and biomedicine. The REU students will participate in developing the next generation of products and processes based on biological molecules and organisms.

The broader impact will be to target recruitment of students to graduate school to offer research experiences to those who may not normally have such an opportunity, particularly women, underrepresented minorities, and first-generation college students, making a significant contribution toward enhancing the diversity of the research and education infrastructure at Texas A&M University and toward increasing the production of African American and Hispanic engineers and scientists. This project creates a model program for diverse populations that addresses the exposure to research by undergraduates as well as awareness of the pathway to graduate school. A more diverse engineering and science professoriate benefits society significantly, not only in Texas but nationally, in promoting a more equitable participation of minorities in the engineering and scientific disciplines.

See more of this Session: Educational Proposal Submissions: Successful Examples
See more of this Group/Topical: Education

# (86b) Ongoing 12-Year Long Research Experiences for Undergraduates On Novel Advanced Materials and Processing

**Christos Takoudis**, Departments of Chemical Engineering and Bioengineering, University of Illinois at Chicago, Chicago, IL

The goal is a stronger and diverse U.S. workforce of scientists, engineers, and technologists as well as discovery across the frontier of science and engineering. This requires significant efforts to attract and prepare U.S. students to be highly qualified members in the global technical workforce, to support innovative research thereby enabling people at the forefront of discovery to make important and significant contributions to science and engineering knowledge, and to increase opportunities for underrepresented individuals to conduct high quality, competitive research activities.

Objectives of our ongoing Research Experiences for Undergraduates (REU) Site for 12 straight years include development and enhancement of the students' creativity and ethics in science and engineering, safety training in research laboratories, fostering the understanding of the research process and intellectual property, giving the students experience at writing and presenting their work, and uniquely preparing them for interdisciplinary collaborations in their future careers. Our REU program has already included a total of over 120 undergraduates from the U.S. with a wide spectrum of scientific backgrounds.

Outcomes, benefits and deliverables of our REU program included professional training on safety in the laboratory along with certificates of laboratory safety and waste management, tutorials on making oral presentations and writing technical papers, one-on-one interactions of the REU students with faculty and graduate students from several disciplines in an exciting scientific environment in novel research in emerging technologies and cutting-edge research, literature study on their research topic, hands on experimentation or modeling, weekly presentations on motivation, objective, status, and future planning of research, intermediate and final reports on research, visits of other research labs in the area, acquired confidence in working independently, application of the students' own ideas to state-of-the-art research, planning and carrying out experiments in laboratories, submission of final reports/papers to the Journal of Undergraduate Research (JUR) at the University of Illinois at Chicago (http://jur.phy.uic.edu/ ) that are anonymously reviewed by faculty, and submission of additional research manuscripts to other refereed journals.

Another important unique aspect of the REU Program was the continuous and intense cross-fertilization of ideas among undergraduates with very rich and diverse (academic, cultural, socioeconomic) backgrounds.

See more of this Session: Educational Proposal Submissions: Successful Examples
See more of this Group/Topical: Education

# (86c) NSF IGERT and REU Programs In Engineered Bioactive Interfaces and Devices: An Integrative Approach

**Kimberly W. Anderson**[1], Dr. J. Zach Hilt[2] and Bruce J. Hinds[1], (1)Chemical and Materials Engineering, University of Kentucky, Lexington, KY, (2)Chemical and Materials Engineering, The University of Kentucky, Lexington, KY

In 2007, the University of Kentucky was awarded an Integrative Graduate Education and Research Training (IGERT) grant from the National Science Foundation. This program supports Ph.D. trainees pursuing research and education focusing on Engineered Bioactive Interfaces and Devices. The program is highly multidisciplinary including 20 investigators and 23 trainees from Chemical Engineering, Materials Engineering, Electrical Engineering, Biomedical Engineering, Chemistry, Pharmaceutical Sciences, Molecular and Cellular Biochemistry, and Anatomy and Neurobiology. The research focus has three general themes including development of synthetic architectures, cell and protein interactions with these architectures, and integration into devices. In 2009, the University was also awarded a NSF grant to support a Research Experience for Undergraduates (REU) Program focusing on the same area. Since its inception, the REU Program has provided support for 43 undergraduate students from all over the country to pursue research for ten weeks during the summer. Because both of these programs focus on the same area and include many of the same faculty members, it has provided us with the unique opportunity to integrate the two programs. To integrate the programs, the IGERT trainees are assigned one or two REU students and serve as their mentors during the program. They also interact during technical activities including a poster session and panel discussions lead by the IGERT trainees. IGERT trainees are also asked to organize social activities for the REU students. The culmination of the REU Program is the Poster Competition and Awards Ceremony, where the REU students present their research and are judged by the IGERT trainees. This event is open to the public and not only recognizes REU students for their efforts during the summer but the IGERT trainees are also recognized for their participation as mentors. The day-to-day interactions between these groups benefit the REU students as they not only interact with their research advisors and lab mentors, but can also seek out advice from their IGERT mentors regarding research and graduate school opportunities. The IGERT trainees benefit from these interactions as they learn the value of outreach and gain experience in advising and mentoring undergraduate students. In many cases, these mentorships have continued beyond the summer as the REU students and their IGERT mentors continue to keep in touch. Details of the research, educational and mentorship activities will be presented and both the successes and the challenges of these activities will be discussed.

See more of this Session: Educational Proposal Submissions: Successful Examples
See more of this Group/Topical: Education

# (86d) (NPT)2: Project and Center Grants Through the NSF Advanced Technology Education Program

**Steve R. Duke**, Chemical Engineering, Auburn University, Auburn, AL, Harry T. Cullinan, Alabama Center for Paper and Bioresource Engineering, Auburn University, AL and T. J. Murphy, National Network for Pulp and Paper Technology Training, Thomasville, AL

This presentation will discuss successful proposals that have supported the National Network for Pulp and Paper Technology Training $(npt)^2$, a National Science Foundation Advanced Technology Education Center. $(npt)^2$ is the culmination of more than 15 years of support for technology education and training in the paper and chemical industries. Notable grants to support this work include three multiyear and multi-institutional project grants and two center grants from the NSF ATE program coupled with USDA, industry, and state support totaling more than $15 million. Best practices will be presented related to faculty development, laboratory development, industry supported scholarships and curricula, and college network partnerships.

The National Network for Pulp & Paper Technology Training (www.npt2.org) was created to provide the pulp and paper sector of the US forest products industry with a technologically advanced workforce (TAW) by offering exciting and effective education and training. The National Network is an alliance of community colleges, universities, industry partners and the National Science Foundation offering a proven program for recruiting, training, placing, and retaining future pulp and paper workers by providing technical curriculum, continuing education, scholarships and internships. The National Network for Pulp & Paper Technology Training's scholarships – internships – jobs model succeeds because its partnerships address the complete cycle of worker training and development, from recruitment of new students to retention of existing workers.

The lead institution for $(npt)^2$ is Alabama Southern Community College (Thomasville AL) and the lead university partner is Auburn University through the Alabama Center for Paper and Bioresource Engineering in the Department of Chemical Engineering.

# (86e) Interdisciplinary Undergraduate Option Network In Nanoscience and Molecular Engineering Education

**Rene M. Overney**, Chemical Engineering, University of Washington, Seattle, WA

The first interdepartmental NanoScience and Molecular Engineering (NME) Option Network has been established at the University of Washington, involving multiple departments from Bioengineering, Chemical Engineering, Materials Science and Engineering, Electrical Engineering, Mechanical Engineering, Chemistry to Physics, with an extension to community colleges (North Seattle Community College and Edmonds Community College). Within the NME network, discipline-linked NME Option programs are offered to undergraduate students providing *hands-on experience* and strong *interdisciplinarity* in NME, and integration in departmental undergraduate core curricula. Since the inception of the NSF NUE supported NME program in 2009,[1] the program has raised great interest from both students and faculty by building on the existing infrastructure, sharing resources responsibly and sensitively, operating in parallel with novel curricula developments, lowering barriers between disciplines, and offering through official departmental Options, credit friendly and opportune NME programs.

[1] National Science Foundation Nanotechnology Undergraduate Education (NSF NUE) program, Award Number 0938558.

See more of this Session: Educational Proposal Submissions: Successful Examples
See more of this Group/Topical: Education

# (86f) AIM AT NANOTEC

**Dr. Srinivas Palanki**, Department of Chemical and Biomolecular Engineering, University of South Alabama, Mobile, AL

In the past 15 years the field of nanotechnology has matured to the level where devices based on nanotechnology are emerging and making an impact to today's society. This field is truly interdisciplinary and requires a synthesis of knowledge from science and engineering principles to develop nanotechnology devices. In this talk, we demonstrate how recent innovations in engineering education, coupled with video creation technology, can be used to develop a set of modules that can be integrated into several sophomore, junior and senior level core courses in three different engineering departments (Chemical & Biomolecular, Mechanical, and Electrical & Computer Engineering). The modules are developed by faculty members who have research experience in nanotechnology and are integrated seamlessly with material covered in core engineering courses. Each of these modules involves problem solving skills at the levels of knowledge, comprehension, application, analysis, synthesis and evaluation. This allows instructors to use parts of the same modules in multiple courses. The specific objectives of the proposal are as follows:
• Develop 12 modules in an asynchronous, on-line learning format that teach science and engineering aspects of nanotechnology at multiple levels
• Implement thesemodules in several core and elective courses in Chemical & Biomolecular, Mechanical, and Electrical & Computer Engineering Departments.
• Assess outcomes and disseminate results to the engineering community.

See more of this Session: Educational Proposal Submissions: Successful Examples
See more of this Group/Topical: Education

# (86g) BioEMB Project - Introducing New Curricular Areas Into Undergraduate Education

**Claire F. Komives**, Department of Chemical and Materials Engineering, San Jose State University, San Jose, CA and Erik J. Fernandez, Department of Chemical Engineering, University of Virginia, Charlottesville, VA

The BioEMB project was funded as an NSF CCLI Phase I initially beginning in 2007. Funding rates at that time were about 10% for Phase I CCLI proposals. A Phase II grant was awarded to begin in 2009. The project was viewed favorably by reviewers because it addressed a clear need for the undergraduate curriculum, it had clear goals and measurable objectives, and it had a specific assessment plan. The successful work in the Phase I project enabled a quick and positive response to the first Phase II project submitted. This presentation will describe the aspects of the project that made it fundable and a brief overview of the project and outcomes will also be presented.

See more of this Session: Educational Proposal Submissions: Successful Examples
See more of this Group/Topical: Education

# (86h) Chemical Engineering Process Design Class As a Venue for Examining Group Dynamics

**Robert R. Beitle**[1], Douglas Behrend[2] and Rachel Schwartz[2], (1)Ralph E. Martin Department of Chemical Engineering, University of Arkansas, Fayetteville, AR, (2)Department of Psychology, University of Arkansas, Fayetteville, AR

I will complete the abstract within the next few days. We will present information on building a grant that looked at interactions between students when a blind person was part of a design team. The study was conducted during senior design class and required adaptive technology to be also developed.

See more of this Session: Educational Proposal Submissions: Successful Examples
See more of this Group/Topical: Education

# TEACHING CHEMICAL ENGINEERING IN EUROPE

*Martin J. Pitt, University of Sheffield, Sheffield, UK*

## Biography

*Martin Pitt worked in industry in R & D and as a project chemical engineer and chemical plant manager before taking a PhD as a mature student at Loughborough University, then becoming a lecturer in chemical engineering at the University of Leeds, then a Senior Lecturer at the University of Sheffield where he is now Coordinator of Design Teaching. He is a former Chairman of the IChemE Education Subject Group and is currently Chairman of the Working Party on Education of the European Federation of Chemical Engineering.*

## Introduction

A remarkable transformation of European higher education has taken place in what is known as the Bologna process. In this, qualifications throughout Europe have been harmonized along the lines of Bachelor, Master and Doctoral qualifications with the idea that (say) a Bachelor degree in one country in chemical engineering should be accepted in another country both for employment and for progression to Master and Doctor. Prior to this, qualifications for chemical engineers had a range of names and varied from 3 to 7 years.

## EFCE Working Party on Education

The European Federation of Chemical Engineering (EFCE) consists of 40 member societies in 30 European countries, who nominate about 1000 representatives to working parties or sections. The Working Party on Education (WPE) currently has members from 24 countries: including 27 academics and 7 industrialists, plus corresponding members from Australia, India and South Africa. The secretariat is jointly held by the Institution of Chemical Engineers (UK), DECHEMA (Germany) and Société Française de Genie de Procedés (France). The WPE organizes conference sessions and other meetings, but its work related to Bologna may be characterized by three parts: analysis, synthesis and development.

### *Analysis*

Firstly, a study was made of (pre-Bologna) degree (or equivalent) programs in chemical engineering and closely related fields in various European countries and the USA. This showed, for example, a strong tradition of industrial chemistry in France and (mechanical) process engineering in Germany, but considerable similarities between UK and US programs.

We also had to try and understand each other's educational systems, including methods of entry to university which varied from highly competitive to open to all. This was done by setting up websites to describe our own country's practice to others.[1]

### *Synthesis*

The second phase involved trying to decide what was and was not chemical engineering, and hence what should be in the education of a chemical engineer, irrespective of the title of the qualification. Should courses focussing on local pharmaceutical or paper industries be excluded if they did not feature design of distillation columns? How much engineering would a chemistry degree require to be called chemical engineering?

As a result, the WPE produced an outline syllabus including what it felt to be a minimum set of topics to be included in a first degree, but deliberately allowing room for other topics according to local conditions and possible specializations, plus incoming subjects. This had the flaws of a committee compromise, but was nevertheless well received.[2]

### *Development*

Over the next few years, the WPE totally revised these recommendations and arranged them in a way which matched both the Bologna agreement and developments in engineering education for bachelor and masters degrees.[3] This document was widely used, and was particularly helpful for those setting up new chemical engineering courses. It particularly focussed on learning outcomes rather than simple course content or length of studies. We believe that we were the first subject to do this in any detail and the latest document has been very well received. We were also cognizant of the accreditation procedures of the IChemE and EUR-ACE.[4] The most recent revision completes the Bologna process by extending the recommendations to Doctoral level.[5]

## References


1. EFCE WPE website http://efce.info/wpe.html

2. EFCE WPE (2000) The Chemical Engineering Core Curriculum Project http://efce.info/efce_media/Downloads/wpe_corecurricula-p-302.pdf

3. EFCE WPE (2005) EFCE Recommendations for Chemical Engineering Education in a Bologna Two Cycle Degree System http://efce.info/efce_media/Downloads/EFCE_Bologna_Recom_0905.pdf

4. European Network for Accreditation of Engineering Education: EUR-ACE http://www.enaee.eu/the-eur-ace-system/

5. EFCE WPE (2010) EFCE Recommendations for Chemical Engineering Education in a Bologna Three Cycle Degree System http://efce.info/Bologna_Recommendation.html

# New Curriculum to Culture Future Chemical Engineers

Baoguo WANG*, Jinli Zhang, uqian Xia, Jinyu Han

School of Chemical Engineering, Tianjin University, Tianjin 300072, P. R. China

**Abstract:** In this paper, chemical versity education philosophy is introduced. Chemical engineering program is used as an example to illustrate the curriculum design and key issues are discussed.

**Keywords**: Undergraduate education, Tianjin University, Chemical Engineering, Curriculum

## Introduction

Tianjin University (TJU), the first institution of higher learning in modern China, was founded on 2 October 1895 and known as Peiyang University in her early days. In September 1951, Peiyang University and Hebei Institute of Technology were merged to form Tianjin University. Since then, TJU has been recognized as one of the key national universities in China. Now, TJU employs about 4600 staff, and 1900 of them are academic faculty. The current student population comprises about 15000 undergraduates, 10000 postgraduates and 1000 international students. The mission of TJU is to advance knowledge and educate students in science, engineering, technology, and other areas of scholarship that will best serve China and the world in the 21st century.

## Education Philosophy

TJU adopts liberal education as a philosophy of education that empowers individuals with broad knowledge and transferable skills as well as a strong sense of values, ethics, and civic engagement. It also provides training in specific areas and the opportunity of personal development through social interaction. In General, a

* Author to whom all correspondence should be addressed. bgwang@tju.edu.cn

liberal education includes a general education curriculum that provides broad exposure to multiple disciplines and ways of knowing, along with more in-depth study in at least one field or area of concentration. It should be pointed out that the university education cannot equip graduates with all the knowledge and skills they will need to deploy over an entire career, thus it is vital that the university education can lay a solid foundation on which further education and training can build.

Tianjin University graduates should have the following qualities and skills:

(1) Profound respect for truth and intellectual integrity;

(2) Highly developed cognitive, analytic and problem-solving skills;

(3) Capacity for critical thinking, rational inquiry and life-long learning;

(4) Extensive knowledge of a particular discipline;

(5) International awareness and openness to the world, based on understanding and appreciation of social and cultural diversity and respect for individual human rights and dignity;

(6) Ability to participate effectively in team work while respecting individual differences;

(7) Leadership capacity and ability to plan work and to use time effectively;

(8) Understanding of the principles of environmental protection and sustainable development.

**Chemical Engineering Program**

There are 51 undergraduate programs at TJU and 37 of them are engineering programs. The chemical engineering program at TJU is one of the best four-year engineering program in China. It equips graduates for successful careers in chemical and related process industries. Students are exposed to the fundamental science underpinning a wide range of process engineering concepts, and there are plenty of opportunities to put that knowledge into practice in design and research exercises.

The aims of the chemical engineering program are to produce graduates who have a coherent understanding of chemical engineering, a sound theoretical grasp of the program combined with practical experience and an awareness of their responsibilities for the society and the environment. The graduates should be able to apply their knowledge and skills to complex, open-ended engineering problems. The graduates should be capable of becoming professional chemical and process engineers in industry or following a postgraduate route into a research, industrial or academic career. In order to achieve these aims, the Chemical Engineering Program has the following objectives:

(1) To recruit good students from a range of geographical, social and academic backgrounds.

(2) To produce graduates who have vision and the ability to address the challenges posed by society through the deployment of the skills and knowledge gained during their courses.

(3) To equip students with appropriate practical skills in information processing, data analysis, problem solving, teamwork, and interpersonal communication.

(4) To encourage students to develop responsible attitudes towards the needs of society and the environment in the application of their engineering and economic knowledge and to ensure that they have particular regard for the importance of safety and sustainable development in their industrial life.

(5) To encourage students to develop appropriate attitudes towards their own future professional development.

(6) To cultivate students the abilities of self-study and life-long learning.

(7) To provide a harmonic environment within the School such that students enjoy the University learning experience sufficiently to want to maintain contact with the School in its future recruiting, teaching, research and social activities.

**Chemical Engineering Curriculum**

University education consists of both curriculum and extra-curriculum activities. The undergraduate curriculum establishes a platform to train students' creativity, critical thinking and analytical skills, produce graduates who can provide innovative solutions to complex engineering problems.

The curriculum for the Chemical Engineering Program is outlined in Table 1. The high standard of research is penetrated directly into the curriculum design and other teaching links, ensuring the graduates well equipped with up-to-date techniques, methods and knowledge basis. The teaching and learning is orientated to the expressed needs of industry and has an excellent reputation. Moreover, the investment in facilities and staff at TJU can provide an excellent experience in a friendly and supportive environment.

In the chemical engineering curriculum, the subjects are divided into two types: compulsory and elective subjects. General electives are normally selected in the first two years, while specialized electives are selected in the last two years. In common with other engineering disciplines the practice of chemical engineering depends upon an understanding of the underlying science. The design of a chemical reactor requires knowledge of the kinetics of chemical reactions and the mathematical techniques which can formulate them into design equations. Not surprisingly, therefore, the semesters contain subjects in mathematics, physics and chemistry as well as subjects which are obviously engineering in nature.

Table 1 Chemical Engineering Curriculum

| Semester | Compulsory Subjects |
|---|---|
| 1 | English(A), Mathematics(A), Inorganic Chemistry(2A)+Lab, Information Technology, Introduction to Chemical Engineering |
| 2 | English(B), Mathematics(B), Physics(2A), Inorganic Chemistry(2B)+Lab, Physical Chemistry(A)+Lab, Introduction to Process Design |
| 3 | English(C), Linear Algebra, Physics(2B), Physics Lab A+B, Organic Chemistry (A), Physical Chemistry(B)+Lab, Physical Chemistry Lab(B), Chemical Process Safety |
| 4 | Organic Chemistry(B)+Lab, Engineering Mathematics, Fluid Mechanics, Heat Transfer, Design Project 1, Chemical Engineering Thermodynamics, Environmental Protection and Sustainable Development, Research Project 1 |
| 5 | Instrumental Analysis & Lab, Chemical Equipment Design, Mass Transfer, Chemical Engineering Lab 1, Chemical Process Flowsheeting, Introduction to Biochemical Engineering, Chemical Reaction Engineering |
| 6 | Design Project 2, Analysis and Synthesis of Chemical Processes, Transport Phenomena, Chemical Engineering Lab 2, Research Project 2, Instrumentation and Control, Separation Processes, Chemical Process Design, Industrial Chemical Processes |
| 7 | Industrial Visit and Practice, Environmental Chemical Engineering, Novel Separation Technology, Molecular Science and Product Engineering, Reactor Design and its Application |
| 8 | Final Design Project, Final Research Project |
| Semester | **General Electives** |
| 2-6 | Chinese-English Translation, Business English, Western Culture, Law Foundation, Symphony Appreciation, Communication and Negotiation, Technical Writing, Introduction to Management, Introduction to Economics |
| Semester | **Specialized Electives** |
| 5-8 | Biochemistry, Microbiology, Introduction to Clean Technology, Introduction to Polymer, Organic Electrochemistry, Metabolic Engineering, Carbon Nanomaterials and "Green Battery" Technologies, Petroleum Refining Technology, Advances in Catalysis Science, Cell and Tissue Engineering, Surfactant Chemistry, Data Processing and Numerical Solutions, Chemical Engineering Data, Environmental Management Systems, Life Cycle Assessment |

Mathematics, physics and chemistry are the basic sciences that underpin chemical engineering. These courses are primarily taught by faculty members at the School of Science in the first two years. In the first year, over one-quarter of the credits are given over to the study of mathematics. Engineering mathematics is taught by faculty members of the School of Chemical Engineering in the fourth semester. Typical chemical engineering problems are solved in this course and the mathematical techniques learned in this course can be used in the study of other courses such as mass transfer and separation processes.

Chemistry (including inorganic, organic, physical and analytical chemistry) covers mainly the first two years. Inorganic chemistry is taught in the first two semesters and physical chemistry is offered in the second and third semester. Organic chemistry is lectured in the third and fourth semester. Analytical chemistry (or instrumental analysis) is given in the fifth semester. The unique characteristic of the study of chemistry is that all courses are accompanied with hands-on laboratory exercises.

**Design Skills**

The purpose of engineering education is to train the future engineers. Thus it is vital to train the students' design skills.

The design courses are structured so that the design aspects are covered in four years. In the first year, a new course "Introduction to Process Design" is developed to give students a general idea about the outlines of process design. Design issues are also mentioned in the course "Introduction to Chemical Engineering". In the second year, following the courses "Fluid Mechanics" and "Heat Transfer", the first design project is introduced. In addition, students have the opportunity to visit several chemical plants and gain first hand experience about the chemical plant design.

In the third year, the second design project is given following the courses "Chemical Equipment Design" and "Mass Transfer". Design aspects include incorporation of the equipment units into a plant flowchart selection of processes according to the technologic, economic and environmental effects; mass and heat balance calculation; design of main equipment; design or selection of affiliated equipments; drawing process flowchart and equipment diagram.

In semester 7, the students spend an average of two weeks in a chemical plant to investigate a chemical process from the plant layout to the operation. In addition, they learn the safe operation procedure of a chemical process and the economic, environmental and sustainable development issues involved in the design of chemical processes.

Furthermore, detailed chemical process design issues are taught in the courses "Chemical Process Design" and"Analysis and Synthesis of Chemical Processes" . In the courses, students are presented with open-ended engineering problems, which are slightly modified for use in the classroom. Students must tackle the problems using the chemical engineering knowledge that they have acquired. The emphasis in these courses is on engineering decision-making, engineering design and improving students' analytical and problem-solving skills.

The final design project starts at the beginning of the 7$^{th}$ semester after the selection of the projects and team members. The designated group then performs the literature survey, reading and data collection. It is expected that the first 4 weeks in the 8$^{th}$ semester be fully occupied by the design project. Students conduct design and research projects in parallel over the next 8 weeks. The final design project is completed in the 13$^{th}$ Week of the final semester.

The final design project requires the students to undertake a major design task utilizing the knowledge gained throughout the chemical engineering program. This comprises the following tasks: design of a process to meet a specified requirement; feasibility study of alternative processes which meet the specification; determination of sequence for investigation of a chemical manufacturing project and preparation of a report; consideration of environmental impacts and sustainability issues; preparation of flowcharts; confirmation of effects derived from market forecasts; economic evaluation; preparation of estimates for the minimization of capital and production costs; specification of equipment; selection of materials; and specification of instrumentation location, staff and labour requirements and safety precautions.

Students who successfully complete the final design project will have mastered the skills necessary to complete a chemical engineering feasibility study and to carry out the integrated process and equipment design for an industrial chemical process.

They will be presented with an initially poorly-defined task for which much of the design data is not available. In completing the design they will apply most of the skills acquired earlier in their course and will learn to function as part of a team and to manage their time effectively.

**Academic Administration**

TJU adopts a 2-semester system. The first semester consists of 22 weeks of teaching and learning and 4 weeks of winter holiday. It commences in late August and ends in January of the next year. The first 4 weeks are designated for intensive engineering practice and the next 16 weeks for teaching, the last 2 weeks are allocated for assessment and examination. In general, new students enter into the university in the $2^{nd}$ week (beginning of September) and start their lectures in the $5^{th}$ week. The second, third and fourth year undergraduates return to the university in the first week and begin their engineering practice programs. The second semester consists of 18 weeks. The first 16 weeks are for teaching and the last 2 weeks for assessment and examination. The summer holiday lasts for 8 weeks. The convocation ceremony is normally held in the first week of July.

It should be distinguished that the credit system in TJU is different from that of the British system. In the TJU credit system, one credit represents 16 hours of lectures or 32 hours of laboratory work. A student is expected to spend 2 additional hours for private study after the one-hour lecture. Thus, one TJU credit represents a total of 48 study hours (lecture + private study hours). A full program load comprises 40 credits in one academic year, and the minimum number of credits needed to graduate is 160 over four years.

**Summary**

TJU is committed to generating, disseminating, and preserving knowledge, and to working with others to bring this knowledge to bear on the world's great challenges. The University motto is "Seeking Truth from Facts". TJU expects its graduates to be well-educated, well-informed responsible citizens, and able to contribute effectively to their communities wherever in the world they choose to reside and work.

# (98c) Building Human Capacity in Chemical Engineering in Kazakhstan

**Stefaan Simons**, School of Engineering, Nazarbayev University, Astana, Kazakhstan

The collapse of the former Soviet Union and the transition of the new independent states to market economies has highlighted the need for a radical overhaul of the higher education system. This is particularly the case in engineering sciences. Due to national requirements, in regards to human development and economic growth, chemical engineering has been highlighted as a key discipline that must be modernised to provide qualified home-grown talent to run the highly technical facilities required to bring about the goals of energy security, increased life expectancy and sustainability. Several universities and organisations across Europe and the CIS are now collaborating to develop modern chemical engineering degree programmes at key institutions in Kazakhstan and Russia. However, in a move that will accelerate the establishment of chemical engineering in Kazakhstan, a new university, Nazarbayev University, has been founded in Astana, the nation's capital city, that is aiming to be the leading university in Central Asia, offering degrees and conducting research to the highest international standards. The School of Engineering, partnered with University College London, will offer degrees in chemical engineering, as well as mechanical, civil and, eventually, electrical engineering.

The ultimate goal is for chemical engineering degree programmes in Kazakhstan to gain international recognition for the quality of the education and of the graduates they produce and for the discipline to take root in the region as a whole.

The author is on secondment from University College London, where he is a Professor of Chemical Engineering and Director of the Centre for CO2 Technology. His email address at UCL is stefaan.simons@ucl.ac.uk.

See more of this Session: International Perspectives On ChE Education
See more of this Group/Topical: Education

# Chemical Engineering Curriculum in Iran

*Farhang Jalali Farahani[1], Jafar Towfighi Darian[2]*

*1-Chem. Eng. Dept., University of Tehran, Tehran, Iran*

*2- Chem. Eng. Dept., University of Tarbiat Moddares, Tehran, Iran*

## Abstract

In this report, initially, comprehensive information about chemical science history in Iran is presented. Then it will be discussed about chemical engineering in the new world in Iran. In the next section, role of private sector in Chemical Engineering (CE) in the country education will be presented. In recent years the Ministry of Higher Education has decided to establish Technology Parks to fill the gap between the universities and industry. Also, international collaboration and international ranking of the Iranian scientific along with the position of CE will be presented. This ranking is according to the reports released by international organizations related to this field.

In the last sections, educational information about chemical engineering major in Iran is presented. Numbers of students, faculties and research fields and division of these subjects have been reviewed. Furthermore, internal journals, which are related with chemical engineering, are presented.

## Introduction

Land of Persia was one of origins of science in earlier times, and their scientists contributed to the current understanding of nature, medicine, mathematics, and philosophy. Khiam and Razi made important contributions to algebra and chemistry, and others, invented the wind-power machine, and the first distillation of alcohol.

A great scientist, Khaje Nasiro-deen Tusi, believed that a body of matter is able to change but is not able to disappear entirely. He said "a body of matter cannot disappear completely, and it only changes its form, condition, composition, color, and other properties, and turns into a different complex or elementary matter". Five hundred years later, Antoine-Laurent Lavoisier (1743–1794) created the law of conservation of mass, setting down this same idea.[1]

Jaber Ibn Hayyan, the famous Iranian chemist who died in 804 Shamsi in Khorasan, was the father of a number of discoveries. These works had a variety of uses including tinctures and their applications in tanning and textiles; distillations of plants and flowers; the origin of perfumes; therapeutic pharmacy. Jabir Ibn Hayyan, is widely regarded as the founder of chemistry, inventing many of the basic processes and equipment still used by chemists today such as distillation.[1]

## Science in modern Iran

Trying to pursue the golden time of Persian science, Iran's scientists cautiously reach out to the world. Many individual Iranian scientists, are involved in this pursuance.

In Iran, the former oil company was trustee in Chemical Engineering education. This company established Chemical Engineering education because it needed to expert in utilization of oil and gas industry.

Since last four decades, expanding of oil, gas and petrochemicals, and several sugar factories, cement, pharmaceutical and so on caused that chemical engineering established in city of Abadan and some of the universities. Graduate development of chemical engineering at other universities became faster after the revolution in 1978. By today, number of the universities that have Chemical Engineering major is 25, all engineering department in Iranian universities. (Table1 is shows Faculty Numbers, Table 2 is shows Current Students in Chemical Engineering at 2010, Table3 is shows Different Fields of Chemical Engineering)

**Table 1.** Faculty Numbers

| Governmental Members | Private Members | Sum |
|---|---|---|
| 300 | 100 | 400 |

**Table 2.** Current Students in Chemical Engineering at 2010

| Degree | NO |
|---|---|
| B.S. | 3106 |
| M.S. | 728 |
| PhD | 328 |

**Table 3.** Different Fields of Chemical Engineering

| | | |
|---|---|---|
| Polymer Eng. | Petroleum Exploration | Petroleum Drilling |
| Nuclear Eng. | Chemical Eng. | Polymer Science & Technology |
| Biotechnology | Hydrocarbon Reservoirs | Arm mental Eng. |
| Environmental Eng. | Underground Gas Storage | Catalyst |
| LNG Eng. | Separation Process | Process Design |
| Gas Eng. | Control and Modeling | Transport Phenomena |
| Pharmaceutical Eng. | Nanotechnology | Paper Industries |
| Reservoir Eng. | Energy Systems | Food Industries |
| Gas Transport | Renewable Energy | Oil Shale |

Iran is an example of a country that has made considerable advances through education and training in almost all aspects of research during the past 30 years. Iran's university population swelled from 100,000 in 1979 to 2 million in 2006. More than 50% of its science and engineering students are female [2]. Iran's scientific progress is reported to be the fastest in the world.[2]

Iran's government has allocated good amounts of funds for research on high technologies such as nanotechnology, biotechnology, stem cell research and information technology (2008).[4] According to UNESCO science report 2010, most of the research in Iran is government funded with the Iranian government providing almost 75% of all research funding.[3] The share of private businesses in total national R&D funding according to the same report is very low being just 14%. The rest of approximately 11% of funding comes from higher education sector and non-profit organizations.

According to the State Registration Organization of Deeds and Properties, a total of 9,570 national inventionswere registered in Iran during 2008. Compared with the previous year, there was a 38-percent increase in the number of inventions registered by the organization.[4]

Theoretical and computational researches in Chemical Engineering are highly developed in Iran, and the researchers regularly publish works in high impact factor journals [4]. Despite the limitations in funds, facilities, and international collaborations, Iranian chemical engineers have been very productive in several experimental fields such as, pharmaceutical chemistry, and organic and polymer chemistry.

## Scientific collaboration

Iranians welcome scientists from all over the world to Iran for a visit and participation in seminars or collaborations. Many Nobel laureates and influential scientists such as Bruce Alberts, F. Sherwood Rowland, Kurt Wüthrich, Stephen Hawking, and Pierre-Gilles de Gennes visited Iran after the revolution. Some universities also hosted American and European scientists as guest lecturers during recent decades.

## Private sector

Parallel to academic research, several companies have been founded in Iran during last few decades and have been productive in CE research. For example CinnaGen, established in 1992, is one of the pioneering biotechnology companies in the region. CinnaGen won Biotechnology Asia 2005 Innovation Awards due to its achievements and innovation in biotechnology research.

## Technology parks

The government of Iran has also plans for the establishment of 50-60 "Science and Technology Parks" by the end of the fifth Five-Year Socioeconomic Development Plan by 2015. Those which are involved in CE are shown in Table 4 :

**Table 4.** List of Parks and their focus area

| Park's name | Focus area | Location |
|---|---|---|
| Guilan Science and Technology Park | Food, Biotechnology, Chemicals, Electronics, Environment, ICT, Tourism. | Guilan |
| Pardis Technology Park | Advanced Engineering (mechanics and automation), Biotechnology, Chemicals, Electronics, ICT, Nano-technology. | 25 km North-East of Tehran |
| Khorasan Science and Technology Park (Ministry of Science, Research and Technology) | Advanced Engineering, Food, Chemicals, Electronics, ICT, Services. | Khorasan |
| Sheikh Bahai Technology Park (Aka "Isfahan Science and Technology Town") | Materials and Metallurgy, Information and Communications Technology, Design & Manufacturing, Automation, Biotechnology, | Isfahan |

***International rankings (CE 3$^{rd}$ place based on statistics)***

Based on the local statistics, among the academic publishing in Iran, most of the papers belong to Pure Science, Medical Science, and Engineering, respectively. And in the Engineering section, Electrical Engineering, Civil Engineering, and Chemical Engineering have the highest publications in Iran.

- According to the Institute for Scientific Information (ISI), Iran improved its academic publishing output nearly tenfold from 1996 to 2004, and has been ranked first globally in terms of output growth rate (followed by China with a 3 fold increase).[5][6]
- Based on Thomson Reuters, Iran has shown a remarkable growth in science and technology since the past one decade. Most of this growth has been in engineering and chemistry producing 1.4% of the world's total output in the period 2004-2008.
- Iranian science and technology output accounted for 1.02% of the world's total output (That is ~340000% growth in 37 years of 1970-2008).[7] 25% of scientific articles published in 2008 by Iran were international coauthor ships. The top five countries coauthoring with Iranian scientists are from US, UK, Canada, Germany and France.[8][4]
- A 2010 report by Canadian research firm Science-Matrix has put Iran in the top rank globally in terms of growth in scientific productivity with a 14.4 growth index.[10]
- Iran with a yearly growth rate of 25% is doubling its total output every three years in science and technology and at this rate may get to the level of Canadian annual output in 2017.[10]
- In US government report titled "Science and Engineering Indicators: 2010" prepared by National Science Foundation, Iran has the world's highest growth rate in Science&Engineering article output with an annual growth rate of 25.7%.
- Iran has been ranked 49th for citations, 42nd for papers, and 135th for citations per paper in 2005.[10] According to a British government study (2002), Iran has been ranked 30th in the world of scientific impact.[11]
- 2008 report of Institute for Scientific Information (ISI), Iran has been ranked 15th in 2009 in the field of nanotechnology as of presenting articles.[12]
- Science Watch reported in 2008 that Iran, during the period 2005-2009, had produced 1.71% of world's total engineering papers.[11]

- The Royal Society in collaboration with Elsevier published in 2011 a study called "Knowledge, networks and nations" surveying global scientific era. According to this study Iran has had the world's fastest growth rate in science and technology. During the period 1996-2008, Iran had increased its scientific output by 18 folds.[14][15][16]

## Iranian journals listed in the Institute for Scientific Information (ISI)

According to the ISI, Iranian researchers have published a total of 60,979 scientific studies in major international journals in the last 19 years (1990–2008). In 2009, Iranian researchers published 15,000 Thomson Reuters-indexed papers.[13]

## Ranking by number of publications (SCImago)

SCImago Research Group's SCImago institutions rankings: 2009 world report ranks all institutions which had more than 100 outputs indexed in the multinational publishing giant Elsevier's Scopus database. The ranking comprises 1,527 higher education institutions, 335 health organizations, 216 government organizations, 29 private bodies and 17 other organizations. (Table 5)

**Table 5.** Ranking by number of publications 2009

| National rank | University | city | Rank in world |
|---|---|---|---|
| 1 | University of Tehran | Tehran | 925 |
| 2 | Sharif University of Technology | Tehran | 1777 |
| 3 | TarbiatModares University | Tehran | 2756 |
| 4 | Shiraz University | Shiraz | 1064 |

The Iranian Journal of Chemical Engineering, (IJChE) accredited by the Ministry of Science, Research and technology, is under supervision of and published quarterly by the Iranian Association of Chemical Engineers (IAChE).

Although for many years the Iranian scientists published their scientific papers in various national general engineering journals or in international journals, this newly established journal is gaining more attractions for publication of high quality papers in chemical engineering and related fields. All submitted papers undergo extensive peer review process, which may then be accepted as full papers, research notes or review articles. (Table 6)

**Table 6.** List of National Journal relating to Chemical Engineering

| |
|---|
| Water & Wastewater Journal<br>Water & Wastewater Consulting Engineers Research Development |
| Petroleum Research<br>Research Institute of Petroleum Industry |
| Iran Energy<br>National Committee of Energy |
| Research of Iranian Water Resources |
| Journal of Separation and Transport Phenomena<br>Ferdowsi University of Mashhad |
| Fuel & Combustion |
| Rubber Industries<br>Rubber Industries Eng. Research Co. |
| Journal of Separation Science & Engineering<br>University of bahonar Kerman |
| Chemical Eng.<br>University of Tehran |
| Iranian Chemical Engineering<br>Iranian Association of Chemical Engineering |
| Chemical, Polymer and Material Science and Technology |
| Iranian Journal of Biotechnology |
| Iranian Journal of Chemistry and Chemical Engineering |

Increasing relationship between Oil, Gas and Petrochemical Industries and Chemical Engineering Departments in Iran has grown substantially. For example there are some Chemical Engineering Departments founded by Ministry of Oil. (Tehran, Shiraz, Sharif, Sahand Tabriz, Amir Kabir University, etc.). There are many joint research projects, majority of M.Sc. and Ph.D Chemical Eng. Student's researches in the universities are requested and sponsored by Oil and Gas industries. Some experts from industry are supervising and teaching the graduate students and there are many short courses offered by faculties to industry and so on.

Oil, Gas and Petrochemical Industries are first priority and the most important sector in Iran. So Chemical Engineering Departments are increasing both quantitatively and qualitatively. And main goal of these departments is to respond to the needs of emerging technologies in this sector by graduating high qualified students and by developing new scientific fields in Chemical Engineering.

## References


1- http://www.wikipedia.com

2- http://aljazeera.com/news/articles/34/Iran-s-science-progress-fastest-in-world-Canadia.html

3- http://www.unesco.int, media-services, Asia leaping forward in science and technology

4- Iran: Huge Investments on Nanotech, Iran daily, 30 Oct 2008

5- http://www.dti.gov.uk/files/file27330.pdf

6- http://www.berr.gov.uk/files/file27330.pdf

7- http://www.timeshighereducation.co.uk/story.asp?storyCode=410623&sectioncode=26

8- http://www.nsf.gov/statistics/seind10/c5/c5s.htm

9- http://www.nsf.gov/statistics/seind10/c5/tt05-b.xls

10- http://www.science-metrix.com

11- http://www.berr.gov.uk/files/file11959.pdf

12- http://www.bernama.com/bernama/v5/bm/newsworld.php?id=453647

13-http://sciencewatch.com/ana/fea/10novdecFea/

14- http://vikno.eu/eng/sciense-and-technology/sciense-and-technology/the-future-science-will-be-set-by-china-and-iran.html

15- http://www.wired.co.uk/news/archive/2011-03/29/china-leads-new-science-giants

16- http://royalsociety.org/policy/reports/knowledge-networks-nations/

# (98e) Chemical Engineering Education in Colombia. the Universidad De Los Andes Experience

**Oscar A. Alvarez**, Chemical Engineering, Universidad de los Andes, Bogota, D. C, Colombia

Though the general framework of the current analysis is chemical engineering education in Colombia, this report is centered in the Universidad de los Andes experience. In Colombia there are 14 universities with chemical engineering programs, from which 12 programs have an educational cycle of 5 years proposed, 1 program has a 4.5 years educational cycle and 1 program has a 4 years educational cycle (Universidad de los Andes). The last one has a similar structure to American BSc engineering programs.

Chemical Engineering education at Universidad de los Andes in the nineties was based merely on the transfer of knowledge (with a proposed duration of 5 years). Since 2003 started a process of curricular reform, officially implemented on 2006, which pursued an education based on the development of abilities coupled to the knowledge and application of chemical engineering fundamentals.

As a result of this process, new learning spaces were created in which the student is allowed to acquire knowledge, to develop abilities and to use engineering tools in the design of products and processes, and also to take the first steps in research. These learning spaces are structured based on a series of courses in the following way:

Process design:

This is the classic and fundamental part of chemical engineering; developed in the courses of mass balance, thermodynamics, transport phenomena, separation processes, unit operations, process simulation and process plant design. Also, searching for an integration course, in which modern engineering tools were used, a required undergraduate level course on optimization of chemical processes is offered.

Product design:

Three spaces were created: the first at the beginning of the educational cycle (Introduction to Chemical Engineering), the second at the middle (Mid-career Project) and the third at the ending (Undergraduate Graduation Project). Each of these spaces faces the student with the challenge of a product design, using chemical engineering tools and within a social and/or economical and/or environmental context, to solve a specific problem or to generate new alternatives for the development of the country or the industry.

Research experience:

The first contact of the student with research, basic or applied in product and/or process design, is achieved in two learning spaces, one elective (Special Project) and the other required (Undergraduate Graduation Project).

See more of this Session: International Perspectives On ChE Education
See more of this Group/Topical: Education

**ID# 221166**

# THE ROLE OF PHYSICAL PROPERTY DATABASES IN CH. E. EDUCATION

*Mordechai Shacham, Ben-Gurion University of the Negev, Beer-Sheva, Israel*

*Michael B. Cutlip, University of Connecticut, Storrs, CT*

*Michael Elly, Intel Corp., Qiryat Gat, Israel*

## Introduction

Numerical problem solving in Chemical Engineering typically requires a mathematical model of the problem along with physical and thermodynamic data and correlations (equations) of the chemical substances involved. The required pure compound properties are usually divided into constant properties and temperature and/or pressure dependent properties. In most ChE textbooks, the preparation of the mathematical model associated with a particular problem is typically emphasized. The property data required for solution of the problems are usually provided in tabular or graphical form in the appendix of the textbook and/or in a CD associated with the book (see for example, Felder and Rousseau, 2000 and Himmelblau and Riggs, 2004).

Recently, various databases that contain extensive physical property data for a large number of compounds have become available. Typical examples are the DIPPR® database (Rowley et al., 2010) and the NIST database (http://webbook.nist.gov/chemistry/) databases. The use of the data available in these databases for problem solving has significant benefits over the use of the data provided in the textbooks. Some of the more important advantages are:

1. *In current engineering practice, the databases are used as a principal source of property data and correlations.* Thus it is important that students become experienced with the application of these sources during their educational programs.
2. *The databases provide consistent sets of correlations for temperature-dependent properties enabling the use of solution techniques independent of the format in which the property data is provided.* If temperature-dependent data are provided in tabular or graphical forms, for example, this prevents the use of standard numerical methods for problem solution, and data format dependent ad hoc solution techniques have to be used.
3. *The data provided in databases are continuously updated with the new data as they become available.* Property data in textbooks are often taken "as is" from references that may be as much as half a century old. The textbook data may be incorrect or even contradictory. It is important that students become accustomed to using reliable data.
4. *The data available in the databases are evaluated and in cases when multiple values are available for the same property, one recommended value is selected by the database professionals.* There may be very substantial differences between property values reported by different investigators. Brauner et al., 2005, for example mention the case of the melting point of 4-methyloctane for which the recommended value is 159.95 K while one reported experimental value is 219.62 K. The students must be made aware of the fact that several property values may be available and learn how to find and utilize the value with the highest confidence in its correctness.
5. *The databases usually provide uncertainty values (upper limit on experimental error) which enable estimation of the uncertainty of the problem solution using error*

*propagation analysis*. An important goal of chemical engineering education is to impart to students that the numerical "solution" of a problem almost always has some associated uncertainty.

A convenient option for chemical engineering educators to incorporate the use of databases into their teaching has been integrated into the POLYMATH package. This widely-used computational tool for numerical problem solving now contains a "sample database" subset of the DIPPR© Database[1]. The "sample database" contains 34 constant properties and 16 temperature dependent property correlations for 112 compounds that are most frequently used in chemical engineering textbooks. A special interface POLYMATH Database Interface (PDI) now enables a convenient search of the sample database for selected compounds. The desired properties can be selected, and the data and correlations outputted in a format that can be copied and pasted directly into a computer code with all the significant figures given in the database. The formats that are currently supported are for POLYMATH[2], MATLAB[3] and Excel[4].

In the following example, the use of the POLYMATH interface to the DIPPR Database will be demonstrated for an adiabatic reactor design problem that will illustrate and highlight the database usage.

**Production of Acetic Anhydride – An Example**

Process models based on material and energy balances usually include some constant and temperature dependent data. A typical example of this type is presented by Fogler (1992). This example involves the vapor phase cracking of acetone to ketene and methane

$$CH_3COCH_3 \rightarrow CH_2CO \quad + \quad CH_4$$
$$\text{Acetone (A)} \rightarrow \text{Ketene (B)} + \text{Methane (C)}$$

as a key step in manufacturing acetic anhydride. The reaction is first order with respect to acetone (A) and the specific reaction rate can be expresses by

$$\ln(k) = 34.34 \quad \frac{34222}{T} \tag{1}$$

where *k* is the specific reaction rate (1/s) and *T* is the temperature (K). It is desired to design a tubular reactor which yields 20% conversion. The case considered here involves adiabatic operation of the reactor, feed flow rate of $F_{A0}$ = 8000 kg/hr acetone, inlet temperature of $T_0$ = 1035 K and pressure $P$ = 162 kPa.

The mole balance equations for this system can be written

$$\frac{dF_A}{dV} = r_A; \frac{dF_B}{dV} = \quad r_A \text{ and } \frac{dF_C}{dV} = \quad r_A \tag{2}$$

[1] The Design Institute for Physical Properties and its acronym DIPPR® are registered trademarks of the American Institute of Chemical Engineers (AIChE®)
[2] POLYMATH is a product of POLYMATH Software, http://www.polymath-software.com
[3] MATLAB is a trademark of The Math Works, Inc. http://www.mathworks.com
[4] Excel is a trademark of the Microsoft Corporation, http://www.microsoft.com

where $F_A$, $F_B$ and $F_C$ are the molar flow rates (in mol/s) of compounds A, B and C, respectively, $V$ is the volume ($m^3$), and $r_A$ is the rate of the reaction mol/($m^3$*s) with respect to A:

$$r_A = kC_A \qquad (3)$$

Here $C_A$ is the concentration of $A$ (mol/$m^3$) and it can be calculated using the ideal gas law.

$$C_A = \frac{F_A}{(F_A + F_B + F_C)} \frac{P}{RT} \qquad (4)$$

The enthalpy balance equation yields

$$\frac{dT}{dV} = \frac{r_A\left\{\left[\Delta H^0_{Rx}(T_R) + \Delta H_Q(T)\right]\right\}}{F_A C_{PA} + F_B C_{PB} + F_C C_{PC}} \qquad (5)$$

where $\Delta H^0_{Rx}(T_R)$ is the heat of the reaction at a reference temperature ($T_R$, usually 25 °C) and $\Delta H_Q(T)$ is the enthalpy change that results when the temperature is raised from $T_R$ to some temperature $T$. $C_{PA}$, $C_{PB}$ and $C_{PC}$ are the heat capacities (J/mol*K) of components $A$, $B$ and $C$, respectively.

Fogler (1992) rewrote these equations in terms of conversion ($X$ = 1-$F_A/F_{A0}$) as the use of conversion enables obtaining graphical solution for the adiabatic case. Rewriting the mole balance equations in terms of the conversion yields:

$$\frac{dX}{dV} = \frac{r_A}{F_{A0}} \qquad (2A)$$

Since the flow-rates of the various compounds not calculated explicitly, Eq. 4 cannot be used for calculation of $C_A$ as it needs to be expressed in terms of the original concentration of A in the reactor feed $C_{A0}$, the conversion in the reactor $X$, and the temperature change in the reactor relative to the feed temperature $T_0$.

$$C_A = \frac{C_{A0}(1 \quad X)T_0}{(1 + X)T} \qquad (4A)$$

The use of the conversion requires revision of the enthalpy balance equation.

$$\frac{dT}{dV} = \frac{r_A\left\{\left[\ H^0_{Rx}(T_R) + \ C_P(T \quad T_R)\right]\right\}}{F_{A0}(C_{PA} + X \ C_P)} \qquad (5A)$$

where $\Delta C_P$ is the overall change in the heat capacity per mol of A reacted.

**Solution of the Example Problem Using Traditional Techniques**

Fogler (1992) used the Simpson graphical technique to find the volume of the reactor and the exit temperature for conversion of 20% of acetone data. Equations (2A), (3), (4A) and (5A) were used in the solution. Fogler (1992) provided standard heats of formation and 2nd order polynomial equations for calculating $C_P$ for the three compounds. As the example is based on Jeffrey's (1964) publication, it is assumed that this publication is the source of the data used by Fogler. The set of equations and data (in POLYMATH format) as presented in the 3rd edition of the Fogler book is shown in Table 1. The row numbers shown in Table 1 are not part of the program; they were added as references for the explanations that follow.

The model is in a compact form as many of the calculations were carried out separately. In line 3 the value of $F_{A0}$ is given in (mol/s). The value of $C_{A0}$ in mol/m$^3$, calculated using the ideal gas law is shown in line 5. The equation for $C_{PA}$ is shown explicitly for compound $A$ (in line 7) however for the rest of the compounds the equations are not shown and only the calculated expression for $\Delta C_P$ is presented in line 8. The heats of formations of the individual compounds are not shown either, only the resultant numerical value of $\Delta H^0_{Rx}(T_R)$ is presented.

Figures 1 and 2 show the change of the conversion and temperature, respectively, in the tubular-flow adiabatic reactor as function of the reactor volume. The curves that were obtained by the model shown in Table 1 are marked as "Jeffrey's data". The conversion of 20% is obtained with the reactor volume $V = 1.191$ m$^3$ and the exit temperature in this case is $T = 937$ K. Fogler (1992) obtained reactor volume of 1.27 m$^3$ with the very same outlet temperature using Simpson graphical integration technique.

**Extracting Physical Properties from the DIPPR Database Using the POLYMATH Database Interface (PDI)**

The physical property data and correlations that needed for the enthalpy balance include heat of formation data and equations for calculating heat capacity and enthalpy values at temperature $T$ for the reactant and the two products. Current, up to date data and equations for these properties can be found in the DIPPR database (Rowley et al., 2010). This database can be accessed through the POLYMATH Database Interface (PDI) software.

The interface program enables searching the database for a particular compound, marking the desired properties and obtaining as output the necessary data and correlations in a format that can be copied and pasted directly into a computer code.

The information that is provided by the interface program in the POLYMATH format is demonstrated in Table 2 for the properties of acetone that are required for the example problem. This table shows most of the information as provided by the interface when the POLYMATH format output is requested. The code that is generated by the interface program includes correlation equations, definition of constant values and comments. Lines 1-11 can be generated at once by selecting the desired "Compounds and Properties" and the "Basic Report." The "Report Level" determines the amount of information that is to be included as comments.

Lines 1 through 4 contain the information related to the heat of formation of acetone. In line 1, the full name of the property and the full name of the compound are shown (as a comment: text that starts with the "#" sign and ends with the end of the line). In line 2, the family of the compound and in line 3 its chemical formula are presented. In line 4, the property value related information is displayed. The units of the property and its uncertainty ( $< 1\%$ in this case) are included as comments. The variable into which the value of the property is entered made up from the symbol of the property (*HFOR* in this case) and the chemical formula of the compound involved.

The information provided for ideal gas heat capacity includes the full name of the property and the compound (line 5), the compound's family (line 6) and its chemical formula (line 7). For temperature dependent properties DIPPR provides the information regarding range of validity for the temperature (line 8) and range of validity for the property (line 9).

The ideal gas enthalpy and heat capacity (lines 10 – 11) deserve special discussion. In most cases DIPPR provides the coefficients for the Aly and Lee (1981) equation for this property. The integrated form of this equation, for calculating ideal gas enthalpy is considerably different than the heat capacity equation. For the benefit of the users, whenever the ideal gas heat capacity is requested by the user, the equation for ideal gas enthalpy is also included, as a comment (see line 10). Note that the enthalpy equation includes an integration constant *Hcon*. This constant depends on the standard state selected for the enthalpy calculation. For example, selecting as standard state pure gaseous component at 25 °C (298.15 K) yields *Hcon* as ideal gas enthalpy at 298.15 K. This value has to be subtracted from the enthalpy of ideal gas at temperature *T*.

**Solution of the Example Problem Using Basic Model Variables and DIPPR Property Data**

The traditional solution technique that includes the use of "conversion" in the differential equations, requires extensive manipulation of the equations. This extra effort is not required when numerical solution techniques are used. Furthermore, in order to provide clear and complete documentation of the problem, it is preferable to include all the original data and variables in the model and carry out the calculations, even the simplest ones, with the computer model. The POLYMATH model for solution of the example problem, using these principles and data from the DIPPR database is shown in Table 3. For most of the variables the units are provided in forms of comments.

The principal model equations are shown in lines 2-13. The mole balance equations (Eq. 2) are in lines 2-4 while the enthalpy balance equation (Eq 5) is in line 5. The notation ICP_C3H6O, ICP_C2H2_O and ICP_CH4 is used for representing $C_{PA}$, $C_{PB}$ and $C_{PC}$, respectively. The division by 1000 is used for converting the $C_P$ units from J/kmol*K (as provided by the DIPPR correlation) to J/mol *K for consistency with the mole balance equations. The variables $r_A$ (Eq. 3), $C_A$ (Eq. 4) and $k$ are calculated in lines 6, 8, and 12, respectively and the term: $\Delta H^0_{Rx}(T_R)+\Delta H_Q(T)$ is calculated in line 13. The conversion $X$ is not required for the solution, using this formulation; however, it is calculated (in line 7) as the reactor volume is specified in terms of the desired conversion.

In lines 15–33, the heat of formation data and the equations for calculating the heat capacity and enthalpy of the reactant and the products are presented. Only the information which is essential for the problem solution is shown. This includes the property values, units, upper limit on the uncertainty of the property values and temperature range of applicability of the correlation equations. The range of applicability is important in order to ensure that the solution of the problem does not require the use of one or more of the equations outside of their range of validity. The uncertainty level of the data is important in order to determine what size of "safety factor" has to be added to the calculated reactor volume, so that the desired conversion can be achieved even in the "worst case" situation.

The problem specific constants, initial and final values are grouped together in lines 35-44. The inclusion of all the data and calculations needed for the problem solution and grouping together the model equations, physical property data and problem specific data provide clear and complete documentation of the model. This in turn enables easy and efficient debugging, modification and extension of the model of the problem. The POLYMATH problem solution is obtained after the problem equations are automatically ordered just prior to the solution.

The graphical solution obtained using the model presented in Table 3 is shown in Figures 1 and 2 (the curves marked as "DIPPR data"). Observe that the temperature and the conversion in the reactor are significantly lower in this case than when using the Jeffreys data. The volume required in order to achieve 20% conversion is 2.052 $m^3$, thus the reactor size is larger by 72% than the size obtained using the Jeffreys data. The model shown in Table 1 cannot be used for locating the causes of difference because most of the original data was omitted, and it is necessary to consult the detailed description provided by Fogler (1992). It turns out that they main source of the difference is the heat of formation value of ketene which is provided by Jeffreys (-61.09 kJ/mol) and the one that provided by DIPPR (-47.5 kJ/mol). The source of the DIPPR heat of formation is Padley et al. (1986) where Jeffrey's publication is from 1964. This demonstrates the importance of the use of the most up to date physical property data from reliable sources in process design calculations.

## Conclusions

The example presented here demonstrates some of the advantages of the use of property databases and mathematical models that contain accurate and documented data typically needed in chemical engineering problem solving. This work demonstrates, in particular, that using physical property information from dated sources may lead to considerable errors in design. The inclusion of the original data in the model is shown to make the debugging, modification and extension of detailed modeling problems more easily accomplished.

It is important for educators to introduce students to available and convenient problem solving tools. Additionally students should be exposed to:

- Use of consistent physical property data with documented uncertainty and range of applicability values, extracted from reliable property databases.
- Documentation of the problem, that includes a complete set of the original equations, property data and correlations and problem specific data, in the computer model, clearly marking the units and uncertainties of the variables and temperature range of applicability of the correlations.

## *References*


1. Aly, F. A., and L. L. Lee, "Self-Consistent Equations for Calculating the Ideal Gas Heat Capacity Enthalpy, and Entropy". Fluid Phase Equilib. 6, 169 (1981)
2. Brauner, N., Shacham, M., Cholakov, G. St. and Stateva, R. P., "Property Prediction by Similarity of Molecular Structures – Practical Application and Consistency Analysis", Chem. Eng. Sci. 60, 5458 – 5471 (2005)
3. Felder, R. M. and Rousseau, R. W., *Elementary Principles of Chemical Processes*,3$^{rd}$ Ed, John Wiley & Sons, Inc, Hoboken, New-Jersey, 2000.
4. Fogler, H. S., *Elements of Chemical Reaction Engineering*, 2$^{nd}$ Ed, Prentice Hall, Englewood Cliffs, New Jersey, 1992.
5. Himmelblau D. M., and Riggs, J. B., Basic principles and Calculations in Chemical Engineering, 7$^{th}$ Ed., Prentice-Hall, Upper Saddle River, New-Jersey, 2004.
6. Jeffreys, G.V., A Problem in Chemical Engineering Design: The Manufacture of Acetic Anhydride, 2$^{nd}$ ed. (London: Institute of Chemical Engineers, 1964)
7. Pedley, J.B.; Naylor, R.D.; Kirby, S.P.; Thermochemical Data of Organic Compounds; Chapman and Hall, London, 1986.
8. Rowley, R. L.; Wilding, W. V.; Oscarson, J. L.; Yang, Y.; Zundel, N. A., DIPPR Data Compilation of Pure Chemical Properties, Design Institute for Physical Properties, (http//dippr.byu.edu), Brigham Young University, Provo, Utah, 2010.

Table 1. POLYMATH Solution of the Example Problem Using the Equations Presented by Fogler (1992)

| No. | Equation/ # Comment |
|---|---|
| 1 | d(X) / d(V) = -ra/Fao |
| 2 | d(T) / d(V) = -ra*(-deltaH)/(Fao*(Cpa+X*delCp)) |
| 3 | Fao = 38.3 |
| 4 | To = 1035 |
| 5 | Cao = 18.8 |
| 6 | Tr = 298 # K |
| 7 | Cpa = 26.63+0.183*T-45.86*10^(-6)*T^2 |
| 8 | delCp = 6.8-11.5*10^(-3)*T-3.81*10^(-6)*T^2 |
| 9 | k = 8.2*10^14*exp(-34222/T) |
| 10 | ra = -k*Cao*(1-X)/(1+X)*To/T |
| 11 | deltaH = 80770+6.8*(T-Tr)-5.75*10^(-3)*(T^2-Tr^2)-1.27*10^(-6)*(T^3-Tr^3) |
| 12 | V(0) = 0 |
| 13 | V(f) =1.191 |
| 14 | X(0) = 0 |
| 15 | T(0) = 1035 |

Table 2. Physical properties of acetone as obtained from the DIPPR database by the POLYMATH interface

| No. | Equation/ # Comment |
|---|---|
| 1 | # Enthalpy of Formation of Ideal gas at 298.15 K and 100000 Pa of ACETONE |
| 2 | # Compounds Family: KETONES |
| 3 | # Compounds Structure: CH3COCH3 |
| 4 | HFOR_C3H6O = -215700000 # J/kmol (Uncertainty < 1%) |
| 5 | # Ideal Gas Heat Capacity of ACETONE |
| 6 | # Compounds Family: KETONES |
| 7 | # Compounds Structure: CH3COCH3 |
| 8 | # Valid Temperature Range: 200K to 1500K |
| 9 | # Valid Value Range: 60487 to 188200 (J/kmol*K) |
| 10 | # HIG_C3H6O = 57040*T+163200*1607*coth(1607/T)-96800*731.5*tanh(731.5/T) +HCON_C3H6O # J/kmol |
| 11 | ICP_C3H6O = 57040 + 163200 * (1607 / T / sinh(1607 / T)) ^ 2 + 96800 * (731.5 / T / cosh(731.5 / T)) ^ 2 # J/kmol*K (Uncertainty < 1%) |

Table 3. POLYMATH Solution of the Example Problem Using DIPPR Physical Properties

| No. | Equation/ # Comment |
|---|---|
| 1 | # Model equations |
| 2 | d(FA)/d(V) = rA |
| 3 | d(FB)/d(V) = -rA |
| 4 | d(FC)/d(V) = -rA |
| 5 | d(T)/d(V) = (-(deltaH)) * (-rA) / ((FA * ICP_C3H6O + FB * ICP_C2H2O + FC * ICP_CH4 )/1000) |
| 6 | rA = -k * CA |
| 7 | XA = (FA0 - FA) / FA0 |
| 8 | CA = yA * P / (8.31 * T) |
| 9 | yA = FA / (FA + FB + FC) |
| 10 | yB = FB / (FA + FB + FC) |
| 11 | yC = FC / (FA + FB + FC) |
| 12 | k = 8.2E14 * exp(-34222 / T) |
| 13 | deltaH = (HFOR_C2H2O +HFOR_CH4 -HFOR_C3H6O+HIG_C2H2O +HIG_CH4 -HIG_C3H6O)/1000 #J/mol |
| 14 | # |
| 15 | # Thermodynamic data and correlations from the DIPPR database |
| 16 | # ACETONE |
| 17 | HFOR_C3H6O = -215700000 # J/kmol (Uncertainty < 1%) |
| 18 | HCON_C3H6O = 57040*Tref+163200*1607*coth(1607/Tref)-96800*731.5*tanh(731.5/Tref) # J/kmol |
| 19 | HIG_C3H6O = 57040*T+163200*1607*coth(1607/T)-96800*731.5*tanh(731.5/T)-HCON_C3H6O # J/kmol |
| 20 | # Valid Temperature Range:200K to 1500K |
| 21 | ICP_C3H6O = 57040 + 163200 * (1607 / T / sinh(1607 / T)) ^ 2 + 96800 * (731.5 / T / cosh(731.5 / T)) ^ 2 # J/kmol*K (Uncertainty < 1%) |
| 22 | # KETENE |
| 23 | HFOR_C2H2O = -47500000 # J/kmol (Uncertainty < 5%) |
| 24 | HCON_C2H2O = 36940*Tref+68650*1490*coth(1490/Tref)-45240*640*tanh(640/Tref) # J/kmol |
| 25 | HIG_C2H2O = 36940*T+68650*1490*coth(1490/T)-45240*640*tanh(640/T)-HCON_C2H2O # J/kmol |
| 26 | # Valid Temperature Range:150K to 1500K |
| 27 | ICP_C2H2O = 36940 + 68650 * (1490 / T / sinh(1490 / T)) ^ 2 + 45240 * (640 / T / cosh(640 / T)) ^ 2 # J/kmol*K (Uncertainty < 1%) |
| 28 | # METHANE |
| 29 | HFOR_CH4 = -74520000 # J/kmol (Uncertainty < 1%) |
| 30 | HCON_CH4 = 33298*Tref+79933*2086.9*coth(2086.9/Tref)-41602*991.96*tanh(991.96/Tref) # J/kmol |
| 31 | HIG_CH4 = 33298*T+79933*2086.9*coth(2086.9/T)-41602*991.96*tanh(991.96/T)-HCON_CH4 # J/kmol |
| 32 | # Valid Temperature Range: 50K to 1500K |
| 33 | ICP_CH4 = 33298 + 79933 * (2086.9 / T / sinh(2086.9 / T)) ^ 2 + 41602 * (991.96 / T / cosh(991.96 / T)) ^ 2 # J/kmol*K (Uncertainty < 1%) |
| 34 | # |
| 35 | # Constants, initial and final values |
| 36 | Tref=298.15 #K |
| 37 | P = 162 * 1000 # Pa |
| 38 | FA0 = 38.3 # mol/m^3 |
| 39 | FB(0) = 0 |
| 40 | FA(0) = 38.3 |
| 41 | FC(0) = 0 |
| 42 | T(0) = 1035 #K |
| 43 | V(0) = 0 |
| 44 | V(f) = 2.052 # m^3 |

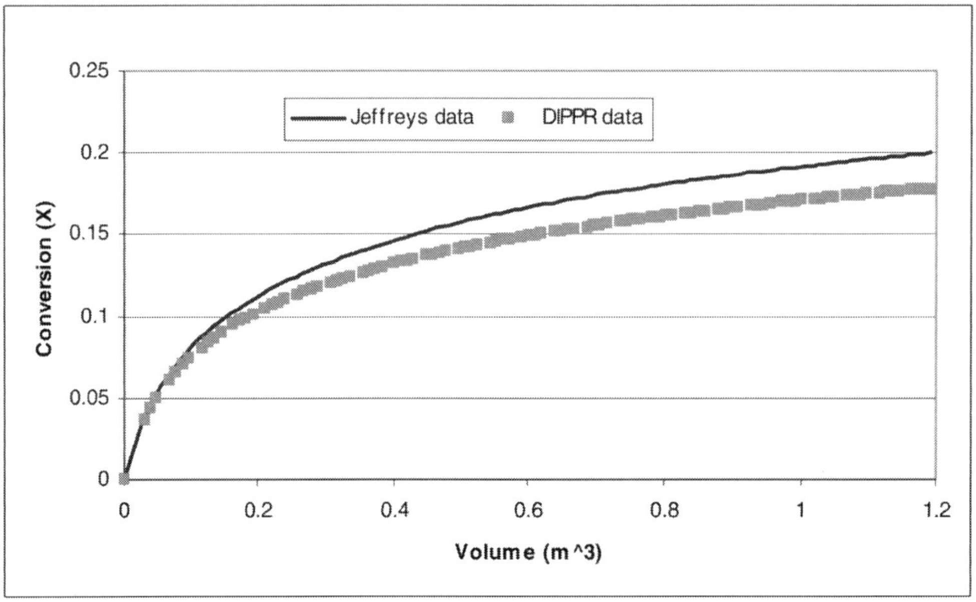


Figure 1 – Conversion of acetone as function of reactor volume

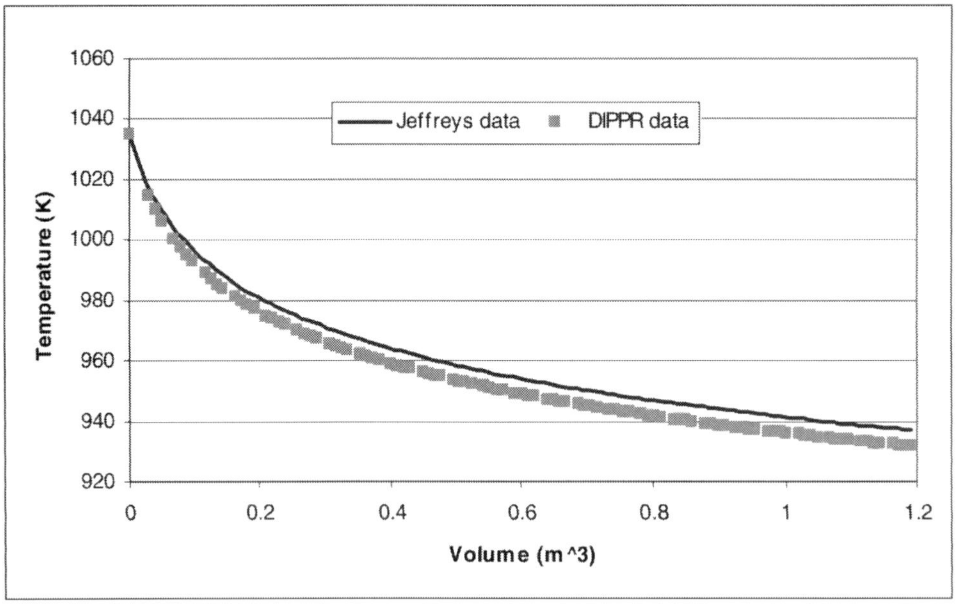


Figure 2 – Temperature change in the reactor as function of volume

# (148b) Molecular Simulation Modules for Instruction In Thermodynamics, Transport, Kinetics, and Materials

**David A. Kofke** and Andrew J. Schultz, Chemical and Biological Engineering, University at Buffalo, The State University of New York, Buffalo, NY

We describe the results of our project to develop and use molecular simulation modules for undergraduate and graduate instruction. These modules have been developed to support instruction of molecular concepts appropriate to a range of courses in the chemical engineering curriculum. The aim is to provide students with tools that permit them to see and manipulate models for the molecular behavior that gives rise to the macroscopic properties they learn about in thermodynamics, transport, and other courses.

Each module consists of an interactive molecular simulation and written supporting material related to it. The molecular simulation is written in the Java programming language, and in most cases it can be run as an applet in a web browser. In all cases the simulation can instead be downloaded and run as an application on the user's machine. The supporting material provides an introduction to the concepts demonstrated by the applet, with some detailed background about what is being modeled. There are also examples describing the use of the simulation, and exercises suitable to use as homework assignments. We have developed a consistent web-based interface that organizes all of the material in each module.

The simulations have been used in classrooms at several institutions, and independent assessments have been performed. Results show that students enjoy the simulations and consider that they help them to learn the material.

See more of this Session: Computing and Technology In Chemical Engineering Education II
See more of this Group/Topical: Computing and Systems Technology Division

# (148c) Teaching Molecular Dynamics and Monte Carlo Simulations: Lessons Learned From the Statistical Thermodynamics Workshops At the School of Advanced Studies In Applied Thermodynamics, Rio De Janeiro, Brazil

**Edward J. Maginn**[1], Frederico W. Tavares[2], Charlles R.A. Abreu[3], Jindal K Shah[4] and Craig Tenney[4], (1)Chemical & Biomolecular Engineering, University of Notre Dame, Notre Dame, IN, (2)Chemical Engineering Program COPPE, Federal University of Rio de Janeiro, Rio de Janeiro, Brazil, (3)Chemical Engineering, Universidade Estadual de Campinas, Campinas, Brazil, (4)Chemical and Biomolecular Engineering, University of Notre Dame, Notre Dame, IN

The authors presented two weeklong workshops on theory and applications of molecular dynamics and Monte Carlo simulations at the Federal University of Rio de Janeiro, Brazil in the summers of 2010 and 2011. The workshops were sponsored by the Chemical Engineering Program of the Alberto Luiz Coimbra Institute (COPPE/UFRJ/Brazil). Participants ranged from new graduate students in a diverse range of fields of study to experienced faculty working in related areas who were interested in learning about molecular simulations. In this talk, we will discuss how the course was designed, how we overcame some of the challenges involved in teaching such a course, and suggestions for how the course could be adopted to other audiences.

See more of this Session: Computing and Technology In Chemical Engineering Education II
See more of this Group/Topical: Computing and Systems Technology Division

**Paper #: 238570**

# LIQUID/VAPOR EQUILIBRIUM VIA EQUATIONS OF STATE FOR FIRST SEMESTER SOPHOMORE STUDENTS

*By*
*Daniel Forciniti*
*Chemical Engineering Department*
*Missouri University of Science and Technology*
*Rolla, MO 65401*

## INTRODUCTION

The availability of powerful computers and of "ready to use" software for the estimation of thermodynamic properties of pure components or mixtures have transformed some basic thermodynamic courses. Many textbooks come with software already developed or with enough information for the students to write simple programs. Are we forgetting concepts by emphasizing calculations? An affirmative answer to this question will definitely weaken the quality of our programs. It is obvious that some compromise is needed between the teaching of concepts and the temptation of overusing computers.

I started to include more and more complex (computer based) calculations in the two thermodynamic classes that I teach as the accessibility of computers increased (I have been doing this for 20 years). Soon I realized that many of the homework problems were not didactic all because the students were becoming just "calculating machines". The fast pace of homework assignments left little room to explore conceptual issues. A few years ago I started with the idea of "mini project" in which the students must use an equation of state to generate a table of thermodynamic properties for a particular compound.

They use an equation of state that I select from the open literature (we use from 6 to 10 different E.O.S. each semester but no two groups have exactly the same problem because they use different compounds). They have to derive expressions for enthalpy and entropies for that particular equation of state and use those values to find the saturation pressure for a given temperature. In addition they have to generate H-S, and P-V plots and compare their calculations against experimental data that they have to find too (an excellent tool to increase the national membership). The parameters of the equation of state are not explicitly given but some references are made available.

Some specific tasks are: 1) Derivation of the expressions for enthalpies and entropies for the equation of state. This is done by hand and they have to check their solution using Maple. 2) Solving the cubic equation of state for a particular temperature (sub-critical) at various pressures. 3) Determining the position of the spinodal and binodal curves by finding points of divergence of the isothermal compressibility and by finding the pressure at which the condition of chemical equilibrium is satisfied. 4) Preparing the required plots and including experimental values.

The "project" has become a staple for the Thermo 1 course and to a lesser extent of Thermo 2. We have incorporated this into our ABET strategy. The work done by the students is used as a performance indicator for the program outcome: "The S&T Chemical Engineering program graduate will have an ability to apply knowledge of mathematics, science, and engineering fundamentals".

**EXAMPLE**

A class of approximately 60 students was divided in groups of 2 according to their cumulative GPA. They were asked to prepare two plots: P vs. V and H vs. S using an equation of state (E.O.S. assignments are below). The P-V plot must have at least three sub-critical isotherms and one supercritical whereas the H-S plot must have at least two isobars. They needed to present their results (calculated) as well as experimental values in those plots. Specific tasks were:

1) Derivation the expressions for enthalpies and entropies for your equation of state.
2) Solution of the cubic equation of state for a particular temperature (sub-critical) at various pressures. They needed to have at least three solutions in the liquid branch, three in the vapor branch and enough points in the two phase region to determine the position of the spinodal and bimodal.
3) Calculation of entropies and enthalpies at two different pressures.
4) Preparation of plots and inclusion of, properly referenced, experimental values.

Table 1. Example of E.O.S. used in the project. No single group has exactly the same project. Cheating is nearly impossible.

| Group | Compound | E.O.S. |
|---|---|---|
| A | Methane | Kedge and Trebble (Ref. 5), non cubic with critical correction |
| B | Decane | Esmaeilzadeh and M. Roshanfekr (Ref. 6) |
| C | Propene | Ishikawa, W.K. Chung (Ref. 7) |
| D | Methane | Kedge and Trebble (Ref. 5), cubic with critical correction |
| E | Dodecane | Esmaeilzadeh and M. Roshanfekr (Ref. 6) |
| F | Propane | Ishikawa, W.K. Chung (Ref. 7) |
| G | Methane | Kedge and Trebble (Ref. 5), non cubic without critical correction |
| H | Carbon dioxide | Ishikawa, W.K. Chung (Ref. 7) |
| I | Benzene | Esmaeilzadeh and M. Roshanfekr (Ref .6) |
| J | Methane | Kedge and Trebble (Ref. 5), cubic without critical correction |
| K | | K.H. Kumar and K.E. Starling. (Ref. 8) |
| L | Carbon Dioxide | Boston-Mathias (Ref 2) |
| M | Carbon Monoxide | Twu with SRK (generic) (Ref 1 and 2) |
| N | Decane | K.H. Kumar and K.E. Starling. (Ref. 8) |
| P | Propylene | Twu with Peng Robinson (generic) (Ref 1 and 2) |
| Q | Ethanol | Twu with TST (individual component paramaters) (Ref 1 and 2) |
| R | Ethanol | K.H. Kumar and K.E. Starling. (Ref. 8) |
| S | Ethylene | Twu with Peng Robinson (generic) (Ref 1 and 2) |
| T | Ethylene | Kozoi (Ref 4) |
| U | n-Octane | Twu with TST (generic) (Ref 1 and 2) |
| V | n-Octane | Harmens (Ref 3) |
| W | Toluene | Harmens [Ref 3) |
| Y | Benzene | K.H. Kumar and K.E. Starling. (Ref. 8) |
| Z | Benzene | Esmaeilzadeh and M. Roshanfekr (ref. 6) |

1) Twu, C.H., Sim, W.D., and Tassone, V. Getting a handle on advanced cubic equations of state. Chem. Eng. Progress, November 2002.
2) Neau et al. The Soave, Twu and boston-Mathias alpha functions in cubic equations of state. Fluid Phase Equilibria, 276:156-164 (2009)
3) Harmens and Knapp. Three-Parameter cubic equation of state for normal substances. Ind. Eng. Chem. Fundam. 19:291-294 (1980)
4) Koziol. Quintic equation of state for pure substances in sub- and supercritical range. Fluid Phase Equilibria 263:18-25 (2008)
5) C.J. Kedge and M. A. Trebble. An empirical near –critical correction for cubic and non-cubic equations of state. Fluid Phase Equilibria, 194-197: 401-409 (2002)
6) F. Esmaeilzadeh and M. Roshanfekr. A new cubic equation of state for reservoir fluids. Fluid Phase Equilibria, 239: 83-90 (2006)
7. T. Ishikawa, W.K. Chung and B.C. –Y. Lu. AIChE J., 26:372 (1980).
8. K.H. Kumar and K.E. Starling. Ind. Eng. Chem. Fundam., 21:255 (1982).

The project was graded according to the rubric shown in Table 2. Approximately 10 % of the students returned superb reports. 20% of the student returned acceptable report. Approximately 30 % of this class failed Thermo 1. The students that passed this class joined me the following semester for a heat transfer class. Their thermodynamics background was excellent.

Table 2. Example of a rubric used to evaluate this class.

Rubric: Project

| Topic | Poor (0 Points) | Fair (40 Points) | Good (80 Points) | Excellent (100 points) | Points |
|---|---|---|---|---|---|
| Experimental Data | Experimental data are missing | Few experimental data are included. Reference is missing | Sufficient experimental data are included. Reference is included. | Experimental data covering of regions of interest are included. References are completed and properly cited. | 0.01x |
| PV calculations | PV calculations are incomplete. The parameters of the equation of state are not calculated properly (if applicable). | PV calculations are incomplete. The parameters in the equation of state are properly calculated. No documentation is included. A sample calculation is not included | PV calculations are complete. The parameters in the EOS are properly calculated. Documentation is incomplete. A sample calculation is included but it is incomplete | PV calculations are complete. The parameters in the EOS are properly calculated. Documentation is complete: Spread Sheets, MathCad pages or equivalent are included and properly explained. A sample calculation is included. | 0.06x |
| Derivation of residual enthalpies and entropies | The derivations were not done or they were incorrectly. Maple solutions may or may not be present | The derivations were done but they are not correct. Maple solutions may or may not be present. | The derivations are correct but details are missing. Maple solutions are present and used to compare own. | The derivations are correct and enough details are included. Maple solutions are present and used to compare own. | 0.02x |
| Calculation of Entropies and Enthalpies | No calculations were done. | Calculations were done but they are incorrect. The ideal gas contribution was not considered. Reference state is not | Calculations were done and they are correct. Calculations include a limited number of points. | Calculations were done and they are correct. Calculations include a large number of points. The ideal gas contribution was included. | 0.06x |

| | | | | | |
|---|---|---|---|---|---|
| | | well defined. | The ideal gas contribution was included.<br>Reference state is not well defined. | Reference state is well defined. | |
| PV and HS plots | Both or one the two plots are missing. | Both plots are present but they are not complete.<br>Experimental data is not included.<br>Plots are of poor quality. | Both plots are present and they are complete.<br>Comparison with experimental data is not satisfactory.<br>Plots are of fair quality.<br>The binodal curve is present but not the spinodal | Both plots are present and they are complete.<br>Comparison with experimental data is satisfactory.<br>Plots are of excellent quality.<br>Both binodal and spinodal curves are present. | 0.06x |
| Presentation of Work | The work done is not explained. | The work done is explained but not in detail.<br>Overall presentation is poor. | The work done is explained in detail.<br>Overall presentation is fair. | The work done is very well explained.<br>The grammar and style are of good quality.<br>Overall presentation is very professional | 0.04x |
| Total percent points (max 25 points) | | | | | ----- |

# (148f) Integrating Computational Transport Phenomena Into the Undergraduate Engineering Curriculum

**Charles A. Petty**[1], Satish Muthu[1] and Andre Benard[2], (1)Chemical Engineering and Materials Science, Michigan State University, East Lansing, MI, (2) Mechanical Engineering, Michigan State University, East Lansing, MI

The field of Computational Transport Phenomena (CTP) or, equivalently, computational fluid dynamics (CFD), has advanced to the point where commercial software can readily be integrated into the undergraduate STEM curriculum. CTP methods provide a powerful and intuitive means to understand the consequences of common assumptions used to analyze the transport of momentum, energy, and mass in complex and simple geometries. Although current CTP technologies can simulate a wide variety of physical problems encountered in engineering and in other disciplines, the use of advanced computational methods in the undergraduate engineering curriculum is not as widespread as it should be.

The proposed presentation will discuss how CTP methods can be used to support student learning and intuition of transport phenomena by simulating the consequences of balance principles, constitutive models, boundary conditions, and initial conditions. Selected examples will be presented where CTP has been employed in the undergraduate curriculum to support specific learning objectives. The presentation will also discuss technical and practical barrier problems that limit the introduction and use of advanced computational methods into the undergraduate curriculum.

See more of this Session: Computing and Technology In Chemical Engineering Education II
See more of this Group/Topical: Computing and Systems Technology Division

# (163a) Intentional Course Design and Outcomes Assessment - Shifting the Curve In Material and Energy Balances Course

Inci Ayranci, Chemical and Materials Engineering, Univeristy of Alberta, Edmonton, AB, Canada and **Suzanne Kresta**, University of Alberta, Edmonton, AB, Canada

**Intentional Course Design and Outcomes Assessment**
***Shifting the Curve in Material and Energy Balances Courses***
**Inci Ayranci and Suzanne Kresta**
**University of Alberta**

The common wisdom is that a bi-modaldistribution in process analysis is "normal". As class sizes in second yearchemical engineering at the University of Alberta have grown to over 100students, we have focused on the process analysis course to consciously uncoverand remove barriers to student learning which result in the "bi-modaldistribution". Our solutions include visual learning, experiential learning,industrial best practices and structured problem solving techniques which are nowembedded in the course. The modified teaching approach progresses in threestages: first, vocabulary building through research on a specific process andthrough flowsheet construction; second, structured visual problem solving toolswhich are also the back bone of industrial best practice; third, activelearning exercises throughout the course to pull out questions and ensure thatstudents are well prepared to tackle problems independently. While there isstill see a tail in our distributions, the lower hump in the curve hasdisappeared. This is a clear indication that we are reaching and helpingstudents who previously were lost and bewildered. These teaching methods takerelatively little effort to implement in the classroom, and create a learningenvironment based on interaction and critical thinking.

Part of this work was supported the Shirleyand Fraser Russell Teaching Fellowship, which was awarded to Inci Ayranci in2010. This fellowship provides a graduate student with the opportunity to gainteaching experience under the mentorship of a faculty member. Inci createdactive learning exercises to improve the understanding and the applications ofthe concepts to the problems. These exercises were very well received and appreciatedby the students.

See more of this Session: In Honor of T.W. Fraser Russell: 2010 W.K. Lewis Award Recipient
See more of this Group/Topical: Education

# (163b) A Web Based Reactor Design Game Used In A Freshman Introduction to Chemical Engineering Course

**Nese Orbey** and Molly Clay, Chemical Engineering, University of Massachusetts Lowell, Lowell, MA

A WEB BASED REACTOR DESIGN GAME USED IN A FRESHMAN INTRODUCTION TO CHEMICAL ENGINEERING COURSE

By Nese Orbey and Molly Clay

This paper describes a project to effectively introduce first year students to chemical engineering. We did this of as a part of an introductory freshman class at University of Massachusetts Lowell using a computer based game developed by Professor TWF Russell (http://www.mht.che.udel.edu). Sixty students were enrolled in the class. Key concepts were introduced during a lecture hour. The students were then split in three two hour lab groups where each student had access to a computer. Two such sessions were required.

The concept of taking a technology from laboratory to manufacturing was demonstrated using this web based game emphasizing the critical need for experimental data, mathematical modeling, and design with uncertainty because of competition. The total market for the product is presented as a demand curve.

The methodology used and the results obtained will be presented and student feedback will be discussed.

See more of this Session: In Honor of T.W. Fraser Russell: 2010 W.K. Lewis Award Recipient
See more of this Group/Topical: Education

# (163c) New Techniques for Just-In-Time Teaching of Chemical Engineering

**J.Will Medlin**, Department of Chemical and Biological Engineering, University of Colorado, Boulder, CO

The use of web-based resources in teaching holds promise for increasing the level of communication between students and instructors and for allowing instructors to focus class time on topics where students need the most help. Recently, several instructors in our department have adopted web-based "just-in-time teaching" (JITT) tools for use in chemical engineering courses. This presentation will describe three key elements of our application of JITT: online quizzes on assigned reading material prior to the start of class, active learning exercises that frequently employ "concept tests" during class, and "screencasts" of problem solutions as a classroom supplement. Content to implement these elements in core chemical engineering classes has been developed and made available by a team at the University of Colorado, as will be described in the presentation.

The online quizzes include a relatively difficult multiple choice portion aimed at ensuring that students read the assigned text and consider it critically. The quizzes also include an open-ended question aimed at promoting discussion and creativity. The instructor is able to review the results of the quiz while designing each classroom meeting, so that activities are planned in part based on the preliminary assessment. Concept tests are generally multiple-choice questions that emphasize a conceptual understanding of course material rather than a detailed derivation or the presentation of a problem solution. Instead, such derivations and problem solutions are provided in online screencasts, which allow students to work along with a narrated problem solution at their own speed. In this presentation, we will illustrate our use of JITT techniques with several examples, and will also discuss student feedback on use of the technique.

See more of this Session: In Honor of T.W. Fraser Russell: 2010 W.K. Lewis Award Recipient
See more of this Group/Topical: Education

# (163d) Classical Thermodynamics and Biochemical Engineering

**Stanley I. Sandler**, Department of Chemical Engineering, University of Delaware, Newark, DE

In accordance with T. W. Fraser Russell's longstanding interest in undergraduate education, this talk will focus several examples of the applications of classical thermodynamics to biochemical engineering that can easily be incorporated into the undergraduate program.

See more of this Session: In Honor of T.W. Fraser Russell: 2010 W.K. Lewis Award Recipient
See more of this Group/Topical: Education

# (163e) OPTIMIZING IN-VITRO FERTILIZATION TREATMENT: A Pedagogical Case Study of Random Phenomena Analysis

**Babatunde A. Ogunnaike**, Chemical Engineering, University of Delaware, Newark, DE

Engineering practice has always required involvement in the acquisition, analysis, interpretation and exploitation of data in one form or another, so that the "well-educated" engineer ought to be truly conversant with how to handle data. This is even more so today, with many engineers now working in non-traditional areas ranging from molecular biology to finance, and the problems to be solved continuing to widen in scope and becoming more interdisciplinary. While most engineers are traditionally well-trained in problem formulation and problem solving when all the entities involved are considered deterministic in character, many remain uncomfortable with problems involving random variations, if such problems cannot be idealized and reduced to the more familiar "deterministic" types. Ironically, while students of generations gone by were trained in basic data analysis, most of today's students are not. And for the very few that receive any form of training in probability and statistics, most tend to perceive such training more as a set of instructions on *what* to do and *how* to do it, than as a training in fundamental principles of random phenomena.

At the University of Delaware, the course in "Random Phenomena" is taught by presenting basic fundamental principles, methods, and tools for formulating and solving engineering problems that involve randomly varying phenomena. The premise is that by emphasizing fundamentals and basic principles, and then illustrating these with examples, the student will be better equipped to deal with a range of problems wider than that explicitly covered in class. In this presentation, we discuss a particular case study that is used in this class as a pedagogical tool to illustrate how to solve complex, real-life problems involving randomly varying phenomena. It demonstrates how the mathematics of probability, probability models, statistical data analysis, model validation, and mathematical optimization can be used effectively to answer a central question in in-vitro fertilization (IVF) treatment: *how many embryos should be implanted to maximize the chances of success while simultaneously minimizing the chances of multiple births*?

# (163f) Using Gas Sparger Models to Determine Mass Transfer Coefficients and Bubble Area for Fermentation

**Anne Skaja Robinson**, Department of Chemical Engineering, University of Delaware, Newark, DE and Twf Russell, University of Delaware, Newark, DE

It is often challenging experimentally to determine the mass transfer area, a, in the constitutive equationfor the rate of mass transfer, $r_A$, where $C_A^{I}$ is the rate of mass transfer ofcomponent A between the phases $C_A^{II}$ and $K_a$ .Almost all experiments are designed to determine KMa ,which is an equipment specific quantity, limiting its utilityfor scaling up. Using an example developed for *Mass and Heat Transfer: Analysis of Mass Contactors and HeatExchangers*, we show that model equations originally developed for commerical-scale gas-liquid contactors(Otero, 1983) can be applied to data obtained for a small scale fermentor (Biostat UD30) using only the gasflow rate and no assumptions to obtain separate values of $a$ and that compare very well with the composite valueobtained from the data of Tobajas, M. et al (2003).Otero, Z (1983) PhD Dissertation, University of DelawareTobajas, M., Garc'a-Calvo, E, Wu, X, Merchuk, J. (2003) *World J. Microbiol. Biotechnol.* 19:391-8.

See more of this Session: In Honor of T.W. Fraser Russell: 2010 W.K. Lewis Award Recipient
See more of this Group/Topical: Education

# (163g) Teaching Mixing to Under-Graduate Chemical Engineers

**Richard K. Grenville**[1], Thomas A. Simpson[1] and Arthur W. Etchells III[2], (1) DuPont Engineering Technology, Wilmington, DE, (2)AWE3 Enterprises, Philadelphia, PA

As Chemical Engineers we are concerned with the transformation of raw materials into valuable products. These transformations may require the mixing and reacting of the raw materials and the separation of the desired products from by-products. Also, as these transformations take place, the materials may need to be heated and/or cooled. All the disciplines required to understand the operation and design of mass and heat transfer equipment are covered in the under-graduate Chemical Engineering curriculum.

Mixing has been identified as a critical operation for the Chemical Industry (see for example, Smith, 1990, *Trans IChemE.* 68, pp. 3 - 6) yet it is not, generally, taught to Chemical Engineering students. There are a number of reasons for this. Among those given are:

It is the University's job to "teach the basics" and the engineer will learn "practical skills" once he/she has started work.

In order to add a course on mixing, some other subject would have to be removed from the curriculum.

But the same comments are not made about heat exchangers or distillation columns.

If we accept that there is no "market" for a full mixing course at most Universities, how can we ensure that the students get some exposure to this technology before they graduate? We believe that a working understanding of mixing can be taught by adding lectures as mixing modules to the classic Chemical Engineering courses. For example:

An agitator impeller is a pump and it can be characterized in the same way as the pump in a piping system that students will have studied in Fluid Mechanics.

Design of agitators for heat and mass transfer can be covered in the traditional courses.

Mixing issues relating to fast, competitive reactions can be covered in the traditional Reaction Engineering course. When does the assumption that a reactor is "perfectly-mixed" fail?

The content of several proposed modules will be presented.

See more of this Session: In Honor of T.W. Fraser Russell: 2010 W.K. Lewis Award Recipient
See more of this Group/Topical: Education

# (163h) Fraser Russell's Influence On Chemical Engineering At Rowan

**Robert P. Hesketh**, Zenaida Gephardt and Mary Staehle, Chemical Engineering, Rowan University, Glassboro, NJ

Fraser Russell has had a substantial influence on the chemical engineering program at Rowan University. This presentation will summarize his contributions through current faculty that completed a PhD under his direction. These faculty members include Zenaida Gephardt, Robert Hesketh and Adjunct faculty member Art Etchells. In addition two faculty members at Rowan have been mentored by through the University of Delaware Fraser & Shirley Russell Teaching Fellow Program: Brian Lefebvre and Mary Staehle. Finally Fraser has presented his educational initiatives at Rowan to students and faculty.

See more of this Session: In Honor of T.W. Fraser Russell: 2010 W.K. Lewis Award Recipient
See more of this Group/Topical: Education

# (166a) Laboratory Activities for Introducing Students to Biomaterials

**Jennifer Vernengo** and Kevin Dahm, Chemical Engineering, Rowan University, Glassboro, NJ

Because of the increasing prevalence of musculoskeletal, cardiovascular, and neurodegenerative diseases, there is a necessity to engineer biomaterials that can be used to treat these painful and debilitating disorders. This paper presents two experiments that introduce freshman and senior-level undergraduate Chemical Engineering students to the design and evaluation of biomaterials. By designing ceramic-polymer composites for bone substitution and hydrogels for cartilage replacement, students develop new knowledge in biomaterials science that also contributes to ABET learning outcomes for undergraduate Chemical Engineering courses. Assessment results will be presented, including those of pre and post-tests, which indicate the effectiveness of the learning activities

See more of this Session: Labs and Experiments In the ChE Curriculum
See more of this Group/Topical: Education

# (166b) Integration of the Unit Operations Laboratory with a Focus On Biofuel Production

**Jim Pfaendtner** and Danilo Pozzo, Chemical Engineering, University of Washington, Seattle, WA

This presentation will highlight our reconfiguration of the senior chemical engineering unit operations (UO) lab (CHEM E 437) at the University of Washington. Using the themes of biotechnology and renewable energy, we have developed a new pedagogical model, which emphasizes 1) collaboration between teams and 2) developing instincts for thinking about the impact of single unit operations on the operation of an entire production facility. The production of renewable energy is one of the most important technological problems that we face today. This challenge also offers us an opportunity to motivate and shape the early careers of Chemical Engineering undergraduate students. The most innovative component of the new laboratory is the complete integration of new and existing experimental stations. The senior UO lab was integrated to model a bio-fuel production plant where student groups work on individual operations that make up a complete process. This full-plant view of the laboratory allows students, for the first time, to evaluate the effects of their decisions on upstream and downstream plant operations. Furthermore, it also provides a common framework to promote active discussion and engagement among student groups. The transformation of the course included the development of completely new modules for fermentation of biomass and the modification of existing equipment and modules for the treatment, separation and extraction of product and waste streams. The new fermentation modules utilize internet-based remote monitoring technologies to track the development of fermentations while students are outside of the laboratory. Fully interconnected units now define a common goal of reducing costs and improving productivity and replace the original independent and unrelated experiments. The new structure also allows us to easily incorporate design concepts, such as cost analysis and environmental compliance, into the laboratory. The objective of the re-designed course is to provide a realistic structure that is congruent with what students will experience after graduation. The new laboratory structure is also designed to foster leadership, creative thinking, composure under uncertainty and the critical review of information. Furthermore, with the new structure, we also continue to meet the original learning objectives of instructing students on the basics of experimental planning and reporting. We will additionally present evaluation and assessment of the student experience from the first round of the new lab and our updated plans for continuing to improve the student experience in our course.

See more of this Session: Labs and Experiments In the ChE Curriculum
See more of this Group/Topical: Education

# (166c) A Drug Delivery Experiment Using Alginate Microspheres

**Stephanie Farrell** and Jennifer Vernengo, Chemical Engineering, Rowan University, Glassboro, NJ

Controlled release drug delivery has recently become a major research and development focus area of today's pharmaceutical industry. New drug delivery systems are continually being designed to dispense a drug at a predetermined rate, to a target area in order to maintain a therapeutic concentration of drug at a desired location in the body. Rowan University is currently working with the Engineering Research Center for Structured Organic Particulate Systems (ERC-SOPS) of Rutgers University to develop educational materials involving drug delivery technology. We describe a drug delivery module in which students investigate factors affecting drug release from hydrogel drug delivery systems. Calcium alginate microspheres and other geometries are produced containing a model drug, limonene (yellow food dye) or bovine serum albumin. Using a guided inquiry approach, students investigate the effects of surface area, degree of cross linking, drug molecular weight, and stirring speed. Students learn introductory concepts related to experimental design and data analysis, mass transfer, and biomaterials. Calculus and chemistry concepts are also reinforced.

See more of this Session: Labs and Experiments In the ChE Curriculum
See more of this Group/Topical: Education

# (166d) Developing Tools for Teaching Chemical Engineering Unit Operation Design

**Rachael H. Elder**, Department of Chemical and Biological Engineering, University of Sheffield, Sheffield, United Kingdom

The teaching of chemical engineering unit operation design in a project week environment is essential for a synoptic course and enhanced student learning. This mini project, funded by the Engineering Subject Group of the Higher Education Academy, has two distinct aims: improving student awareness of resources available for Chemical Engineering Design and development of a pilot Staff Toolkit to assist lecturers new to Chemical Engineering Design. The second year design week is used as an example of how different resources can be introduced and used in a design exercise.

Experience from supervising third year undergraduate design projects over the last couple of years has highlighted the significant amount of time spent on directing students to resources where they can find information. This not only detracts from supervision time but also reduces the useful learning content of the project. Students should be aware of a wide range of resources and not just those relevant for the design they complete during their design projects. The rationale behind this work is that by providing a more thorough grounding in the second year design week, students will be better prepared for their IChemE accredited design project and for life as a Chemical Engineer.

Second year design week involves a task based design, focussing on prescribed unit operations in a defined process, so is ideal for directing the students to the wealth of new online design resources relating to both underlying theory and practical design of unit operations. The synoptic nature of the project week makes it a vital part of the curriculum to ensure student understanding. This paper discusses some of the different resources available and the way they have been implemented into the design week.

Student awareness of online resources before and after design week was evaluated via questionnaires and focus groups. The impact of the online resources introduced in the design week on the cohort's third year design project was also evaluated. The results show that the students found the design week and online resources extremely useful.

Providing a teaching resource in the form of a pilot staff toolkit is another key aim of the project as it will be an invaluable teaching aid, especially given the increasing number of academics lacking real industrial design experience. The pilot of the staff toolkit will be demonstrated and discussed.

See more of this Session: Labs and Experiments In the ChE Curriculum
See more of this Group/Topical: Education

# (166e) Alternative Lab Reports - Engineering Effective Communication

**Daniel Lepek**, Department of Chemical Engineering, The Cooper Union, New York, NY and Richard Stock, Chemical Engineering, The Cooper Union, New York, NY

For many chemical engineering undergraduate programs, the laboratory sequence allows students to experience the application of chemical engineering outside of the classroom. Typically, the results of these experiments are analyzed and written up in a classic laboratory report format. Although writing laboratory reports is a valuable skill, today's engineers are expected to report their results and express their findings in different forms of oral, written, and visual communication.

We have introduced a new initiative at The Cooper Union in the senior chemical engineering laboratory sequence that provides the students with a range of skill-sets that prepares them to communicate successfully on a professional level in different contexts and environments. As a baseline, students were expected to submit 2 laboratory reports a semester, written on the level of a scholarly journal article. To develop and stretch their communication skills, students used their laboratory experiences to submit executive memos to company CEOs, formulate grant proposals to government agencies. and created posters that convey their results in a visual setting. Students also made oral presentations tailored to a variety of key audiences of different scientific and corporate backgrounds.

The Cooper Union's CONNECT (Cooper's Own No Nonsense Engineering Communication Training) Program, also initiated as part of this laboratory sequence in 1996, partnered this initiative, providing students with feedback and specialized training in both written and oral communication as well as graphics.

In this presentation we will illustrate student work using these report alternatives, indicate how we assessed their effectiveness, present examples of the feedback and specialized training stimulated by the students' work and outline the lessons learned as we move on to our next year.

See more of this Session: Labs and Experiments In the ChE Curriculum
See more of this Group/Topical: Education

# (166f) Wiki Implementation In a Senior Unit Operations Laboratory

**James P. Abulencia**, Department of Chemical Engineering, Manhattan College, Riverdale, NY

Laboratory reports have traditionally been the primary deliverable for students enrolled in the unit operations laboratory course. These reports are typically comprised of several sections that emulate the format of a traditional scientific publication. Because these documents can be very lengthy, it is time consuming for educators teaching this course to read and assess the work. To this end, a wiki system was implemented wherein students can collaboratively contribute to the body of knowledge pertaining to a particular experiment. This wiki is comprised of four parts: 1) "Background", which focuses on industrials applications, 2) "Theory and Equations", which highlights important definitions, principles, and equations pertaining to the experiment, 3) "Technical Notes", which provides a venue for sharing experimental details to subsequent groups, and 4) "Other" for facts/details that do not directly belong under the three aforementioned headings. Students are assessed on the impact of their contribution, as well as a final exam based on the material in the wiki. The benefits are as follows: 1) students are placed in an environment that encourages collaborative learning, 2) a final exam ensures that all students understand the experiments (not only the group leaders), and 3) the jigsaw approach provides student satisfaction on the personal level for contributing to the class' knowledge.

Note to reviewer: A survey will be distributed at the conclusion of the course, recording student opinion on their experience with the wiki. Thus, the results are not currently available. However, students have articulated a positive experience overall.

See more of this Session: Labs and Experiments In the ChE Curriculum
See more of this Group/Topical: Education

# (166g) Web-Lab: Enhancing the Undergraduate Engineering Experience

**Cynthia Collins**, Department of Chemical and Biological Engineering, Rensselaer Polytechnic Institute, Troy, NY and **Kristala Jones Prather**, Chemical Engineering, Massachusetts Institute of Technology, Cambridge, MA

Web-Lab is an innovative SBE program that web enables laboratory experiments. A collection of bioreactors are located miles away in a central facility, and students, regardless of their physical location, can access and remotely interact with the bioreactor in real time over the internet. Multiple schools can access these web-based experiments and incorporate them into their curriculum, either as a lab or to supplement a lecture course. This is the objective of the SBE Web-Lab Initiative—to advance engineering education by breaking beyond the confines of the classroom and enhancing the typical classroom experience.

The upside of the web-based laboratories is to reduce the cost incurred per student to deliver state-of-the-art laboratory exercises and thereby enable more students to participate in a modern laboratory experience. Web-based laboratories will improve undergraduates' education by better preparing engineers with the latest skills and tools to compete in the global workforce.

See more of this Session: Labs and Experiments In the ChE Curriculum
See more of this Group/Topical: Education

# (166h) An Integrated Curricula Approach to a Great Capstone Lab Experience and Its Potential Implementation In South America

**Rocio Tijaro**, Ingenieria, Universidad Arturo Prat, Iquique, Chile, Mario Oyanader, Chemical Engineering, Tenneessee Tech University, Cookeville, TN and Joseph Biernacki, Chemical Engineering, Tennessee Technological University, Cookeville, TN

Traditional Unit Operation Labs (UOL) consist of a series of well define experiments that are run every semester with very little modification. As a matter of fact, a steady operation of the many units (Absorption Column, CSRT, Heat Exchanger, etc…) is very much preferred so the students have the chance to validate performance of unit the set up. In addition, a traditional UOL course usually have 3 to 4 credits hours, whereas in a Lecture-Lab integrated curricula the number of credit drops significantly to 1 to 2 credit hours. In this case, making a UOL core course an actual learning tool can be an opportunity for creativity and enhancement. In this paper, it is described how Chemical Engineering core courses converge into one single project with an open-ended problem. Particular attention is given to the strategic planning of the course, logistic associated, ABET requirement and external assessments. Other aspects are also includes as for example the details of the expected deliverables of the course, final presentations, and statistical analysis performed on the data collected. After five years of implementation at Tennessee TECH University, the pros and cons of the approach are evaluated for testing in a different environmental setting as it may be the North to the South American culture. The author's particular intentions are to document potential barriers as well as opportunities in a more global engineering education. The base case for the analysis is the new UOL concept as of a progressive program.

See more of this Session: Labs and Experiments In the ChE Curriculum
See more of this Group/Topical: Education

# (232a) Process Hazards Emphasis In the New ABET Chemical Engineering Program Criteria

**Jeffrey J. Siirola**, AIChE, Kingsport, TN

This presentation will outline the US Chemical Safety Board concerns and recommendations that lead to the drafting of new ABET chemical engineering program criteria, the wording that is currently working its way through the ABET approval process, other ABET "editorial" changes being proposed for the chemical engineering program criteria, likely ultimate wording of the program criteria, and expected effective date of implementation. The intent of the Education and Accreditation Committee in drafting the new criteria wording will also be summarized.

See more of this Session: Integrating Safety Into the Chemical Engineering Curriculum - Where, What, and How (Part 1)
See more of this Group/Topical: Education

# (232b) Promoting Acceptance of Process Safety Curriculum Requirements

**Scott Berger**, American Institute of Chemical Engineers, New York, NY

In 2009, the US Chemical Safety and Hazard Investigation Board (CSB) recommended to AIChE and ABET to develop requirements related to understanding safety hazards and how to prevent them. The resulting new ABET requirements are anticipated to take effect in 2012. It would not be unusual if some administrations of universities offering chemical engineering degrees were concerned or even upset about the new requirements.

In this short presentation, I will attempt to address the concern over the new requirements by discussing a broad range of benefits to the quality of the students' chemical engineering education and students' future employability, as well as some additional benefits that may accrue to the overall campus experience based on implementing the new requirements.

See more of this Session: Integrating Safety Into the Chemical Engineering Curriculum - Where, What, and How (Part 1)
See more of this Group/Topical: Education

# (232c) Resources, Recommendations and Overcoming Challenges for Integrating Safety Into the Chemical Engineering Curriculum

**Amy Theis**, Fauske & Associates, LLC, Burr Ridge, IL

Interviews were performed with more than ten instructors and professors who are currently incorporating safety into the chemical engineering curriculum. These individuals described their methodology, resources and challenges in this effort. Excellent programs exist with courses dedicated to chemical process safety; some curricula also include a lab safety component. Implementation varies between schools; best practices and tips for effective teaching methods will be summarized. Examples of methods to overcome both explicit and perceived challenges will be discussed. Implementation strategies will be provided for universities with minimal safety experience as well as those that want to achieve a higher level of excellence. A relationship between industry and universities should exist to reinforce this commitment to safety; examples of this type of partnership will be presented.

See more of this Session: Integrating Safety Into the Chemical Engineering Curriculum - Where, What, and How (Part 1)
See more of this Group/Topical: Education

# (232d) Educational Resources On Process Safety At the SaChE Website

**Thomas O. Spicer**, Ralph E. Martin Department of Chemical Engineering, University of Arkansas, Fayetteville, AR

Started in 1992, the Safety and Chemical Engineering Education Program (SaChE) is a cooperative effort between CCPS and engineering educators to provide teaching materials and programs for the enhancement of undergraduate and graduate students studying chemical and biochemical products and processes.

The SaChE Committee is comprised of representatives from academe, industry, and government in addition to AIChE staff. The SaChE Committee develops, reviews, and distributes classroom and training materials (SaChE Products) with the ultimate aim of safer operations. At its inception, funding for SaChE was provided from CCPS as well as membership fees from participating universities. As part of the Safety Scale-Up Partnership between CCPS and corporate sponsors, SaChE membership has been provided for all US and Canadian chemical engineering departments with AIChE student chapters.

Starting in 2003, SaChE Products were distributed by a web site hosted by the Ralph E. Martin Department of Chemical Engineering at the University of Arkansas (http://www.sache.org). Since the original development of the site, features have been added to further the goal of safety education including: Process Safety Beacon archive; Student Safety Certificate Program (largely based on the SaChE Products); Materials from Faculty Workshops sponsored by SaChE/CCPS; and links to related materials (such as the CSB Video Room).

The site also serves an administrative function by listing the SaChE Committee members in addition to a messaging service that facilitates email communication with Committee members and university and corporate contacts. Much of the material on the site can be viewed by all visitors. Some material on the site is restricted to registered users including the SaChE Products and printable editions of the Process Safety Beacon archive. This paper summarizes the resources available on the site and discusses statistics on worldwide usage gathered by Google Analytics.

See more of this Session: Integrating Safety Into the Chemical Engineering

# Continuing Our Journey to Bridge the Process Safety Gap Between Academia and Industry: Meeting the New ABET Process Safety Expectations

*Bruce K. Vaughen, Cabot Corporation, Tuscola, IL*

*Thomas O. Spicer, Ralph E. Martin Department of Chemical Engineering, University of Arkansas, Fayetteville, AR*

*D. "Trey" Morrison, Exponent, Lisle, IL*

*James A. Klein, DuPont, Minneapolis, MN*

*David A. Rockstraw, Department of Chemical Engineering, New Mexico State University, Las Cruces, NM*

## ABSTRACT

This paper discusses a proposed organizational framework for SAChE products and reference materials that will help meet the new Chemical Process Safety (CPS)-related instructional requirements. The goals of this paper include:

1) To show how the current CPS products, certificates and references correspond to industry's Process Safety Management (PSM) efforts,
2) To provide a summary of the latest efforts for organizing and displaying the products, safety certificates and CPS-related resources into a curriculum-based framework,
3) To describe a matrix that links available SAChE materials to a typical chemical engineering curriculum, and
4) To show a couple of examples of how these CPS products (including the SAChE Process Safety Certification Program) can be incorporated into a typical chemical engineering curriculum to meet the proposed changes from ABET.

The paper concludes with a brief description of future SAChE-related efforts to help us reduce our process safety-related risks.

## BACKGROUND

The Safety and Chemical Engineering Education (SAChE) committee was created in 1992, in part, to help provide instructors with practical Chemical Process Safety (CPS)-related teaching materials (aka "products") from industry [1]. One of SAChE's goals is to help instructors who don't have extensive process safety experience prepare and teach CPS lessons in their classes. A stronger emphasis on CPS in the chemical engineering curriculum will help reduce industry's process safety risk helping prevent major incidents such as the explosion and fatalities that occurred at T2 Laboratories in 2007 [2]. The process safety-related gap between education and application can be visualized in Figure 1 with a broken "bridge over troubled waters" of potential fires, explosions, runaway reactions and toxic releases.

Currently, the SAChE Committee has overseen the creation of more than 90 different products, the development of the safety certificate program (six certificates are currently available for students to complete), and provided a clearinghouse for many useful links to other resources [1]. Now that SAChE has established its strong resource base of teaching materials, we believe it is time to enhance the SAChE product's organizational framework to help instructors more easily locate materials that meet the AIChE/CCPS Process Safety Expectation Guidelines. The objective of making SAChE materials more accessible to faculty will facilitate the response of academia to the proposed 2011-12 ABET Chemical Engineering Program-Specific Criteria to "address the hazards associated with these processes," as is recommended by AIChE and the CSB [2, 3, 4].

## A PROPOSED SACHE FRAMEWORK – THE SACHE LIBRARY

The proposed SAChE Library – an organizational framework - is designed to help chemical engineering instructors find course-specific materials for their use that incorporate the elements of process safety. This section of the paper describes the development of the "SAChE Library" in four parts:

1) SAChE's Link to Process Safety Management (PSM)
2) SAChE's CPS Materials - A New Organizational Scheme
3) A Matrix Linking the SAChE CPS Materials to the ChE Curriculum
4) Examples for Integrating SAChE CPS Materials in a ChE Curriculum.

As is noted above, the SAChE CPS-related materials have been created to help instructors prepare lessons for their particular chemical engineering course (e.g., thermodynamics, kinetics, design, etc.). The SAChE Library – its organizational scheme - is designed to help instructors find their course-specific materials quickly.

Part 1) SAChE's Link to Process Safety Management (PSM)

A useful image that helps describe a "PSM System" is shown as a PSM Wheel in Figure 2 [5]. The PSM System's three parts are as follows:

| | | |
|---|---|---|
| I. | PSM Programs (its 12 PSM Elements) | The Spokes |
| II. | Information Technology Systems (ITS) | The Hub |
| III. | Operational Discipline (OD) | The Tire |

This system is designed to reduce the risks associated with fires, explosions, runaway reactions and toxic releases when handling or processing hazardous materials. By preventing the Loss of Containment (LOC) of flammable or toxic materials, PSM Programs help protect the safety and health of people, the environment, the facilities and the business.

As is shown in Figure 2, Information Technology Systems (ITS) represent the hub of the PSM Wheel. Information Systems include both paper and electronic systems that help communicate PSM-related information between the spokes. The communication may be between equipment, between people or be interfaces between equipment and people. These systems may be simple or complex, local or corporate. They may help document corporate or legal requirements and may require different levels of authorization. PSM risks are minimized when effective communication exists between PSM Elements, when learnings are shared, and when people document, store, maintain and use the process information.

The benefits for effectively applying PSM Programs focus on this concept: with fewer leaks, we reduce the potential for injuries, process safety events, environmental harm, equipment failure, poor yields and poor quality. Hence, PSM Programs help reduce potential disability costs, help reduce potential air, land and water regulatory permit violations and fines, help reduce unexpected production shutdowns and capital equipment replacement costs. A list of the twelve PSM Elements is shown in Figure 3. The details for each PSM Element are beyond the scope of this paper and are described elsewhere [5]. The goal of a robust PSM System is to keep us and our environment healthy so that our neighbors, our customers and our stockholders are happy.

Part 2) SAChE's CPS Materials - A New Organizational Scheme

Currently, the SAChE Committee has overseen the creation of more than 90 different products [1]. Now that there are many educational materials and references, discussions have focused on how best to "frame" these efforts for an efficient "search" for instructors. Prior work has noted reorganizing the existing products within "capsules" designed for specific chemical engineering courses [6]. Hopefully, the efforts noted in this paper capture subsequent discussions about SAChE's Modules and its website.

Note that the synergistic effects of many educators and representatives from industry work together, helping the student better understand different aspects of chemical engineering during their educational experience. The SAChE Committee conducts workshops, in conjunction with representatives from industry, where faculty can visit an industrial facility and get hands-on experience with the application of process safety principles and practices. Note that the approach to incorporate industrial-specific educational materials and examples into existing courses is not unique to the process safety field, as other "modules" designed by experts for use by instructors exist in other fields like sustainability and green engineering [7], financial risk analysis [8] and the use of computers [9], as well.

Part 3) A Matrix Linking the SAChE CPS Materials to the ChE Curriculum

A Matrix linking courses in the Chemical Engineering Curriculum to the PSM System is shown in Figure 4. The rows of this matrix list typical courses in a chemical engineering curriculum, including both fundamental science and mathematics courses as well as chemical engineering specific courses. The columns of this matrix capture the parts of the PSM System shown in Figure 2 and listed in Figure 3. The SAChE Link correlates to the "Design" aspects of a robust PSM System: Process Safety Information (PSI), Process Hazards Analyses (PHA) and Management of Change (MOC). The specific link at the "Design" stage for process safety between education (SAChE) and application (PSM) is noted in Figure 3, shaded with respect to the other PSM Elements in Figure 4 and shown schematically as a "bridge" in Figure 5.

Part 4) Examples for Integrating SAChE CPS Materials in a ChE Curriculum

There are many ways to incorporate the elements of process safety into existing chemical engineering courses without having to create a dedicated process safety-related course within an already overcrowded curriculum. The objective of the new program-specific ABET criteria for chemical engineering can be met, in part, by integrating SAChE CPS Modules into existing courses. The following examples show how these CPS educational materials can be incorporated in the senior design course and how these materials are integrated across an engineering curriculum at New Mexico State University.

Example 1: One approach to meet the ABET expectations has been described in detail in prior work [10]. A SAChE Module, designed to help meet the new ABET expectations, helps bridge the gap between academia and industry by incorporating existing SAChE Modules into the senior design course: "Safety Guidance for Design Projects [11]". The Senior Design Project Block Diagram - a design course "Road Map" - is shown schematically in Figure 6. This Road Map can be used to help the students meet and document the new ABET process safety-related expectations.

Example 2: At New Mexico State University, students must complete seven SAChE Modules to meet their graduation requirements – it is required as a part of their degree. The SAChE Modules are assigned as stand-alone exercises, with "certification" in the assigned modules accounting for approximately 5% of course grade. The course examinations/quizzes include questions drawn from SAChE Module content. In addition, the AIChE Student Chapter offers viewings of the CPS-related videos followed by Q&A sessions with pizza.

The learning modules serve as independent, stand-alone educational assignments for students to complete outside of the classroom and independent of the lecture. These modules contain a bulleted list of learning objectives that permit an instructor to quickly identify relevant content suitable to in-class assessment of student learnings. The Modules are targeted for different education levels, from freshmen through seniors, to allow for easy integration of content across the entire student experience. Hence, over time, the students are immersed in the chemical industry's corporate safety philosophy.

## OUR PATH FORWARD

Where we are today: Process Safety Management (PSM) is defined as the principles and systems applied to identify, understand and control process hazards to reduce process safety risks and protect the safety and health of people, neighbors and the environment [see reference 12 for a recent CCPS overview of PSM]. As risk managers, we focus on the hazards of our processes that may lead to a major Loss Of Containment (LOC), resulting in fires, explosions, runaway reactions and/or toxic releases. Our risk reduction equation, shown in Figure 7 and described in detail elsewhere [13, 14], is used as the basis for reducing our process safety risks. And as engineers across the globe, we "hold paramount the safety health and welfare of the public and protect the environment [15]," we "must be dedicated to the protection of the public health, safety, and welfare [16]," and we "work to enhance the welfare, health and safety of all whilst paying due regard to the environment and the sustainability of resources [17]."

For decades, experts across the globe have pooled and shared their experiences to help apply the principles of process safety to reduce our process safety risks [1, 9, 18, and 19]. There are visible risk reduction efforts, such as meeting the criteria of Responsible Care® [20], and invisible efforts simply performed and undocumented since they are "the right thing to do."

Where we may be going: The authors hope that these visible SAChE-related efforts contribute to bridging the process safety-related gap between academia (education) and industry (application), as is shown in Figure 8, helping reduce our process safety-related risks. We hope that the SAChE Modules are used to help ensure adequate education – combining both training and skills-based testing - meeting our specific regulatory process safety-related requirements to transform raw materials safely and efficiently into products, as well [e.g., 21, 22, 23, 24, and 25]. Future SAChE Committee efforts will include identifying which new teaching modules need to be created or which of the existing modules will need to be reviewed and updated. Future efforts will include the SAChE Website and system "scalability" – i.e., its sustainability – to ensure that accurate and up-to-date information is readily available to instructors.

## REFERENCES


1. The Safety And Chemical Engineering Education (SAChE) Website, An AIChE Alliance, http://www.sache.org/ (accessed 21-August-2011).

2. The US Chemical Safety Board (CSB) T2 Laboratories Report, http://www.csb.gov/ (accessed 21-August-2011): completed investigations, T2 Laboratories Inc. Reactive Chemical Explosion, December 19, 2007.

3. Gunaratnam, Rachael T., "CSB Recommendations From the T2 Laboratories Inc. Investigation," paper presented in Session 339, "Integrating New Safety Requirements Into the Chemical Engineering Curriculum" Session, abstract 339a published in the Proceedings from the 2010 AIChE Annual Meeting, Salt Lake City, UT, November, 2010.

4. Willey, Ronald J., H. Scott Fogler and Michael B. Cutlip, "The Integration of Process Safety Into a Chemical Reaction Engineering Course: The Review of the T-2 Incident," paper presented in Session 339, "Integrating New Safety Requirements Into the Chemical Engineering Curriculum" Session, abstract 339b published in the Proceedings from the 2010 AIChE Annual Meeting, Salt Lake City, UT, November, 2010.

5. Vaughen, Bruce K., "A Process Safety Management (PSM) Overview," Safety and Chemical Engineering Education (SaChE), SaChE Products (accessed 21-August-2011), http://sache.org/products.asp, 2012.

6. Morrison, Delmar (Trey) R. and Russell A. Ogle, "Developing Process Safety Capsules for the Chemical Engineering Classroom," paper presented in Session 339 "Integrating New Safety Requirements Into the Chemical Engineering Curriculum" Session, published in the Proceedings from the 2010 AIChE Annual Meeting, Salt Lake City, UT, November, 2010.

7. Seay, Jeffery, Richard Rezek and Luke Richardson, "Developing Sustainable Engineering Modules for the Chemical Engineering Curriculum," paper presented in Session 405 "Incorporating Green Engineering and Sustainability Into the Curriculum," abstract 405e published in the Proceedings from the 2010 AIChE Annual Meeting, Salt Lake City, UT, November, 2010.

8. Papavassiliou, Dimitrios V., "Introducing Risk Analysis in a Design I Course," paper presented in Session 247 "Steal This Course!," abstract 247a published in the Proceedings from the 2010 AIChE Annual Meeting, Salt Lake City, UT, November, 2010.

9. Computer Aids for Chemical Engineering Education (CACHE), http://cache.org/ (accessed 21-August-2011).

10. Vaughen, Bruce K., "A SAChE Module Designed to Bridge Process Safety's Troubled Waters: Meeting the New Academic Process Safety Requirements," paper presented in Session 339 "Integrating New Safety Requirements Into the Chemical Engineering Curriculum" Session, published in the Proceedings from the 2010 AIChE Annual Meeting, Salt Lake City, UT, November, 2010. Submitted for review to *Chemical Engineering Education*, July 2011.

11. Vaughen, Bruce K., "Safety Guidance for Design Projects," Safety and Chemical Engineering Education (SaChE), SaChE Products, http://sache.org/products.asp, 2011 (accessed 21-August-2011).

12. Sepeda, A. L., "Understanding Process Safety Management," Chemical Engineering Progress, August 2010, Vol 106, No. 6, pp. 26-33.

13. Klein, James A., and Bruce K. Vaughen, "Implementing an Operational Discipline Program to Improve Plant Process Safety," Chemical Engineering Progress (CEP), The American Institute of Chemical Engineers (AIChE), June 2011, pp. 48-52.

14. Vaughen, Bruce K., and James A. Klein, "Improving Operational Discipline to Prevent Loss of Containment Incidents," Process Safety Progress (PSP), The American Institute of Chemical Engineers (AIChE), in press.

15. The American Institute of Chemical Engineers (AIChE), Code of Ethics, http://www.aiche.org/About/Code.aspx (accessed 21-August-2011).

16. The National Society of Professional Engineers (NSPE), Code of Ethics, http://www.nspe.org/Ethics/CodeofEthics/index.html (accessed 21-August-2011).

17. The Royal Academy of Engineering, in collaboration with Engineering Council (UK) and other societies, including the Institution of Chemical Engineers (IChemE): "Statement of Ethical Principles," (accessed 21-August-2011), http://www.raeng.org.uk/societygov/engineeringethics/principles.htm

18. The Center for Chemical Process Safety (CCPS), American Institute of Chemical Engineers (AIChE), http://www.aiche.org/CCPS/index.aspx (accessed 21-August-2011).

19. European Process Safety Centre (EPSC), http://www.epsc.org/ (accessed 21-August-2011).

20. American Chemistry Council (ACC), Responsible Care ® Management Systems and Certification, http://responsiblecare.americanchemistry.com/Responsible-Care-Program-Elements/Management-System-and-Certification/default.aspx (Accessed 21-August-2011).

21. Occupational Safety & Health Administration (OSHA), Standard, OSHA 29 CFR 1910.119, Process safety management of highly hazardous chemicals, http://www.osha.gov/pls/oshaweb/owadisp.show_document?p_table=STANDARDS&p_id=9760 (Accessed 21-August-2011).

22. Environmental Protection Agency (EPA), The Risk Management Plan (RMP) Rule, http://www.epa.gov/osweroe1/content/rmp/index.htm (Accessed 21-August-2011).

23. Health and Safety Executive (HSE), Control of major accident hazards (COMAH), http://www.hse.gov.uk/comah/index.htm , (Accessed 21-August-2011).

24. European Commission (EC), Environment, Chemical Accidents (Seveso II) - Prevention, Preparedness and Response http://ec.europa.eu/environment/seveso/index.htm, (Accessed 21-August-2011).

25. Guidelines for Process Safety Management, AQ/T3034-2010, Effective: 2011-May -01. The Guideline is the first standard regarding the Process Safety Management (PSM) in China (accessed 21-August-2011) "translate this page!" http://www.chemicalsafety.org.cn/News_Minute.aspx?newsid=8068

Figure 1

The Bridge Over Process Safety's Troubled Waters:

The Gap Between Education and Application

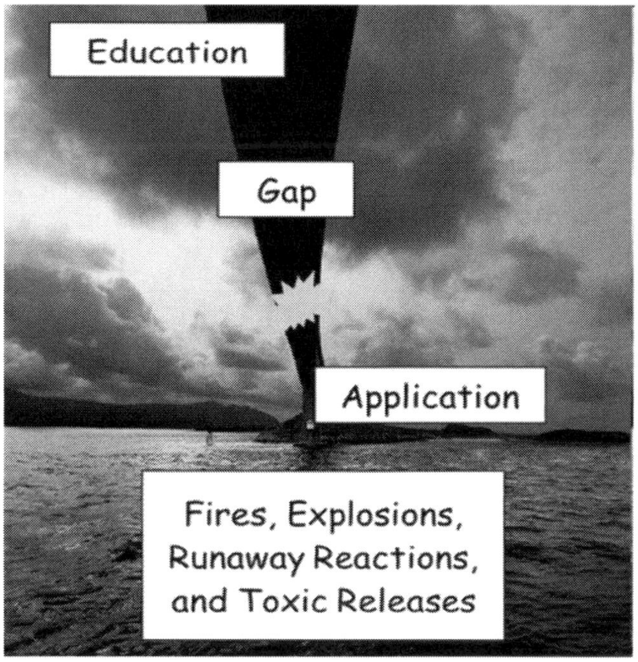


Figure 2

The Process Safety Management (PSM) System:

Image of a Wheel

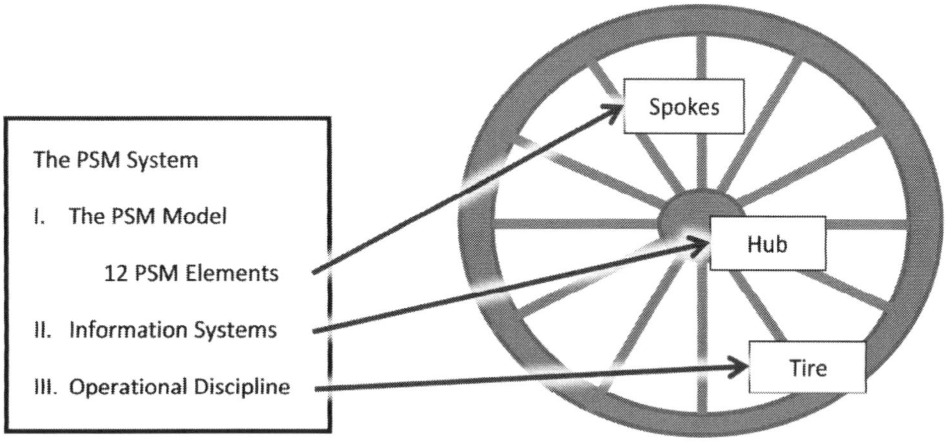

Figure 3

The Process Safety Management (PSM) System

| | | |
|---|---|---|
| I | The PSM Model | PSM |
| 01 | Process Safety Information | PSI |
| 02 | Process Hazards Analysis | PHA |
| 03 | Management of Change | MOC |
| 04 | Pre-Startup Safety Reviews | PSSR |
| 05 | Plant Integrity and Reliability | PIR |
| 06 | Written Procedures | WP |
| 07 | Safe Work Practices | SWP |
| 08 | Contractors and Visitors | C&V |
| 09 | Emergency Management & Planning | EMP |
| 10 | Training and Performance | T&P |
| 11 | Auditing | AUD |
| 12 | Investigations | INV |
| II | Information Systems | IS |
| III | Operational Discipline | OD |

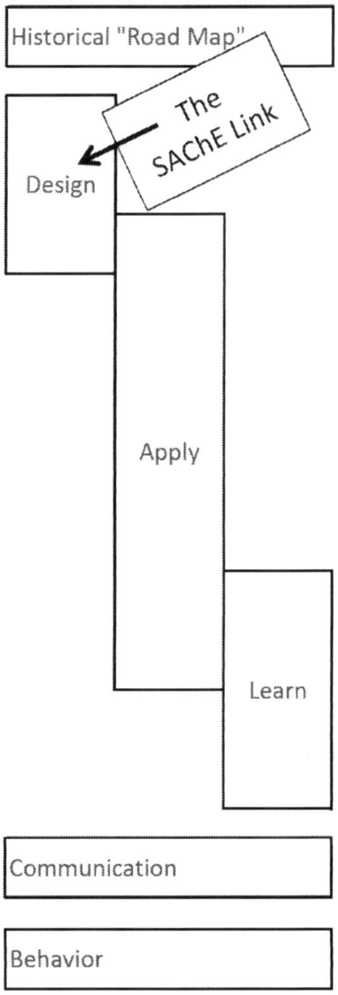

Figure 4

A Matrix Linking The Chemical Engineering Curriculum to the PSM System

| | Chemical Engineering Courses | The PSM System | | | | | | | | | | | | | |
|---|---|---|---|---|---|---|---|---|---|---|---|---|---|---|---|
| | | I. PSM Model | | | | | | | | | | | | II. | III. |
| | | Design | | Apply | | | | | | | | Learn | | | |
| | | 01 | 02 | 03 | 04 | 05 | 06 | 07 | 08 | 09 | 10 | 11 | 12 | ITS | OD |
| | | PSI | PHA | MOC | PSSR | PIR | WP | SWP | C&V | EMP | T&P | AUD | INV | | |
| Science and Mathematics Foundation | Mathematics | √ | √ | √ | | √ | | | | | | √ | √ | √ | |
| | Engineering: Basic, Computers | √ | | | | | | | | | | | | √ | |
| | Chemistry: Basic | √ | | | | | | | | | | | | | |
| | Chemistry: Structure and Reactivity and Lab | √ | √ | √ | √ | √ | √ | √ | √ | √ | √ | √ | √ | | √ |
| | Chemistry: Intro to Quantum | √ | | | | | | | | | | | | | |
| | Physics with Lab | √ | √ | √ | √ | √ | √ | √ | √ | √ | √ | √ | √ | | √ |
| | Biology/Life Science | √ | | | | | | | | | | | | | |
| | Material Science | √ | | | | | | | | | | | | | |
| | Other | √ | | | | | | | | | | | | | |
| Chemical Engineering Specific | Material and Energy Balances | √ | | | | √ | | | | | | | | | |
| | Thermodynamics | √ | | | | | | | | | | | | | |
| | Fluid Mechanics | √ | | | | | | | | | | | | | |
| | Mass and Heat Transfer | √ | | | | | | | | | | | | | |
| | Separation Processes | √ | | | | | | | | | | | | √ | |
| | Reaction Engineering and Design | √ | | √ | | | | | | | | | | | |
| | Chemical Engineering Laboratory | √ | √ | √ | √ | √ | √ | √ | √ | √ | √ | √ | √ | √ | √ |
| | Process Dynamics and Control | √ | | | | √ | | | | | | | | √ | |
| | Chemical Engineering Process Economics | √ | √ | √ | | √ | | √ | √ | √ | √ | | | | |
| | Chemical Process Simulation and Design | √ | √ | √ | | | | | | | | √ | | √ | |

Note: Process Safety Information is required for all PSM Elements!

Figure 5

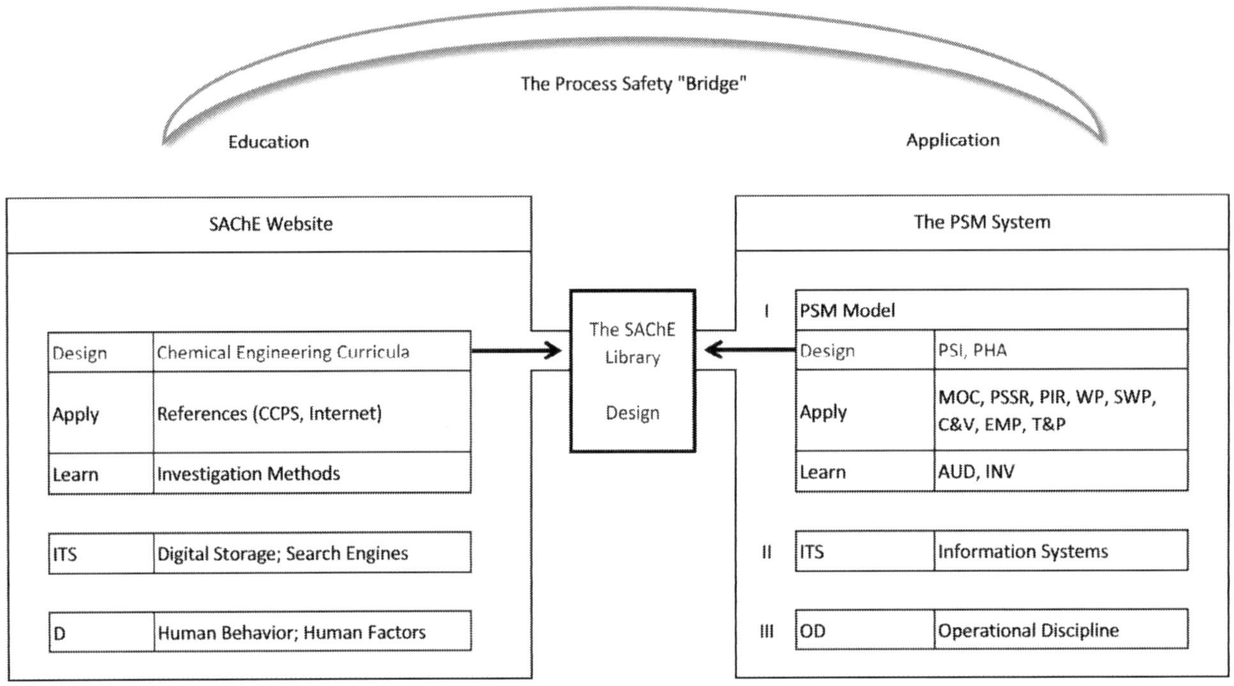


Figure 6

The Design Team Project Road Map

The steps addressing and documenting process safety within a design project

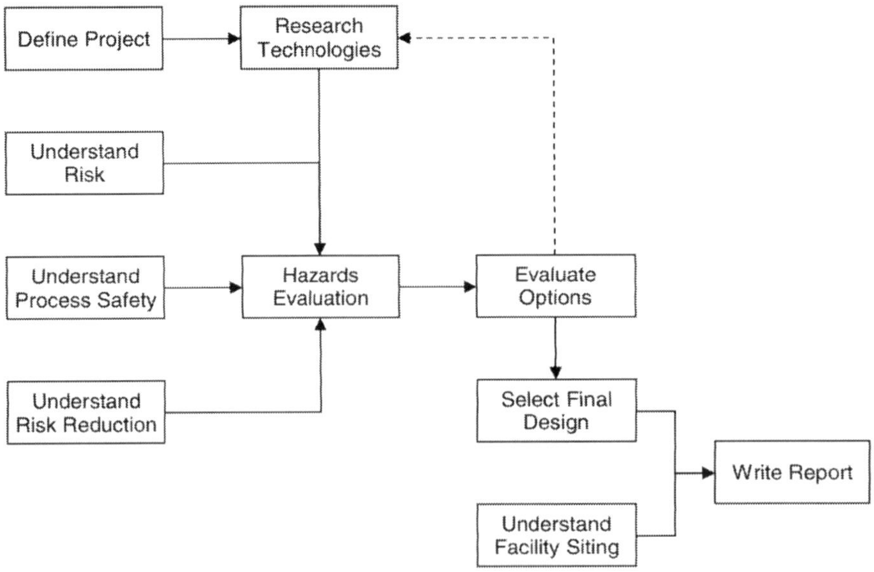

Figure 7

An Approach For Reducing Process Safety Risk

$$R = \frac{F \times C}{D}$$

| | | |
|---|---|---|
| **R** | **R**isk<br>What consequence (event) we are trying to reduce | |
| **F** | **F**requency<br>How often the event may occur | |
| **C** | **C**onsequence<br>How severe the event may be | Process Safety: Fires, Explosions, Toxic Releases and Runaway Reactions |
| **D** | **D**iscipline<br>How we apply our safety principles | Process Safety: Operational Discipline, Conduct of Operations, Human Factors |

Figure 8

The Bridge Over Process Safety's Troubled Waters:

Eliminating the Gap With SAChE

# (232f) Exposing the Blurry Lines Between Personal Safety and Process Safety Education: Contrasting NIOSH Prevention Through Design (PtD) with CCPS Sache

**Delmar R. Morrison**[1], Ryan J. Hart[1] and Pamela Heckel[2], (1)Thermal Sciences, Exponent, Inc., Lisle, IL, (2)Prevention through Design Initiative, National Institute for Occupational Safety and Health, Cincinnati, OH

Chemical engineering curricula have long been focused on instilling the fundamentals of chemical engineering science in undergraduates and advancing the state of the art through postgraduate study. Unfortunately, a comprehensive focus on safety has never been in the core of the chemical engineering curricula; however, since the T2 incident, industry stakeholders have pushed successfully for additional emphasis on process safety education in U.S. curricula. The SAChE program was created as a collaborative effort between CCPS and engineering schools to develop support materials to teach the elements of process safety to chemical engineering undergraduate and graduate students. Recently, NIOSH has led a national initiative called Prevention through Design (PtD) to address occupational safety and health needs in engineering design to "prevent or minimize the work-related hazards and risks associated with the construction, manufacture, use, maintenance, and disposal of facilities, materials, and equipment." We have found that there is a blurry line between the concepts of process safety and personal (i.e., occupational) safety in both industry and in academia. Not only is there a hole in the chemical engineering curriculum, but also the distinction between the two sets of complementary safety objectives is little appreciated. This extended abstract aims to identify the core concepts of PtD as applicable to the chemical engineering curriculum and contrast those to the typical elements of process safety. The objective of the abstract is to provide an outline for creating a PtD module for the chemical engineering curriculum that complements the SAChE modules on process safety. We anticipate significant audience feedback during the session that will assist in the development of this module and aid in its implementation.

See more of this Session: Integrating Safety Into the Chemical Engineering

# (232g) The 3rd Edition of Chemical Process Safety, Fundamentals with Applications

**Daniel Crowl**, Chemical Engineering, Michigan Technological University, Houghton, MI and Joseph F. Louvar, Chemical Engineering, Wayne State University, Detroit, MI

The 3rd edition of Chemical Process Safety, Fundamentals with Applications was published in the early summer of 2011. The third edition has been rewritten to include new process safety technology, new references, and updated data that have appeared since the first edition was published in 1990 and the second edition in 2002.

The third edition contains two new chapters. Chapter 8, "Chemical Reactivity," was added due to the recommendations from the U.S. Chemical Safety Board (CSB) as a result of the T2 Laboratories accident investigation. Chapter 13, "Safety Procedures and Designs," was added to consolidate some material that was scattered throughout the previous editions and to present a more complete and detailed discussion.

This paper will present an historical perspective on this textbook, present our original objectives for this textbook, and discuss some of the difficulties in publishing a first-time text in this area. It also includes our combined experiences of teaching process safety in both industry and academia during the past 20 years.

See more of this Session: Integrating Safety Into the Chemical Engineering Curriculum - Where, What, and How (Part 1)
See more of this Group/Topical: Education

# (273a) Chemical Engineering At the University of Dayton: 2011 ABET Visit

Amy R. Ciric and **Robert J. Wilkens**, Chemical and Materials Engineering, University of Dayton, Dayton, OH

The Chemical Engineering program at the University of Dayton was reviewed by ABET during the 2011-2012 academic year and received a full 6-year accreditation. This presentation will describe our data analysis and continuous improvement processes, our self-study report and on-site documentation, and our preparation for the on-site visit. The presentation will also review the feedback we received from the ABET visit and our continuing assessment processes.

See more of this Session: ABET Accreditation: Best Practices
See more of this Group/Topical: Education

# (273b) ABET Accreditation of Miami University's Chemical Engineering Program

**Shashi B. Lalvani**, Chemical and Paper Engineering, Miami University, Oxford, OH

Chemical engineering program at Miami University grew out of its nearly 50 year old program in Paper Science and Engineering. The first cohort of students graduated from this program in year 2008, when this program was first accredited. The paper will describe the relationship between the curriculum, ABET outcomes and objectives. In addition, the development of various methodologies for measurement of the achievement of the learning outcomes and program objectives will be discussed. A short summary of the best practices for successful accreditation of the major will be addressed as well.

See more of this Session: ABET Accreditation: Best Practices
See more of this Group/Topical: Education

# (273c) Lessons Learned From a Recent Successful Accreditation Visit

**Faith A. Morrison** and Katie Torrey, Department of Chemical Engineering, Michigan Technological University, Houghton, MI

In 2010/11 Michigan Tech came through ABET accreditation with flying colors, but our success was a result of concentrated effort over the course of nearly two years. In this presentation we discuss the lessons we learned and the significant positive benefits the program enjoys as a result of its ABET experience.

See more of this Session: ABET Accreditation: Best Practices
See more of this Group/Topical: Education

## Experience and Perspectives of the CBE Department at NC State University through Its 2010 ABET Accreditation Visit

P. K. Lim[†], Hubert Winston, Lisa Bullard, and Peter S. Fedkiw
Department of Chemical and Biomolecular Engineering
North Carolina State University
Raleigh, NC 27695-7905
[†]To whom correspondence should be addressed

**Extended Abstract**

This presentation will cover the strategy, methodology, and protocol used by the Department of Chemical & Biomolecular Engineering (CBE) at NC State University (NCSU) in addressing ABET Criteria 2–5, the principal criteria in the ABET Self-Study Report. Elements deemed effective in preparing CBE for its 2010 accreditation visit will be presented and discussed.

CBE's Program Educational Objectives (PEOs)–ABET Criterion 2–were formulated and revised with input from faculty, students, alumni, and NCSU administrators. They were assessed using survey data collected–through paper ballots and web-based mechanisms–from CBE's Alumni Advisory Board, NCSU alumni, and employers and professional schools of graduates.

CBE's Program Outcomes–ABET Criterion 3–are based on the learning objectives of 13 core undergraduate courses that are mapped to ABET's Outcomes *a* through *k*. Annual assessment of CBE's Program Outcomes starts with the collection of detailed performance data in core courses that are analyzed by the instructors at the end of each semester to determine if the learning objectives are met and whether any changes need to be made the next time the courses are offered. Students' performance data in each core course are tracked according to the learning objectives using an *Excel*® program that records performance scores problem-by-problem on examinations, quizzes, and homework, category-by-category on reports, presentations, and projects. Each instructor presents the data analysis and assessment of how well the course meets the learning objectives in a course analysis report that is reviewed independently by an ABET Subcommittee consisting of 3 (typically) faculty members. Recommendations for improvement are also a component of the reports. A total of six ABET Subcommittees annually review the course analysis reports of the 13 core courses and the subcommittees present their findings and recommendations in reports to the faculty, which may take specific actions.

CBE's efforts aimed at program improvements–ABET Criterion 4–may be classified in two categories, episodic and sustained. The episodic efforts consisted of lobbying efforts over the 2003-2005 period that culminated in the 2006 establishment of a \$33.5 million, 85,000-$ft^2$ Biomanufacturing and Training Education Center in affiliation and close proximity to the department. The center made possible a new program concentration, the Biomanufacturing Sciences concentration, and strengthened an existing concentration, the Biomolecular

Engineering concentration. It has provided new training and career opportunities to many students in the expanding field of biomanufacturing.

Sustained improvement efforts included development of a new data base and a more direct methodology for PEOs assessment, a better tracking of the implementations and outcomes of recommended actions for course improvements, introduction of new courses and experiments, upgrade of course contents, facilities and experiments, and development of ASPEN, Matlab and Visual Basic programs of generic utility to promote the use of advanced computing tools in the curriculum.

CBE's efforts on curriculum–ABET Criterion 5–focused on a full accounting of the syllabi and credit hours of all courses offered in the curriculum. An important lesson learned is a careful scrutiny of the course content of every engineering elective to justify its engineering credit designation.

# (273e) Preparing for An Interim Visit and Report

**David L. Silverstein**, University of Kentucky, Paducah, KY

The paper will discuss specifics of handling an interim visit for a non-traditional chemical engineering program with two weaknesses with respect to outcomes assessment and faculty development.

See more of this Session: ABET Accreditation: Best Practices
See more of this Group/Topical: Education

# (273f) Time Flies: Keeping up with Preparations for An Accreditation Visit

**Daina Briedis**, Chemical Engineering and Materials Science, Michigan State University, East Lansing, MI

This paper discusses the preparation for and experiences during our Fall, 2010 ABET accreditation visit. Expectations for compliance with criteria are clearly increasing, as revised and harmonized criteria now contain specific definitions, and accreditation teams with newly-trained evaluators appear to be better aligned with criteria requirements. The author will share lessons learned in maintaining a sustainable continuous improvement process and minimizing faculty effort in visit preparation.

See more of this Session: ABET Accreditation: Best Practices
See more of this Group/Topical: Education

# (273g) Beyond the Requirements

**Valerie L. Young**, Chemical and Biomolecular Engineering, Ohio University, Athens, OH

Assessment for accreditation purposes in our department has focused on required aspects of the program that impact all of our BS ChE students. However, over the same time period, we made a number of changes and additions to options that our students have. Arguably, these options also substantially improve our program. In this talk, some strategies will be presented for tracking these types of program improvements and incorporating them into the self-study.

See more of this Session: ABET Accreditation: Best Practices
See more of this Group/Topical: Education

# (299a) Separation Processes Course: A Vehicle for Introducing Basic Principles of Particle Science and Technology and Needs of Various Value-Added Product Industries

**Ecevit Bilgili**, Chemical, Biological and Pharmaceutical Engineering, New Jersey Institute of Technology, Newark, NJ, Rajesh Dave, Department of Chemical, Biological and Pharmaceutical Engineering, New Jersey Center for Engineered Particulates, Newark, NJ and Norman W. Loney, NJIT, Newark, NJ

A great majority of the chemicals and value-added products is multi-phase in nature containing two or more phases. Usually, there exist a continuous phase and dispersed phase in the form of powders, granules, droplets, aerosols, suspensions, etc. Interestingly, most undergraduate engineering curricula including chemical engineering focus on single-phase, single or multi-component systems, largely ignoring the multi-phase systems that are of interest to a multitude of industries such as pharmaceuticals, inks, paints, detergents, and ceramics. Most chemical engineering students graduate with little to no knowledge of such systems. Such basic dispersed-phase concepts as particle size distribution, morphology, porosity, bulk density, surface area, etc. must be learned in industry through on-the-job or external training programs. Particle change mechanisms such as nucleation, growth/shrinkage, breakage and agglomeration are commonly observed in almost all industrial particulate processes; yet, such mechanistic understanding is largely lacking in industry due to the fact that chemical engineers have usually not been exposed to these types of rate processes in their formal education and training.

Incorporation of fundamental concepts of Particle Science and Technology can be introduced either as a stand-alone, elective undergraduate course or in the scope of a core chemical engineering course. In this paper, we illustrate our efforts to incorporate basic concepts of Particle Science and Technology into the undergraduate core course titled “Separation Processes II”. Specific examples will be given in the context of several unit operations such as drying and crystallization as applied to Pharmaceutical Industry, where concepts of dispersed phase properties and particle change mechanisms are introduced. The presentation will also discuss some pedagogical tools such as conceptual questions in assessment, active learning, and lab visits that enhance the student learning process.

See more of this Session: Incorporating Industry Needs Into the ChE Curriculum
See more of this Group/Topical: Education

# (299b) Application of Chemical Engineering Education In Automotive Thermal Management

**Alaa E. El-Sharkawy**, Aero-Thermal Vehicle Integration and Validation, Chrysler Group LLC, Auburn Hills, MI

**Application of Chemical Engineering Education in Automotive Thermal Management**

Automotive industry has been widely understood to be a dominant field for mechanical and manufacturing engineers. However, the wide scope of chemical engineering education provides chemical engineering students with all essential technical skills required for certain areas of automotive industry. It has been an issue that some chemical engineers usually define what they cannot do, not what they can do. Therefore, this paper identifies non-traditional technical areas, in automotive industry, that chemical engineers can excel at. Throughout the years, chemical engineers had significant contributions in areas such as fuels, lubricants and emission control. However, due to the stringent requirements for emissions and for fuel economy improvements, new technical areas such as "automotive thermal management" have become more challenging. This requires temperature management of vehicle components, vehicle systems, materials and fluids. During the course of chemical engineering education, most of the emphasis is focused on chemical process industries. However, several examples from automotive applications can be presented to the students as typical chemical engineering problems. Since transportation is a part of our daily activities, it will be much easier and more interesting for students to realize the significance of these applications since they interact with our daily life. Therefore, student's attention is easily captured during the lecture or exercise. A high level of focus and interest is usually observed when reference is made to real life examples that students can relate to. In addition, these examples point the students toward a wider range of technical fields and career opportunities in automotive industry.

In this paper, examples will be given in the areas of vehicle instrumentation and testing, exhaust after-treatment systems, fuel system thermal management and kinetics of material thermal degradation.

- The basic energy balance around a thermocouple junction and the assumptions made to determine thermocouple dynamic response are usually covered in undergraduate programs. However, this problem can be also presented to the students in case if these assumptions are violated. These assumptions may lead to significant errors in thermocouple readings during actual vehicle testing unless certain installation procedures are considered. Examples will be presented and estimates of potential errors will be evaluated.

- The after-treatments systems including catalytic converters, Diesel Particulate Filters (DPF) and the Selective Catalytic Reduction systems (SCR) will be presented in relationship to chemical engineering fundamentals. This problem can be analyzed, as a series of reactors, using any of the commercial flow-sheet s simulation tools.
- Distillation problems which are taught in chemical engineering undergraduate programs are typically steady state. While distillation process is very important for chemical and petroleum industries, it has also significant applications in the automotive world. Thermal analysis of gasoline fuel tank is often required in order to estimate fuel temperature and fuel vapor emissions. This problem becomes more interesting since it is transient in nature. The fuel mass is continuously decreasing as long as the engine is running. The more volatile hydrocarbons will evaporate first and therefore the fuel composition will be always changing. The rate of heat exchange between the fuel tank and the surrounding environment is also transient. Analysis of this problem requires application of heat transfer models, thermodynamics of hydrocarbon mixtures, flash calculations and mass transfer.
- Basic principles of chemical kinetics are also used to determine the rate of thermal degradation of rubber and polymeric materials. The rate of thermal degradation can be related to the material temperature through the activation energy of each material. Therefore, an estimate of material performance under different driving conditions and at various ambient temperatures can be estimated.

To serve automotive industry needs, it is recommended that an undergraduate course in the area of "automotive thermal management" be developed. The course should be offered as an elective course for senior undergraduate and graduate students. Due to the potential increase in production of hybrid and electric vehicles, the proposed course should also include battery thermal management, fuel cells, and light weight materials.

See more of this Session: Incorporating Industry Needs Into the ChE Curriculum
See more of this Group/Topical: Education

# (299c) Comparison of Expert and Novice Solution Approaches to An Industrially Situated Process Development Project

Ben Sherrett[1], Erick Nefcy[2], Debra Gilbuena[2], Edith Gummer[3] and **Milo D. Koretsky**[2], (1)College of Mechanical, Industrial, and Manufacturing Engineering, Oregon State University, Corvallis, OR, (2)Chemical, Biological, and Environmental Engineering, Oregon State University, Corvallis, OR, (3)Education Northwest, Portland, OR

Engineeringeducators must be mindful of the diverse range of careers their students will pursueand the rapid integration of new technologies into those careers. Such aperspective naturally leads to a shift of focus from teaching students "what tothink" to teaching them "how to think" and, even more specifically, "how toapproach novel problems." Within this context, a greater emphasis isbeing placed on developing students' abilities to transfer knowledge andunderstanding from the classroom to professional practice. The researchreported in this paper seeks to identify and compare methods that expert andnovice engineers use to solve the same novel process development project. Through this study, we seek to contribute to the understanding of transfer inill-structured engineering problems.

Participantsin the study were tasked with developing an optimal "recipe" for a Low PressureChemical Vapor Deposition (LPCVD) reactor that deposits thin films of siliconnitride on polished silicon wafers, an initial step in the manufacture oftransistors. The optimization problem is completed using a virtual laboratoryand involves iteratively developing, testing, and refining solutions based onadjusting input parameters and measuring outputs. The computer simulationgenerates data representative of an industrial reactor and includes both randomand systematic process and measurement error. Users are also encouragedto apply sound engineering methods since they are charged virtual money foreach reactor run and each measurement. This problem is complex and atypical team spends approximately 15-25 hours to complete it.

Formany years, the virtual LPCVD project has been delivered in a capstone chemicalengineering course. The solution process of one student cohort is used torepresent different possible novice solution approaches. Additionally, threeindustrially accomplished experts led teams complete the project. Two ofthe experts had no direct experience with CVD reactors. This choice wasdeliberate as we seek to characterize how experts will solve the problem bytransferring their core engineering knowledge and skills to a novel problem.One expert was a mechanical engineer who lacked domain specific knowledge, butis an internationally recognized leader in engineeringdesign.

Thevirtual LPCVD project contains the features of an ill-structured problem. It isauthentic, ambiguous, and has many possible solutions and solution paths. However, it has more constraints than design problems found "in the wild." Thisaspect reduces the degrees of freedom and facilitates comparison of different solutionprocesses. Additionally, the virtual environment enables a more thoroughassessment of a team's proposed "recipe." Since the error can be removed andfilm thickness "measured everywhere," an absolute metric of performance can beobtained and used for comparison.

Solutionprocesses were documented using talk-aloud protocol analysis, collection oflaboratory notebooks and written reports, and data acquired by the computerinterface. Previously we have reported a graphical method, termed modelrepresentation and usage maps, which allows us to characterize participants'model development as they solve the problem. These maps show the natureof each model component (quantitative, qualitative, graphical, empirical, statistical), its correctness, and its use in the solution process (did itdirect the values of input parameters for a future run, was a run used toquantify model parameters, was the model qualitatively verified, etc.). Themodel maps are the primary artifact of analysis and form the basis to compareparticipants' solution processes. Additionally, semi-structured interviews areused to validate the results of the study and provide further insight.

Solution characteristics andapproaches of the expert and the student teams are identified. Compared to thestudent teams, the expert engineers devote a higher proportion of time toinformation gathering and problem scoping. They also tend to access large°Chunks of information during the problem solving process. Onceactivated, these chunks tend to remain relatively constant in form and are usedthroughout the solution. However, regarding performance, the final recipes ofthe experts were no better than a large fraction of the student teams. Specificways that the participants' prior knowledge and previous experiences wereapplied to this new situation are identified and will be discussed.

See more of this Session: Incorporating Industry Needs Into the ChE Curriculum

# (299d) Adventures with Incorporating Process Safety Into the Chemical Engineering Curriculum

**Daniel Crowl**, Chemical Engineering, Michigan Technological University, Houghton, MI

Until the mid 1980's process safety was primarily an industrial issue, with almost all of the technology being developed and used by industry.

Due to an industrial summer job in the early 1980's, the author was introduced to process safety and was convinced that this material must be incorporated into the undergraduate chemical engineering curriculum. This eventually resulted in a textbook on process safety, several faculty workshops, many instructional modules, a video series, a student certificate program, a ChemE car safety program, and other materials and projects related to process safety.

This paper presents the history of how this happened and also the challenges of bringing a new industrial area into the chemical engineering curriculum.

See more of this Session: Incorporating Industry Needs Into the ChE Curriculum
See more of this Group/Topical: Education

# (299e) Sustainable Integration of Sustainability Into Senior Design for Chemical Engineers

**Paul Blowers**, Chemical and Environmental Engineering, The University of Arizona, Tucson, AZ, Armin Sorooshian, Chemical and Environmental Engineering, University of Arizona, Tucson, AZ and Kimberly Ogden, University of Arizona, Tucson, AZ

This work discusses how to sustainably introduce topics, particularly those regarding sustainability, into a senior design course series for chemical engineers. Historically, a life cycle assessment (LCA)/sustainability senior/graduate student elective had been created and offered to chemical engineering students at the University of Arizona. This elective course introduced LCA, green chemistry, atom economy, and inherent technology trade-off evaluation in an open discussion format with guided projects to evaluate technologies. The bulk of this material was subsequently folded into a "traditional" chemical engineering senior design course without a lot of other content through incorporation of the topics into existing projects and a re-balancing of course materials. The materials were offered for five years before a change of instructors involved in the courses could have impacted the continuation of offering the sustainability topics.

The use of archival recorded lectures, along with clear communication of topical content on sustainability allowed the new instructor to seamlessly pick up and offer the content that was outside of their prior knowledge domain. In particular, concept mapping strategies allowed students to quickly build an approach for handling the complexities of the design process while allowing them to have time to incorporate sustainability topics. In the second course in the design series, students are required to evaluate their designed process for sustainability regarding environmental constraints. Student evaluations and comments favorably indicate that the approaches used and the technical sustainability content are valued. Additionally, employer surveys show that students have a strong awareness of environmental topics and how to approach sustainability issues.

See more of this Session: Incorporating Industry Needs Into the ChE Curriculum
See more of this Group/Topical: Education

# (299f) An Academic Perspective On Developing a Course Framework On Inspiring Students to Be "Innovative"

**Asad H. Sahir**[1], JoAnn S. Lighty[1], Terry Ring[1] and Beverly A. Brehl[2], (1) Chemical Engineering, University of Utah, Salt Lake City, UT, (2)Center for Teaching and Learning Excellence, University of Utah, Salt Lake City, UT

The subject of innovation traditionally has been a concern of business and technology leaders in an organization and is extensively discussed and dealt with in business school curricula. A course framework has been conceptualized to inspire undergraduate engineering students who contribute significantly to the workforce driving innovation in an economy.

The central pedagogical idea of the proposed course framework aims to inform the interested undergraduate student about the important interconnections behind an innovative process or product, formed by an amalgamation of scientific and technological breakthroughs, novel marketing strategies and societal and public policy issues.

This talk will be about the development of a course framework on inspring students to learn about innovation by the Department of Chemical Engineering, University of Utah in conjunction with the University of Utah's Center for Teaching and Learning Excellence. In the development it was revealed that there is a need to investigate incorporation of pedagogical instructional techniques like case-based learning, discussion based learning, reflective writing in addition to the conventional lecture and project based learning practiced in engineering curricula.

See more of this Session: Incorporating Industry Needs Into the ChE Curriculum
See more of this Group/Topical: Education

# (299g) The Evolution of a Curriculum Reform Process: Vertical Integration of Computational Thinking by Incremental Steps

**Daina Briedis**[1], Robert Ofoli[2], Jon Sticklen[3], Mark Urban-Lurain[3], Claudia Vergara[3] and Louise Paquette[4], (1)Chemical Engineering and Materials Science, Michigan State University, East Lansing, MI, (2)Chemical Engineering & Materials Science, Michigan State University, East Lansing, MI, (3)Center for Engineering Education Research, Michigan State University, East Lansing, MI, (4) Mathematics and Computer Science Department, Lansing Community College, Lansing, MI

In the engineering workplace, engineering graduates are expected to be facile in formulating well-defined problems and in selecting an appropriate computational tool with which to develop a solution once a problem is well posed. In the engineering academic community, we have historically considered the analytical tools of traditional mathematics as the primary solution tools our students must systematically master. The language of mathematics is indeed indispensable for representing engineering problems. But solution tools bundled into modern computational environments provide a second and complementary solution capability. As we tread further into the 21st century, globalization, international competition, an increasingly diverse population, and a rapid growth in computational capabilities and infrastructure are some of the challenges that will test the boundaries of engineering ingenuity. It is important that our students become competent in solving engineering problems using computational skills in a broader sense than simply selecting the "right tool."

With its origin in our assessment process to its current status in the curricular integration of computational thinking principles, this project has evolved from the first steps of helping students learn more effectively in particular computing environments (ASPEN and MATLAB) to the development of a process by which a wide variety of stakeholder computing needs and perspectives are analyzed and integrated into engineering curricula. Our objective is to use stakeholder input to infuse computational problem-solving competencies not only in chemical engineering, but across engineering in our college. These competencies are aligned with industry needs and enable students to integrate conceptual knowledge, technical skills, and professional practice.

In the process we first identified and attempted to remedy a lack of curricular continuity in student learning of computational tools and their application to problem solving. The undertaking was then developed as an NSF-funded project involving a thorough survey of engineering corporate stakeholders to identify computational competencies needed in the engineering workplace. These findings were analyzed and categorized based on the framework of computing concepts based on the Fluency with Information Technology (FITness) report. The fundamental nature of these concepts notwithstanding, they are instantiated in practical technologies and applications that allowed us to move from the computational competencies identified in our industrial data to computing concepts that could be integrated in the curricula. These concepts have now been introduced into two engineering curricula in the College of Engineering, Michigan State University and in the pre-engineering curricula of our collaborating partner, Lansing Community College. A cornerstone of our approach lies in using industrially authentic problems that require students to apply various computational concepts for their solution. The link to industry has the effect to more fully engage students in their own problem solving and to bring to problem solving a stronger, more enthusiastic attitude. Early results of assessment of student performance in the application of computational competencies in these courses will be presented.

See more of this Session: Incorporating Industry Needs Into the ChE Curriculum
See more of this Group/Topical: Education

# (372a) Women: Women's Outreach In Materials, Energy, and Nanobiotechnology

**Susan Daniel**, Chemical and Biomolecular Engineering, Cornell University, Ithaca, NY

**WOMEN: Women's Outreach in Materials, Energy, and Nanobiotechnology**

**Susan Daniel**

School ofChemical and Biomolecular Engineering

CornellUniversity, Ithaca, NY 14853

For the past two years,the School of Chemical and Biomolecular Engineering at Cornell Universityorganized and hosted a successful pilot outreach event, called WOMEN (Women'sOutreach in Materials, Energy, and Nanobiotechnology),to foster rural tenth-grade girls interest in STEM and introduce their parentsto the university and career options for their daughters. Twenty familiesvisited our campus for the day. Students ran short experiments or activities designedand organized by graduate students (Figure 1) while their parents learned aboutengineering careers and planning for college. In the second year, we alsointegrated a student-parent laboratory module, based on feedback from parentsurveys in the prior year. We received overwhelmingly positive feedback on theprogram from parents and student participants in both years. All parents saidthey would attend this kind of event again and felt that it broadened theirview of engineering. Parents commented that the information provided gave theman important early start in planning for college and knowing how to encouragetheir daughters to plan for a technical career. All 20 students felt that theprogram broadened their view of engineering and 17/20 stated they wouldconsider an engineering career after the program. In this paper, I will providean overview of the entire program, highlight several experiments we ran withthe students and parents, and discuss detailed feedback from students andparents.

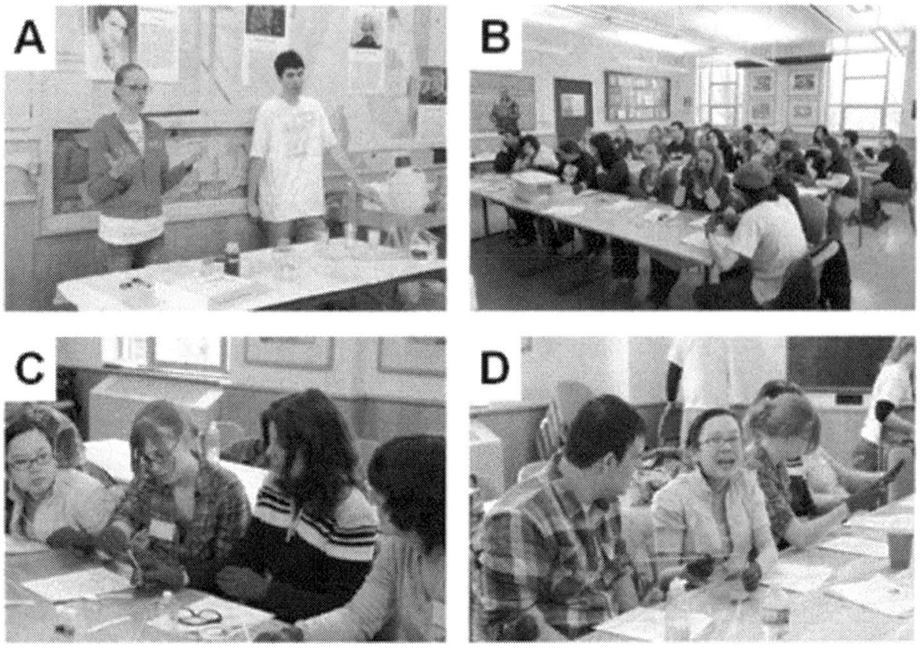


**Figure 1. WOMEN outreach program hosted by the Cornell School of Chemicaland Biomolecular Engineering. (A) Graduate students present the objectives ofthe parent-student lab to the families. (B) Parents and students engaging inthe joint activity. (C) A mother-daughter team at work. (D) A student enjoyingthe activity.**

See more of this Session: Free Forum On Chemical Engineering Education (Part 1)
See more of this Group/Topical: Education

# (372b) Process Technology Learning Activities and Experiments Used In Workshops for K-12 Educators and Students

**Steve R. Duke**, Chemical Engineering, Auburn University, Auburn, AL, Harry T. Cullinan, Alabama Center for Paper and Bioresource Engineering, Auburn University, AL, R. Dale Smith, Alabama Southern Community College, Thomasville, AL and T. J. Murphy, National Network for Pulp and Paper Technology Training, Thomasville, AL

More than 30 workshops and summer institutes have been conducted during the past seven years for K-12 students and K-12 and community college educators as part of outreach and recruiting strategies to generate interest in chemical and paper technology and engineering degrees. The workshops and institutes have been offered at a variety of venues (college campuses, high schools, and rural industrial sites) and a variety of durations from half-day to week-long. We have used a number of hands-on and group learning activities during the workshops, including classroom small-scale experiments, pilot-scale experiments conducted in technology or engineering laboratories, and industry site tours. Best practices and assessment results for several learning activities used at the workshops will be presented.

The summer institutes were conducted through a partnership between Auburn University, Alabama Southern Community College (Thomasville AL), and Lower Columbia College (Longview WA). Faculty and industry partners from these institutions developed and delivered the workshops as part of the National Network for Pulp and Paper Technology Training $(npt)^2$, a National Science Foundation Advanced Technology Education Center. Engineering workshops were also conducted through the Alabama Center for Paper and Bioresource Engineering at Auburn University.

See more of this Session: Free Forum On Chemical Engineering Education (Part 1)
See more of this Group/Topical: Education

# (372c) Does a STEM Researcher's Role Orientation Affect His or Her Ethical Sensitivity to Responsible Conduct of Research?

**Joseph Holles**[1], Michael Bowler[2], Susan Amato-Henderson[3], Jingfang Ren[2], Ted Lockhart[2], Joanna Schreiber[2] and Thomas Drummer[4], (1)Chemical and Petroleum Engineering, University of Wyoming, Laramie, WY, (2)Humanities, Michigan Technological University, Houghton, MI, (3)Cognitive and learning Sciences, Michigan Technological University, Houghton, MI, (4)Mathematical Sciences, Michigan Technological University, Houghton, MI

A role orientation inventory provides a record of the way in which an individual orients himself or herself with respect to a particular role that he or she plays within society. Role orientation inventories were originally developed to measure an individual's professional role orientation, for instance, with regard to being a dentist, doctor, social worker, etc. A role orientation inventory can answer questions such as: Does a doctor see her role as primarily that of providing her patient with a service in exchange for payment? Does a dentist understand his role as fundamentally consisting of serving the best interests of his patients? Does a social worker see her role as an authority figure?

A test of ethical sensitivity measures an individual's ability to recognize and take account of the ethically relevant elements of ethically-charged situations, e.g., the rights and obligations of those involved, applicable principles and ethical guidelines, and the consequences of particular courses of action, etc. Also originally developed for the field of professional ethics, ethical sensitivity tests provide information about an individual's ability to map-out and size-up a situation involving ethics prior to making an ethical judgment. In fact, one's sensitivity to an ethical situation provides the material basis upon which one makes ethical judgments and chooses courses of action.

As part of an NSF sponsored project, we have developed and validated a role orientation inventory for science, technology, engineering, and mathematics (STEM) researchers and a test of an individual's ethical sensitivity to situations involving the responsible conduct of research (RCR). Our role orientation inventory instrument is an adaptation and extension of previous instruments that were originally designed with professionals in mind. We have created an instrument appropriate to the role of a STEM researcher. Our ethical sensitivity instrument embodies an entirely new approach toward measuring ethical sensitivity that relies upon comparing an individual's ethical sensitivity to situations involving RCR with his or her ethical sensitivity to common, everyday ethical situations.

In order to prepare for the final study that will be the culmination of our NSF grant project and to further validate the instruments that we have designed, we have conducted a study designed to examine the relationships that exist between a STEM researcher's role orientation and his or her ethical sensitivity to RCR. As ethical sensitivity is considered to be the bedrock of ethical behavior, discovering the relationships between a STEM researcher's role orientation and ethical sensitivity to RCR is an important step in understanding the psychology behind STEM researcher's behavior with respect to RCR. But it also illuminates aspects of ethical behavior that ethics educators should take into account and suggests possibilities of new approaches that ethics educators can take toward forming responsible researchers. For instance, if there is a correlation between a STEM researcher's role orientation and his or her ethical sensitivity to RCR, then it is important that ethics educators explore the possibility of developing pedagogical methods for influencing researcher's role orientation in order to enhance their ethical sensitivity to RCR.

See more of this Session: Free Forum On Chemical Engineering Education (Part 1)
See more of this Group/Topical: Education

# (372d) Celebrating the Individual: Encouraging Engineering Students to Find Their Voice

**Lisa G. Bullard**, Carol K. Hall and David F. Ollis, Chemical and Biomolecular Engineering, North Carolina State University, Raleigh, NC

**Celebrating theIndividual: Encouraging EngineeringStudents to Find Their Voice**

**Lisa G. Bullard, CarolK. Hall, and David F. Ollis**

**Department of Chemicaland Biomolecular Engineering**

**North Carolina State University**

Studies on retention of freshman engineering students have acommon theme; students who persist in engineering report a feeling of belongingand a "fit" with the organizational culture.In addition, they make connections with faculty and other students. In order to facilitate this, the authors haveimplemented several mechanisms throughout the curriculum to encourage studentsto reflect on their personal journey and express their individuality.

In several classes the students' first assignment is tosubmit a one-page autobiography, using autobiographies of the instructors asmodels. Our autobiographies included information about our families andpersonal interests as well as our academic interests, and we encouraged thestudents to do the same in theirs. We compile a composite portrait of the classfrom the autobiographies and share it as a memo to the students. Anotherapproach to this (for smaller classes) is to form pairs and ask each pair tointerview the other and introduce their partner to the rest of the class, witha few introductions occurring in each class over the course of several weeks. The results of the brief interviews arecompiled and distributed to the class. Ourgoals in these exercises are to give the students a sense of their instructorsas somewhat normal and approachable human beings and to help them start todevelop a sense of community as a cohort.

An initial challenge in our junior professional developmentcourse is to open the student up to writing reflectively about some keypositive factors (parents, mentors, teachers, etc) or negative experiences(roadblocks and resolutions) important in the student's life. To invite such atypical introspection, weask students to first view Randy Pausch's "Last lecture" video to demonstratehis response to an insurmountable personal challenge: his diagnosis of terminalpancreatic cancer in his mid-forties. We have been impressed over time with thedepth and openness demonstrated by our student writers in this first writingopportunity, and their candor in discussing their roadblock and barriers. This writing also sets the stage for asubsequent mock job interview, during which students are asked about theirresiliency, e.g., behavior inchallenging and difficult personal, work, or school situations, and whatthey learned from (attempted) resolutions.

Several instructors in our department invite students tosubmit as extra credit a creative expression of their experience in thecourse. Typical submissions range frompoems to artwork to music videos to edible creations. Thisassignment encourages students to reflect on their experience in the course andto attempt to express that experience in a tangible way. Hearing other studentsbemoan the difficulty of the class or the fear with which they approached ithelps incoming students understand that they are not alone in theirapprehension. It provides an outlet forthose students with a creative bent to express their individuality, hopefullydispelling the stereotype that engineers aren't creative. Finally, sharing theextra credit submissions on the last day of class ends the semester on apositive and often humorous note and serves to create an indelible memory forboth the students and the instructor.

See more of this Session: Free Forum On Chemical Engineering Education (Part 1)
See more of this Group/Topical: Education

# (372e) Gender Similarities and Differences In Belonging Among Engineering Graduating Seniors At Two Universities

**Tamara Floyd-Smith**, Chemical Engineering, Tuskegee University, Tuskegee, AL, Denise Wilson, Electrical Engineering, University of Washington, Seattle, WA, Diane Jones, University of Washington, Seattle, WA, Melani Plett, Electrical Engineering, Seattle Pacific University, Seattle, WA, Rebecca Bates, Minnesota State University at Mankato, Mankato, MN and Nanette Veilleux, Simmons College, Boston, MA

One primary goal of engineering education, as with any educational enterprise, is to achieve particular learning outcomes. For engineering education, one format for articulation of student learning outcomes is the Accreditation Board for Engineering and Technology (ABET) outcomes a-k. Continuous improvement as it relates to engineering education and ABET outcomes a-k traditionally has been approached from the standpoint of (1) faculty professional development to become better teachers, (2) increasing student exposure to the practice of engineering through corporate internships and research experiences and (3) incorporation of the latest technology to assist students. However, it is important to realize that the ABET ladder of outcomes which represents all levels of Bloom's taxonomy is overlaid with the student life experience represented by Maslow's hierarchy of needs. Thus, if a student has not successfully advanced beyond the third level of Maslow's hierarchy, belonging, it may become the bottleneck in the goal to achieve learning outcomes.

Recognizing the importance of student belonging in engineering education, researchers at multiple institutions began a multi-year study of belonging among engineering and computer science students. The main goal of this larger study is to test the linkage between STEM student's sense of belonging while in school and their engagement in their studies for different types of institutions. One potential outcome of the larger study is a list of interventions or best practices to ensure that students have achieved a sense of belonging at their respective institutions. The instrument for the larger study is a 200+ item survey that, in addition to addressing the central research question, lends itself to testing additional linkages.

In this paper, the focus is on engineering students, gender similarities and differences in belonging and the linkage between sense of belonging and academic grades. During spring 2010, students completed surveys at University of Washington (UW), Tuskegee University (TU), Simmons College, Seattle Pacific University, and Minnesota State University at Mankato. Simmons College, Seattle Pacific University and Minnesota State University at Mankato do not have chemical engineering programs. Consequently, data from those institutions was not included for this work. Students from Tuskegee University (TU) and University of Washington (UW) were recruited primarily from engineering disciplines. Based on their self-reports, the students included in these analyses were advanced or graduating undergraduates (seniors or $5^{th}$-year seniors). Thirty six engineering students from TU participated and ninety-nine engineering students from UW participated. The chemical engineering participants totaled eight or 22% and thirty five or 35% from TU and UW respectively.

Data from TU and UW were analyzed separately. Then, *t*-tests were conducted to determine if the means for the males and females were significantly different. The means alone suggest that females at both schools were less likely to feel that they belong to their academic major but more likely to feel connected to the university community. However, the differences were not statistically different and lead to the conclusion that male and female graduating seniors in engineering report the same level of belonging for all levels at both UW and TU. Finally, examination of Pearson correlations between GPA in major and the levels of belonging revealed notable findings for female students in the engineering majors at Tuskegee University. The significant correlation (0.67) between the sense of belonging in class and GPA in major indicates that females who reported more positive sense of belonging in their classes also reported higher GPAs in their major. A similar finding occurred for the correlation (0.55) between belonging at the university level and GPA in the major. The correlations for the other groups ranged from low to moderate for these variables.

See more of this Session: Free Forum On Chemical Engineering Education (Part 1)
See more of this Group/Topical: Education

# (372f) Role of Collaboration In Enhancing Creativity and Innovation In Engineering Education: Examples From Fluid Mechanics and Biotransport Courses

**Chinyere P. Mbachu**, Robby J. Sanders and Pedro E. Arce, Chemical Engineering, Tennessee Technological University, Cookeville, TN

Global problems in engineering education include the method teaching and learning strategies are developed and applied in engineering curricula. Efforts have been made on innovation and creativity in Science and Engineering Education for effective practical and successful experiences and outcomes. There are several methods assessing creativity and innovation in engineering students that integrate different techniques in the curriculum with the provision of a safe environment for experimentation, learning, and student enrichment in these critical topics;  however several aspects needs additional attention including for example, implementation protocols, strategies for designing efficient environments, etc. Furthermore, the impacts of creativity and innovation in Engineering Education are important if the factors and traits of modifying the engineering classroom experience are identified and assessed. One very important and somewhat forgotten aspect is the role of collaboration in promoting creativity and innovation in Engineering Education. The preliminary studies will be presented based on the examples gathered in two courses in Chemical Engineering, i.e. a junior level Fluid Dynamics course and a senior level Bio-transport of the Bio-molecular concentration. The finding will be related to the Linear Engineering Sequence (L.E.S) of the High Performance Learning Environment (Hi- PeLE) that is being used in several courses in Chemical Engineering at Tennessee Technological University Cookeville, TN USA.

See more of this Session: Free Forum On Chemical Engineering Education (Part 1)
See more of this Group/Topical: Education

# (372g) The Research Proposition and Professional Development: First Year Graduate Student Preparation

**David F. Ollis**, Chemical and Biomolecular Engineering, North Carolina State University, Raleigh, NC

**The Research Proposition and Professional Development: Update on First Year Graduate Student Preparation**

Twenty years ago, our department initiated an independent research proposition course for all first year PhD candidates. Student performance in this spring semester three unit course was treated as a graduate qualifier exam, and both students and faculty have been supportive of this requirement, as summarized earlier[1].

Over the last decade, our first year approach to research education has broadened. Peter Kilpatrick added a one unit fall course, Introduction to Research, a professional development course including research ethics, presentations, and publications. While these two courses were satisfying as stand-alone efforts, recent faculty and graduate student sentiment pushed for an earlier engagement of student with research advisor, PhD committee, and research itself.

In response, we have developed a yet broader first year experience encompassing a pair of two unit courses, one each in fall and spring. In the first, professional development topics are followed by creation of an independent, ten page research proposal. The second, spring semester effort requires the student, in consultation with her new advisor, to develop a NSF length proposal for the prospective PhD effort, and present it to her nascent PhD committee and course instructor. Additionally, earlier engagement with the PhD committee is now achieved through a January, second year oral report. The customary university Preliminary Exam occurs at the beginning of year three, and includes both a document (progress and plans) and an oral presentation.

In summary, we now have the following early research requirements:

| Semester | Activity | Deliverables |
| --- | --- | --- |
| Fall (1st) | Intro to research | 10 p proposal: independent |
| Spring(2nd) | PhD research proposal | 15 p. collab. proposal: PhD plan |

| | | |
|---|---|---|
| Spring (4th) | Progress report | Oral presentation to PhD committee |
| Fall (5th) | Preliminary exam | PhD progress & plans (document) and presentation to PhD committee |

---

Taken together, these activities constitute a broad and continuing "Introduction to Research" including considerable practice opportunities in writing proposals (3) and delivering oral presentations (4). These formal structures guarantee that all topics central to setting the stage for a successful research PhD experience are encountered early in what is typically a five year effort.

Our formal approach is consistent with recent studies of "How People Learn" (2), in which Donovan, Bransford and Pellegrino argue that "To develop competence in an area of inquiry, students must (a) have a deep foundation of factual knowledge, (b) understand facts and ideas in the context of a conceptual framework, and (c) organize knowledge in ways that facilitate retrieval and application"

We believe that our early introduction of literature searching and reading reviews and original articles centered around a simple hypothesis provides opportunity to initiate foundation knowledge construction, that the conceptual framework of writing in proposal format provides a focus for the student to demonstrate "understanding of facts and ideas in the (research) context", and that the written proposals and oral presentations repeatedly force the student to "organize knowledge in ways that facilitate retrieval and application."

Evaluation by graduate students

Formal course evaluations for the first three semesters of the independent, fallproposal and the first two of the collaborative spring proposal appear in Table 2 .These two writing courses fair well compared to our other department graduate courses (including reactors, transport, thermodynamics, and applied mathematics) and to the 1-5 absolute standard of our evaluation scale. Given the apparent initial hostility of new graduate students toward technical writing, these end-of-semester evaluations represent a significant achievement.

Table 2

Graduate student course evaluations: (Grad writing course / grad course dept average)

| Course: (5.0 max) | Intro to Research | PhD proposition |
|---|---|---|

| | Fall semester | | | Spring semester | |
|---|---|---|---|---|---|
| (% student participation)* | (84) | (65) | (46) | (67) | (47) |
| The instructor… | F 08 | F 09 | F 10 | S 09 | S 10 |
| 1. …stated course objectives | 4.4/4.6 | 4.5/4.5 | 4.6/4.6 | 4.6/4.3 | 4.8/4.5 |
| 2. was receptive outside class | 4.6/4.9 | 4.2/4.3 | 4.5/4.3 | 4.4/4.3 | 4.6/4.4 |
| 3. explained difficult materials | 4.3/4.3 | 3.9/4.0 | 4.6/4.2 | 4.4/3.9 | 4.6/4.3 |
| 4. was enthusiastic re/teaching | 4.6/4.5 | 4.2/4.4 | 4.6/4.5 | 4.6/4.3 | 4.4/4.4 |
| 5 was prepared for class | 4.2/4.6 | 3.8/4.4 | 4.5/4.6 | 4.6/4.3 | 4.0/4.4 |
| 6. gave prompt, useful feedback | 4.4/4.3 | 4.2/4.0 | 4.5/4.3 | 4.3/4.0 | 4.2/4.2 |
| 7. used instructional technology | 4.1/4.5 | 3.3/4.2 | 4.0/4.3 | 4.3/4.3 | 4.2/4.4 |
| 8 treated students with respect | 4.6/4.6 | 4.4/4.5 | 4.8/4.6 | 4.5/4.5 | 4.7/4.5 |
| 9. was an effective teacher | 4.5/4.5 | 4.1/4.2 | 4.7/4.3 | 4.4/4.1 | 4.3/4.4 |
| The course … | | | | | |
| 10. …readings were valuable aids | 4.4/4.5 | 3.6/4.2 | 4.3/4.2 | 4.6/4.2 | 4.3/4.3 |
| 11. assignments aided learning | 4.6/4.6 | 4.1/4.3 | 4.5/4.4 | 4.8/4.3 | 4.1/4.2 |
| 12. was intellectually challenging | 4.4/4.7 | 3.9/4.4 | 4.3/4.6 | 4.8/4.4 | 4.6/4.5 |
| 13. improved subject knowledge | 4.4/4.6 | 4.1/4.5 | 4.4/4.6 | 4.9/4.4 | |

4.6/4.5

14. was excellent 4.3/4.4 3.9/4.2 4.3/4.3 4.4/4.2 4.3/4.3

---

* Student completion of online university survey is optional, (unfortunately) not mandatory.

.

Instructor time commitment

Scheduled instructor-student discussions are important to the student, and also constitute a substantial and necessary part of the instructor's time commitment to the course. For example, the fall two unit course includes two 20 minute sessions with each student, so for our two most recent years, class size was 20-24 students, hence requiring 7-8 hours of discussions twice a semester. The final presentation at 15 min(fall) and 30 min (spring practice) and 1 hr.(spring final) /student add 5-6 and 30-36 hours of instructor contact per semester, respectively. These times commitments are partially offset by moving from two classes/wk to one/week as the semesters progress.

Evaluations: Faculty

After the first two academic years (08/09 and 09/10) of the present format, our faculty who had accepted new students in these years were surveyed to ascertain their assessment of the new spring student-advisor collaborative format for producing a PhD research proposal. Our questionnaire asked about changes in the speed and depth of advisor engagement, integration into advisor lab group, and encounters with the PhD committee.

The results in Table 3 (next page) show that faculty are overwhelmingly positive about the new format. The faculty have, necessarily, a longer experience with current and former proposal formats for the first year experience than do the grads. The table shows that the new format results in faster engagement with a PhD research topic, advisor conversations, integration into lab groups, and conversations with the PhD committees. We conclude that the new format fruitfully addresses the prior concerns of both our faculty and graduate students.

Table 3

Faculty Survey: Spring research proposition course

AS-agree strongly, A-agree, N-neutral, D-disagree, DS-disagree strongly

| | AS | A | N | D | DS |
|---|---|---|---|---|---|
| SPRING PROPOSITION | | | | | |
| The spring proposition … | | | | | |
| 1. …increased speed of student engagement with (Ph D) research topic | 8 | 7 | 1 | 0 | 0 |
| 2. …increased speed/depth of engagement with research advisor | 5 | 9 | 2 | 0 | 0 |
| 3. … increased speed of integration into lab group | 5 | 6 | 4 | 0 | 1 |
| 4. …led to earlier formation of PhD committee | 6 | 6 | 3 | 0 | 1 |
| 5. …led to earlier engagement with (some of) PhD committee | 5 | 9 | 1 | 1 | 0 |
| 6. … allowed earlier advising/counseling of student by advisor | 4 | 9 | 2 | 1 | 0 |

---

Whether such spring effort is productive can also be judged by student evaluations. In Table 2, responses to statements 10-14 illustrate graduate student satisfaction with what was learned from the spring proposals as the responses were among the highest in the table. No student has complained that the second, spring collaborative proposition was a duplicative or repetitive version of the fall independent proposal !

Conclusions

Research is the dominant activity of graduate programs. Consequently, formal training in aspects of research should be a logical part of first year graduate study. The present article illustrates how to smoothly engage new graduate students with their research topics and committees through construction, presentation, and defense of several written propositions. The courses described here have been taught in one form or another for twenty years, and have been strongly accepted and endorsed by both new PhD graduate students and their faculty advisors as the surveys reported here indicate.

These two courses could easily be taught elsewhere. Graduate student opportunity to write creative papers about research has been repeatedly shown to be productive via our twenty year history of these offerings, now to more than 300 graduate PhD candidates. The total teaching load for such a course is similar to that for any three unity traditional offering. What is most different and required is the presence of an instructor dedicated to enhancing each student's storytelling skill in the research domain.

References

1. Ollis, D. F., The Research Proposition, Chemical Engineering Education, Fall 1995, p. 222.

2. Donovan, M. S., Bransford, J. D, and Pellegrino, J. W., (eds) How People Learn: Bridging Research and Practice, National Academy Press, Washington, DC 1999, p. 12.

See more of this Session: Free Forum On Chemical Engineering Education (Part 1)
See more of this Group/Topical: Education

**Time on Task as an Assessment Tool for Student Learning**

*Paper #372h*

*Benjamin J. Davis, The Cooper Union (New York, NY)*

## Introduction and Motivation

Based on data from the National Survey of Student Engagement[1], it's clear that undergraduate students in the United States aren't attending class and studying as much as they could. Many works on higher education have been gleaned from this NSSE data, including most famously Academically Adrift[2], but as an educator the one statistic that caught my eye the most was the breakdown of the number of hours spent "Preparing for class (studying, reading, writing, doing homework or lab work, analyzing data, rehearsing, and other academic activities)" by engineering seniors (Table 1) below:

**Table 1: Data from 2011 NSSE of Hours Spent on Academic Activities**

| Hours | # of responses | % |
|---|---|---|
| 0 | 28 | 0.2% |
| 1-5 | 993 | 7.2% |
| 6-10 | 2163 | 15.8% |
| 11-15 | 2333 | 17.0% |
| 16-20 | 2586 | 18.9% |
| 21-25 | 1954 | 14.3% |
| 26-30 | 1317 | 9.6% |
| 30+ | 2330 | 17.0% |
| **TOTAL** | 13704 | 100.0% |

Ignoring part-time students, a full credit load for an engineering program is around 12 credits. If you assume that a student taking 12 credits should be working on material for their courses for about 2 hours per credit, that's 24 hours per week outside of class on the low end. According to this study, only 41% of students are spending that much time per week on their studies. Is that enough?

Malcolm Gladwell has famously suggested in his book Outliers[3] that the key to success in a field is practice. Specifically, 10,000 hours of practice is a rule of thumb for the time it takes for a person to become proficient at something. Even at 30 hours per week over 8 semesters of college, a student will not even approach half that number of total hours on their coursework. But the difference between a student spending 10 hours a week on their schoolwork and 30 is enormous when compiled over an academic career. Right now we make no distinction between the former student and the latter if and when they graduate (aside from the obvious impact of that extra work on the student's GPA). Should we start?

## Literature Review

[1] NSSE Annual Results for 2011. Question 9a for Engineering majors. http://nsse.iub.edu/html/annual_results.cfm (accessed May 2, 2011)

[2] Arum, Richard and Roksa, Josipa. Academically Adrift: Limited Learning on College Campuses. Chicago: University of Chicago Press. 2011. http://highered.ssrc.org/ (accessed May 2, 2011)

[3] Gladwell, Malcolm. Outliers: The Story of Success. New York: Little, Brown and Company. 2008. http://www.gladwell.com/outliers/index.html (accessed May 2, 2011)

Research on "time on task" – the amount of time a student spends actively engaged in learning – is extensive. According to Chickering and Gamson, as of 1991 "most of the research on the variable of engaged time or time on task [had] been conducted at the elementary and high school levels"[4] (the author of that work cites Cross[5] for that insight.) An extensive review of educational research devoted to time on task in K-12 classrooms was published in 1982 by Nancy Karweit[6] and she summarizes the results in a later 1984 article.[7] Karweit states that research to that point shows a positive correlation between in-class instructional time and learning, but that variance in student achievement cannot be consistently explained by these differences. She states that instructors should not necessarily assume that more instructional time will produce more learning, since this time may also vary with other factors that have a more pronounced effect on learning, but she does insist that:

> "Time is a necessary, but not sufficient, condition for learning. Learning takes time, but providing time does not in itself ensure that learning will take place. More time *may* result in more learning – if adequate time was the major cause of the problem in the first place. If other factors were the real cause, then providing more time will not be an effective strategy."

Chickering and Gamson cite time on task as one of their seven principles for good practice in higher education. They state that "the general consensus of research is that the more time students are engaged in learning, the greater the amount of their learning" but that "we must continue to explore the nature of time on task in the college classroom and its influence on student learning." In summary:

> "Time plus energy equals learning. There is no substitute for time on task. Learning to use one's time well is critical for students and professionals alike. Students need help in learning effective time management. Allocating realistic amounts of time means effective learning for students and effective teaching for faculty. How an institution defines time expectations for students, faculty, administrators, and other professional staff can establish the basis of high performance for all."[8]

Alexander Astin, who published two important books on student outcomes and how they are affected by college environments – Four Critical Years[9] in 1977 and What Matters in College?[10] in 1992 – surveyed around 25,000 undergraduate students as first-years in 1985 then gave a follow-up survey four years later. In his summary, he states that:[11]

> "The follow-up questionnaire asked students to indicate how much time they spent studying and doing homework. This form of involvement had significant effects on more than two-thirds of the eighty-two outcome measures: Basically, hours spent studying was positively related to nearly all academic out-comes: retention, graduating with honors, enrollment in graduate school, standardized test scores, and all self-reported increases in cognitive and affective skills. The

---

[4] Chickering, Arthur, and Gamson, Zelda. Applying the Seven Principles for Good Practice in Undergraduate Education. From: New Directions for Teaching and Learning, Volume 47. San Francisco: Jossey-Bass Inc. 1991.
[5] Cross, K. "Teaching for Learning". *AAHE Bulletin*, 1987, 39 (8), 3-7.
[6] Karweit, Nancy. "Time on Task: A Research Review". Report 332 from Meeting of the National Commission on Excellence in Education, 1982. Published by the Center for Social Organization of Schools, January, 1983.
[7] Karweit, Nancy. "Time-on-Task Reconsidered: Synthesis of Research on Time and Learning" Educational Leadership, 1984, 41 (8), 32-35.
[8] Chickering, Arthur, and Gamson, Zelda. "Seven Principles for Good Practice in Undergraduate Education". *AAHE Bulletin*, March, 1987, 3-7.
[9] Astin, Alexander. Four Critical Years. San Francisco: Jossey-Bass Inc. 1977.
[10] Astin, Alexander. What Matters in College?. San Francisco: Jossey-Bass Inc. 1993.
[11] Astin, Alexander. "What Matters in College?". *Liberal Education*, 1993, 79 (4), 4-15.

strongest effects were on overall academic development and preparation for graduate school. This measure also had significant associations with all measures of satisfaction and with the personality characteristics of scholarship and social activism. It also had positive correlations with commitment to the goals of promoting racial understanding, cleaning up the environment, and making a theoretical contribution to science and increasing the student's chances of choosing careers in science, engineering, and college teaching. The only outcomes negatively associated with hours per week spent studying were hedonism, alcohol consumption, smoking cigarettes, the view that the chief benefit of a college education is to increase one's earning power, and the goal of being very well-off financially."

Kuh et al.[12] studied the data from the NSSE from 2000 to 2003 on 6,193 first-year undergraduate students of all majors at four-year institutions and found that students who reported studying more (in terms of hours per week) also had a higher GPA. The level of this correlation varied depending on the students' pre-college preparation (as measured by their ACT score) but was consistently positive across all groups. Meng and Heijke[13] collected extensive data in 2005 on students of all majors in 9 European countries and found that students there spend roughly 32 hours per week on their coursework.

Recent work on time on task in engineering education includes studies at Stanford and Cornell, a 2008 Finnish study on 54 electrical and environmental engineering students by Kolari et al.[14], a study in May of this year on 231 civil and mechanical engineering students at Cal State Los Angeles[15], and another 2011 study on 30 chemical engineering juniors at Montana State[16]. I will discuss the results of the first few surveys separately in the next section, as they include data on hours spent by students on all of their classes together. The study by Guillaume and Khachikian didn't give enough information to find time on task data for the students in each class. The study by Heys and my data are only for single courses and will be compared directly.

## Review of Relevant Data

Overall, the Finnish students from Kolari et al. spent an average of 31 hours per week on task, including both time in class and outside studying / homework. This represented 63% of the 50 hours they were expected to spend based on the number of credits they were receiving and their institutional guidelines. The authors noted that it is "difficult to imagine that the learning outcomes are optimal and that deeper understanding could be achieved" given the low amount of time that the students there spent on task. Undergraduate students surveyed at Stanford[17] in 2000 spent 40.3 hours per week on class time and study (all majors, N = 384), though it isn't clear what their expected time spent should be. Engineering and computer science majors (N = 67) reported spending about two hours more (42.1 hours) per week. The report authors noted:

---

[12] Kuh, George; Cruce, Ty; Shoup, Rick; Kinzie, Jillian; Gonyea, Robert. "Unmasking the Effects of Student Engagement on First-Year College Grades and Persistence". *Journal of Higher Education*, 2008, 79 (5), 540-563.
[13] Meng, C. and Heijke, H., 2005. Student time allocation, the learning environment and the acquisition of competencies. ROA-RM-2005/1E, Maastricht: Research Centre for Education and the Labour Market.
[14] Kolari, S.; Savander-Ranne, C.; Viskari, E.-L. "Learning need time and effort: a time-use study of engineering students". *European Journal of Engineering Education*. 2008, 33 (5-6), 483-498.
[15] Guillaume, D.; Khachikian, C. "The effect of time-on-task on student grades and grade expectations". *Assessment & Evaluation in Higher Education*, 2011, 36 (3), 251-261.
[16] Heys, Jeffrey. "Chemical Engineering Problem Solving: How Important is Persistence?". *Proceedings of the Annual Meeting of the American Society for Engineering Education*, 2011, Paper #AC 2011-327.
[17] Wais, M. and Light, K. 2000. Stanford Student time-use study results.

"that the average number of hours spent in the classroom and labs by science and engineering/computer science majors is actually fewer than that for other students, but they spent nearly five hours more during the week studying and doing homework. These data seem to confirm stereotypes about hardworking engineers and scientists, even if they cannot explain why this is so."

The same author did another survey in 2008 of NYU undergraduates[18] (N = 1212) with a very similar result of 41.0 hours per week spent in class or studying, but this was for all majors and not for engineering. Cornell[19] performs a similar survey every other year which is broken down by college, such that engineers in particular can be analyzed; approximate hours per week (assuming that each bin of hours reported is evenly distributed) are 39.1, 41.3, 35.3, 35.1, 36.7 in 2011, 2009, 2007, 2005, and 2003, respectively. Assuming a full load of 15.5 to 16.8 credit hours at the low end (the minimum number of credits required for graduation in the college of engineering at Cornell is 124 to 134 spread over 8 semesters), this is well below the amount of time these students are expected to be working, which is consistent with the results on Finnish students from Kolari et al. in 2006.[20]

Heys reported that his students in the junior-level Mass Tranfer class at Montana state in 2009 spent an average of 4.3 hours per week on problem sets, with an estimated 1 hour extra spent on studying, reading the text, etc. This is less than the expected amount of time for the 3 credit class of 6 hours per week, but not significantly less.

My students in the junior-level Numerical Methods course (ChE 151) in Spring 2011 averaged 5.7 hours per week studying and working on problem sets. Including the three hours of in-class time gives 8.7 hours of time on task per week. I averaged about one student absent per hour of lecture, which reduces the number to 8.6 hours per week. This number likely represents an underestimate, since some students did not report their hours in some weeks (see Table 2). A three contact-hour course at Cooper Union should mean that the student must spend six hours per week outside of class on work for that course, giving an expected total of 9 hours, which is fairly similar to the amount of time students actually spent.

---

[18] Wais, M. 2008. NYU Undergraduate Student Time-Use Study.
[19] Cornell University. 2003, 2005, 2007, 2009, 2011. Enrolled student survey. http://www.ipr.cornell.edu/IP_E_Enrolled_Student_Survey.htm (accessed September 2, 2011)
[20] Kolari, S., Savander-Ranne, C., and Viskari, E.-L., 2006. Do our engineering students spend enough time studying? *European Journal of Engineering Education*, 31, 499–508.

**Table 2: Hours Spent on Homework for ChE 151 by Week**

| StudentID# | Week3 | Week4 | Week5 | Week6 | Week7 | Week11 | Week13 | Week15 | TotalHours |
|---|---|---|---|---|---|---|---|---|---|
| 1 | 3 | 5 | 5 | 7 | 4 | 5 | 2 | 4 | 35 |
| 2 | 4 | 5 | 6 | 5 | 5 | | 4 | 2 | 31 |
| 3 | 5 | 7 | 4 | 8 | 6 | 8 | 6 | 7 | 51 |
| 4 | 3 | 3 | 7 | | 6 | 5 | 5 | | 29 |
| 5 | 4 | 6 | 8.5 | 9 | 10 | 12 | 3 | 6 | 58.5 |
| 6 | 3 | 4 | 9 | 6 | 3.5 | 10 | 4 | 4 | 43.5 |
| 7 | 7.5 | 8 | 8 | 10 | 6 | 7 | 4 | 6 | 56.5 |
| 8 | | 17 | 10 | 14 | 6.5 | 15 | 4.5 | 5 | 72 |
| 9 | 5 | 5 | 6 | 8 | 5.5 | 8 | 4 | 4 | 45.5 |
| 10 | 6 | 7 | 7 | 7 | 7 | | 3.5 | | 37.5 |
| 11 | 2.5 | 3.5 | 6 | 8 | 8 | | 3 | 9.5 | 40.5 |
| 12 | 8 | 8 | 10 | | 6 | 8 | 4 | 6 | 50 |
| 13 | 4 | 6.5 | 4.5 | 5 | 4 | 6.5 | 3.5 | 7 | 41 |
| 14 | 2 | 2 | 2 | 6 | 2 | 7 | 5 | 5 | 31 |
| 15 | 2 | 6 | 9.5 | 6 | 5 | 4 | | 2 | 34.5 |
| 16 | 4.5 | 6 | 7 | 9 | | 8 | 8 | 5 | 47.5 |
| 17 | 7.5 | 8 | 16 | 8.5 | 10 | | 5.5 | 8 | 63.5 |
| 18 | 6.5 | 5 | 9 | 9 | 8 | 12 | 2 | 4.5 | 56 |
| 19 | 5 | 6 | 5 | 6 | 5 | 6 | 4 | 6 | 43 |
| 20 | 2.5 | 3.5 | 5 | 5 | 5 | 5 | 4 | 2 | 32 |
| 21 | 9 | 10 | 9 | 8 | 7 | | 5 | 4 | 52 |
| 22 | 2.5 | 7 | 13 | 8 | 4 | 7 | 2 | 5 | 48.5 |
| 23 | 4.5 | 7 | 7 | 8 | 8 | 7 | 3 | 4 | 48.5 |
| 24 | 2.5 | 3 | 6.5 | 4 | 7.5 | 6.5 | 4 | 7 | 41 |
| 25 | 4 | 6 | 12 | 6 | 5 | 6 | 6 | 14 | 59 |
| 26 | 2 | 4 | 10 | | 6 | 10 | 5 | 4 | 41 |
| 27 | 3 | 5 | 10 | 8 | 8 | 8 | 4 | 3.5 | 49.5 |
| 28 | | 4 | 5 | | 5 | 7 | 3 | 5 | 29 |
| **PER-WEEK AVERAGE:** | | | **5.66** | | | | **AVERAGE:** | | **45.3** |

In the senior-level process design class, ChE 161.2, the numbers were much different (Table 3); this was a four contact hour class but since the students had been assigned their design projects already there was no in-class time. All time reported was for groups preparing reports, memos, and for their weekly 20 minute presentations. Note that the students were assigned groups for their design projects and often they reported their hours as a group. The time reported was significantly higher than for ChE 151, 17.1 hours per week. This was a four contact-hour course, so the students should have spent 12 hours per week on the course; on average they spent significantly more than that (17.1 hours). This was likely due to the open-endedness of the design project.

**Table 3: Hours Spent on Design Project for ChE 161.2 by Week**

| StudentID# | Group | Week10 | Week11 | Week12 | Week13 | Week14 | Week15 | Week16 | TotalHours |
|---|---|---|---|---|---|---|---|---|---|
| 1 | 4 | 25.0 | 30 | 46.7 | 30 | 30 | 25 | 15 | 201.7 |
| 2 | 4 | 25.0 | 30 | 46.7 | 30 | 30 | 25 | 15 | 201.7 |
| 3 | 4 | 25.0 | 30 | 46.7 | 30 | 30 | 25 | 15 | 201.7 |
| 4 | 4 | 25.0 | 30 | 46.7 | 30 | 30 | 25 | 15 | 201.7 |
| 5 | 6 | 11 | 12.3 | 17.3 | 22 | 22 | 19.3 | 20 | 124.0 |
| 6 | 6 | 11 | 12.3 | 17.3 | 22 | 22 | 19.3 | 20 | 124.0 |
| 7 | 6 | 11 | 12.3 | 17.3 | 22 | 22 | 19.3 | 20 | 124.0 |
| 8 | 5 | 7 | 24 | 7 | 9.3 | 14.5 | 10.75 | 12.5 | 85.0 |
| 9 | 5 | 5 | 20 | 7 | 9.3 | 14.5 | 10.75 | 12.5 | 79.0 |
| 10 | 5 | 5 | 20 | 7 | 9.3 | 14.5 | 10.75 | 12.5 | 79.0 |
| 11 | 5 | 4 | 20 | 7 | 9.3 | 14.5 | 10.75 | 12.5 | 78.0 |
| 12 | 1 | 6 | 14 | 37 | 23 | 38 | 33 | 31 | 182.0 |
| 13 | 1 | 6 | 14 | 35 | 26 | 38 | 31 | 31.5 | 181.5 |
| 14 | 1 | 6 | 14 | 37.5 | 25.5 | 15 | 39 | 37.5 | 174.5 |
| 15 | 2 | 7.7 | 12.7 | 10.7 | 28.7 | 31.7 | 26.7 | 38.7 | 156.7 |
| 16 | 2 | 7.7 | 12.7 | 10.7 | 28.7 | 31.7 | 26.7 | 38.7 | 156.7 |
| 17 | 2 | 7.7 | 12.7 | 10.7 | 28.7 | 31.7 | 26.7 | 38.7 | 156.7 |
| 18 | 7 | 0.5 | 6 | 10 | 26 | 39 | 35 | 38 | 154.5 |
| 19 | 7 | 0.5 | 6 | 11 | 27 | 36 | 35 | 38 | 153.5 |
| 20 | 7 | 0.5 | 6 | 0 | 12 | 34 | 30 | 35 | 117.5 |
| 21 | 3 | 13.7 | 9.0 | 17 | 9.7 | 16 | 14.7 | 27 | 107.0 |
| 22 | 3 | 13.7 | 9.0 | 17 | 9.7 | 16 | 14.7 | 27 | 107.0 |
| 23 | 3 | 13.7 | 9.0 | 17 | 9.7 | 16 | 14.7 | 27 | 107.0 |
| 24 | 8 | 10.7 | 16 | 20 | 20.3 | 14 | 14.3 | 8.7 | 104.0 |
| 25 | 8 | 10.7 | 16 | 20 | 20.3 | 14 | 14.3 | 8.7 | 104.0 |
| 26 | 8 | 10.7 | 16 | 20 | 20.3 | 14 | 14.3 | 8.7 | 104.0 |
| **PER-WEEK AVERAGE:** | | | **17.1** | | | | **AVERAGE:** | | **137.2** |

## Discussion

Some of my notes and feelings on this data:

- Overall, collecting this data was a valuable way for me to assess whether or not my students are spending enough time on their coursework. It is a bit troubling that they are spending a significantly larger amount of time on their process design project than they should, but in my anecdotal experience that is very typical.
- Incorporating this data collection into weekly assignments was very simple; I included an extra (0 point) question on each homework assignment asking them to list their hours spent that week on the course for ChE 151 and recorded their responses as I recorded grades. For ChE 161.2, I had the students submit an invoice each week billing me (their "Manager") with their hours for that week.
- I'd like to see other chemical engineering educators collect similar data to see how well we are doing at keeping our students engaged and working.

- Collecting this data may subject you to review by your institutions internal review board (IRB); be sure to check with your college's IRB before you collect data like this.
- Overall, I do not think data like this is significantly predictive of student future success, nor do I think that the students who worked more learned more. However, performing this check on the number of hours my students worked has made me satisfied that the necessary condition for their learning (that they are spending enough time on the subject) is satisfied.

# (382a) Affecting Change: Creating a Culture of Safety within a Chemical Engineering Education Program

**John F. Sandell**, David W. Caspary and Anton J. Pintar, Chemical Engineering, Michigan Technological University, Houghton, MI

The goal of every safety program is to prevent accidents. However, a safety program designed for use in Chemical Engineering education should have much broader goals. It should help our students develop safety awareness and hazards recognition skills; teach safe design and operation of chemical processing activities; empower our students to establish a culture of safety within their working environment; and, improve their safety awareness in their private lives and in the communities in which they live.

The Chemical Engineering department at Michigan Technological University has 30 years of experience integrating safety education into the undergraduate curriculum. Initially, our safety program was restricted to the senior-level Unit Operations Laboratory and was an autocratic, "follow-the-rules", management-driven safety program. In the late-1980's the safety program was remodeled to reflect several successful industrial safety programs of the time and named Prevent Accidents With Safety (PAWS). In 1990, a required, senior-level course in Chemical Process Safety was added to the Chemical Engineering curriculum. In 2000, under the new ABET program evaluation criteria, the laboratory safety program was included as one of the curriculum's primary assessment tools. By the mid-2000's the PAWS safety program was implemented into the junior-level Fundamentals of Chemical Engineering Laboratory course. And in 2011, at the request of students in Unit Operations Laboratory, our regularly-scheduled safety meetings include a "safety outside the workplace" topic.

The required safety course and comprehensive PAWS laboratory safety program continue to evolve and improve. This presentation will cover important aspects of the evolution and the current scope of the Chemical Engineering safety program at Michigan Technological University.

See more of this Session: Integrating Safety Into the Chemical Engineering Curriculum - Where, What, and How (Part 2)

# (382b) Safety Education throughout the Undergraduate and Graduate Experience At Michigan Tech

**Faith A. Morrison** and Adrienne Minerick, Department of Chemical Engineering, Michigan Technological University, Houghton, MI

Michigan Tech has a mature safety program in its capstone Unit Operations and Plant Operations courses (senior required laboratories; see accompanying abstract by Sandell et al.). The capstone experience builds on a comprehensive safety program that is spread throughout the chemical engineering curriculum. In this presentation, the authors describe Michigan Tech's introductory safety program, safety in the bioengineering electives, and auxiliary safety program. The introductory safety program is part of our Fundamentals of Chemical Engineering laboratory (required junior class) where we focus on establishing the seriousness of plant and laboratory safety, shared responsibility for safety, and learning to identify hazards. Specific skills addressed include appropriate use of personal protective equipment, ladder safety, use of safety eye wash and safety shower, and chemical labeling. In the bio-focused coursework laboratories, we address the special safety issues that impact biochemical engineering education. In addition to all of the chemical safety skills, students learn about routes of biocontamination and practice using engineering controls that isolate biological samples from equipment or people. Students also learn specific protocols for biosafety level 1 and 2 laboratories by working with graduate students who are trained via our graduate safety program and job safety assessment (JSA) reviews prior to setting up and conducting an experiment. This presentation will highlight the comprehensive efforts in our program to infuse students with a safety mentality.

See more of this Session: Integrating Safety Into the Chemical Engineering Curriculum Where, What, and How (Part 2)
See more of this Group/Topical: Education

# A Treasure Hunt towards Process Safety In the Unit Operations Laboratory

**Tuesday, October 18, 2011: 3:55 PM**
Marquette I (Hilton Minneapolis)

**Ronald J. Willey, Kathleen Ziemer and Tracy Carter**
Northeastern University, Boston

# Management buy-in helps!

- I took over the unit operations laboratory in 1986
- Ralph Buonopane was the Departmental Chair
- He supported and heavily encouraged me to integrate process safety into our unit operations laboratory

# Staff must belief in safety

- Not only was I encouraged from the top
- I was pushed from the bottom
- Our technician, Al Bina, everyday, in my office asking:
  - How can we make it better?
- I make a slight suggestion, next you know…

# Our laboratory utility lines are labeled

- 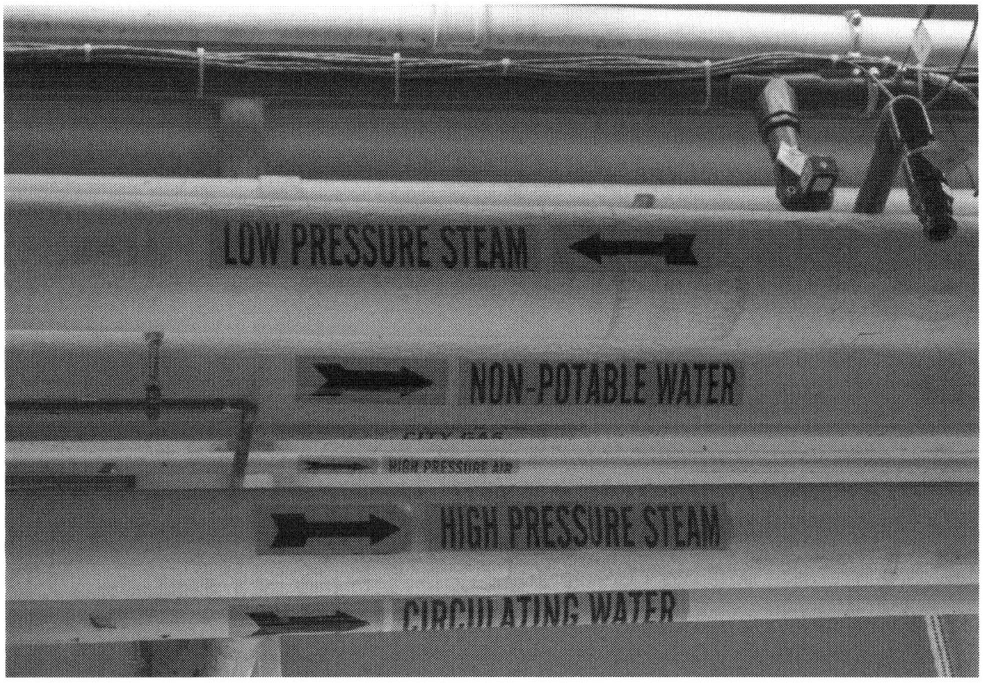

## Or new signage is added to throughout the Laboratory

## However the biggest incremental step towards a safe culture: Our Initial Treasure hunt

- I inherited it from Ralph, who inherited it from the previous instructors
- A process that happens the very first day students are in the laboratory
- Many alumni can recall stories, and their first day experience with this process
- Continues today under Profs. Ziemer and Carter
- Prof. Carter was one of my students in the 90's

# Directions to Students

- Identify equipment used in laboratory experiments (must be submitted at end of the period). Locate the following items in the Chemical Engineering Unit Operations (Baker) Laboratory. Record on this sheet the name of the item, its label [e.g. V21], and its location. Turn in one sheet per group.

# Opening page

Department of Chemical Engineering

CHEM 2311 Transport 1 Laboratory

Identify equipment used in laboratory experiments (must be submitted at end of the period). Locate the following items in the Chemical Engineering Unit Operations (Baker) Laboratory. Record on this sheet the name of the item, its label [e.g. V21], and its location. Turn in one sheet per group.

1. Three types of manual valves

2. Two types of flow measurement instruments

3. Two types of level measurement instruments

4. Two types of automatic regulating valves or control valves

# Portion dedicated to Safety

b. → Eye wash¶

c. → Safety manual¶

d. → $CO_2$ fire extinguisher¶

e. → Dry-type fire extinguisher¶

f. → Lock out tag¶

g. → Emergency first aid box¶

h. → Fire blanket¶

i. → Fire alarm¶
¶
j. → Relief valve¶

k. → MSDS Safety sheets¶

# Photo 1 students looking in general

- To be added before the presentation from the Sept 2011 class

# Photo 2 Students looking

- to be added

# Our method and the whys

- After the treasure hunt all students report back to the key instructor, generally, in the center of the laboratory.
- He or she works down the list item by item asking a student to tell where they located the particular "treasure"
- Not unusual to have even a common instrument such as pressure gauges misidentified.

# An example

- Show me a check valve
- Student identifies one in the laboratory
- If incorrectly identified, we'll explain the correct term, and if safety is involved, tie this into the explanation
- When the check valve is found, we'll explain it's function with a cut away

# Cutaways – to many students their first experience with valves and fittings

## Collectively, we review each laboratory safety item at the end with the whole group

- One key safety training item is the emergency number
- Inside Northeastern it is X 3333 (NOT 911)
- Via Cell phone it is 617-373-3333
- All students must know where fire extinguishers are located

## Emergency Number Posted

# Fire extinguishers

- All students must know the location of laboratory fire extinguishers.
- We also use the time to cover the different types of fire extinguishers A, B, & C
- Generally there is a student who knows the differences and we have them explain these to the students.

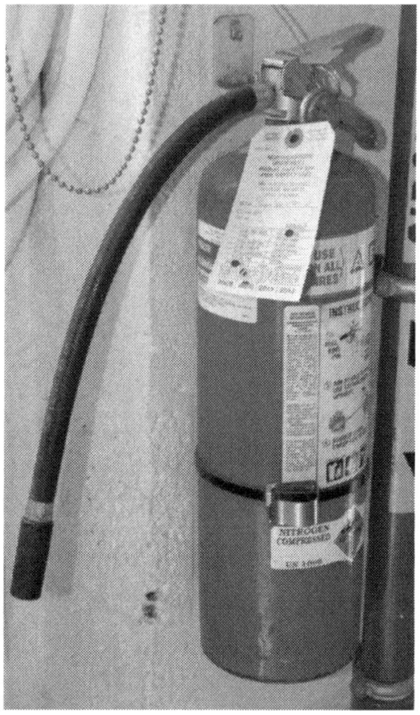

# **Our success? Over 25 years:**

- No major incidents as defined by loss time or hospitalization
- One burn due to hot water when disconnecting a hose
- One foot ran over with a heavy piece of equipment
- One shock due to an ungrounded unit

# Other ideas we stole

- Texas Tech, Robert Bethea, SACHE module on UO Lab Safety?
- Permission to start experiment – Initiated in 1989 and continues today.
- Done correctly, is extremely helpful in detecting issues beforehand and creating an awareness
- Michigan Tech University PAWS program – had about a 2 year lifetime in the early 90's one or two good suggestions evolved in that time.

# Permission to Start Experiment Form

- A check list – Teaching Assistant Dependent
- AND

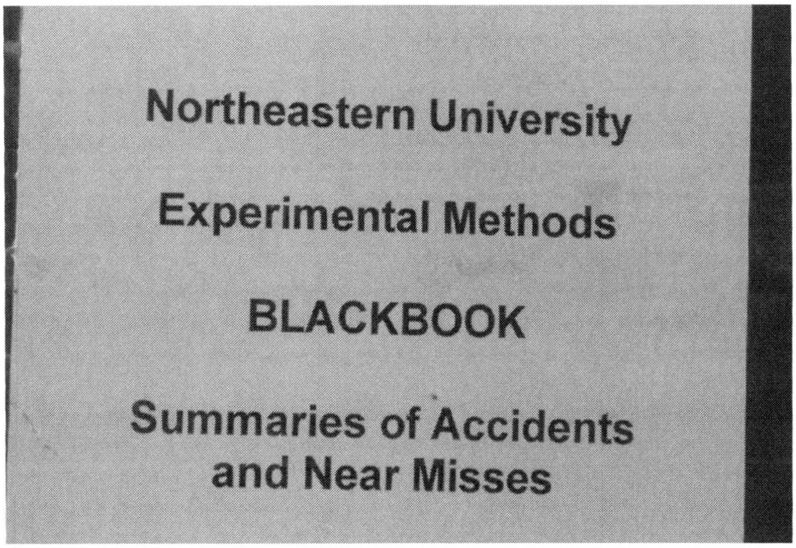

# PAWS website from MTU

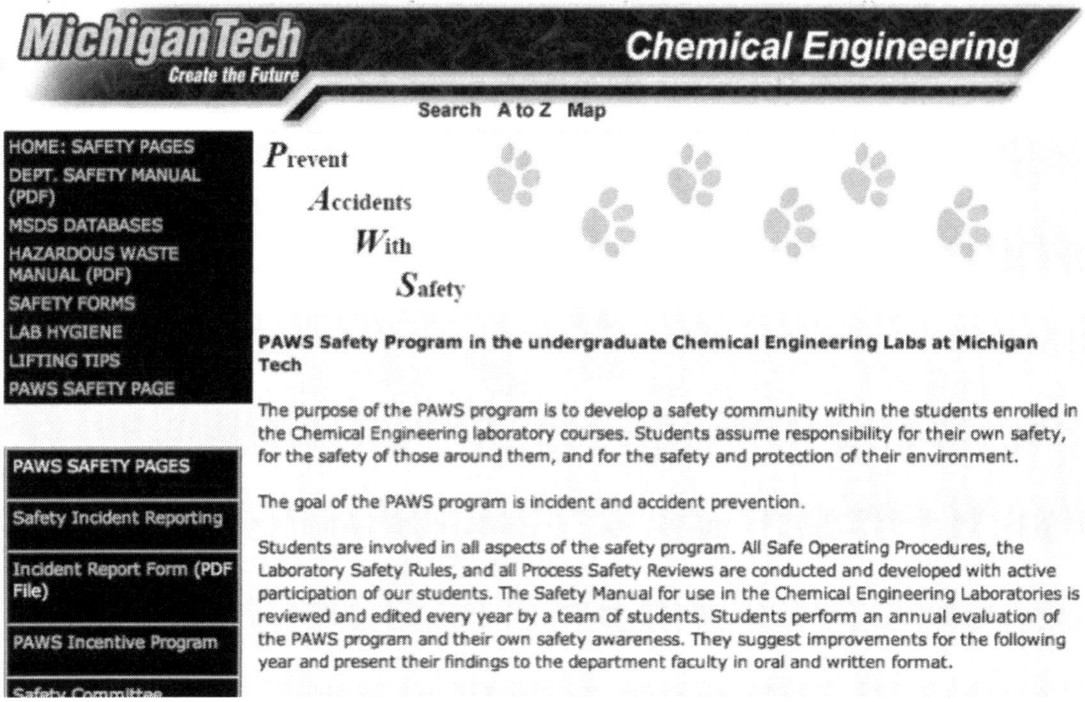


# MTU PAW Form Students fill oou

Prevent Accidents With Safety

Unsafe Acts:

_____Chemical Safety problem
_____Improper PPE
_____Unsafe Act in Lab or Improper Use of Equipment
_____Other: ______________________________

Equipment or Facility Problems:

_____Violation of Hazard Communication Standard
_____Leaks
_____Safety Equipment problem
_____Electrical problem
_____Faulty Equipment
_____Hot Surfaces
_____Odors
_____Missing Guards

# Question to the audience

- Can you share a UO laboratory safety practice with our audience?

# THE OHIO STATE UNIVERSITY PROCESS SAFETY EDUCATION

*Robert W. Johnson*
*Department of Chemical and Biomolecular Engineering*
*The Ohio State University*
*Columbus, Ohio*

## Biography

Robert W. (Bob) Johnson (B.S.Ch.E, M.S.Ch.E, Purdue University) is a Fellow of AIChE and past chair of AIChE's Safety & Health Division. In addition to being on the faculty of Ohio State part-time, he is President of Unwin Company, a process safety consultancy. He teaches the AIChE continuing education course on HAZOP Studies, and was primary author of CCPS' *Guidelines for Hazard Evaluation Procedures, Third Edition* as well as two books on chemical reactivity hazards.

## Abstract

This presentation describes how process safety has been taught at The Ohio State University as a dedicated elective course, including use of the Safety and Chemical Engineering Education (SAChE) safety certificate program and other available resources.

## Introduction

In addition to safety-related topics being included in the core chemical engineering curriculum at Ohio State, most notably in the Unit Ops Lab and in the senior design course, Ohio State offers a 3 credit hour Chemical Process Safety elective to Juniors, Seniors and graduate students. This elective course, originally developed and taught for two years by John E. Corn, instructor at Ohio State (now retired) and past chair of AIChE's Process Development Division, has now been taught for five years, with expanded content each year.

## Text

The primary textbook for this course has been Crowl and Louvar, *Chemical Process Safety: Fundamentals with Applications* (Prentice-Hall). The process safety course used the third (2011) edition of the text for the first time last summer. The text is supplemented by a reading assignment related to Ethics and Professionalism, as well as many on-line resources.

## Topics

In-class lectures are given on the following topics:

1 - Course Introduction

2 - Regulations and Ethics
3 - Safety in Process Design
4 - Fires and Explosions
5 - Fire & Explosion Prevention
6 - Dust Explosions
7 - Chemical Reactivity Hazards
8 - Introduction to Reliefs
9 - Relief Sizing
10 - Hazard and Risk Analysis
11 - Chemical Plant Security
12 - Inherently Safer Design

The course introduction is fairly extensive, covering the anatomy of process safety incidents as well as the spectrum of process hazards shown by way of videos and calculations the potential severity of consequences.

## SAChE Products

Portions of several products of AIChE's Safety and Chemical Engineering Education Committee are worked into the lectures at various points, including the "Introduction to Reactive and Explosive Materials" video in the course introduction. Students are required to complete a minimum of four Safety Certificate Modules, with completion of additional modules being given extra credit. The Piper Alpha video has been shown most years, also.

## Grading

Students can earn a base maximum of 100 points by the following:

| | |
|---|---|
| 10 | Class attendance and participation |
| 20 | Completion of four Safety Certificate Modules |
| 10 | In-class quizzes |
| 10 | Midterm exam |
| 21 | Homework problems |
| 14 | Case study reports and presentations |
| 15 | Final exam |

Extra credit points can be earned by attending more than ten lectures, completing more than four Safety Certificate Modules, and completing extensions to one of the homework problems.

## Future

The Chemical Process Safety will be changing in the coming year, as Ohio State transitions from quarters to semesters. The new ABET requirements are likely to also have an impact. However, it is not known at this time what form that change will take.

# IMPLEMENTING CONSERVATION OF LIFE ACROSS THE CURRICULUM

*Richard A. Davis, Department of Chemical Engineering, University of Minnesota, Duluth, MN*
*James A. Klein' DuPont, North America Operations, Wilmington, DE*

## Introduction

At the University of Minnesota Duluth, the Chemical Engineering Department faculty has significant industry experience that includes industry training in the practice of chemical process safety (CPS). As a faculty, we value CPS instruction in our curriculum foremost because we do not want our graduates to get hurt or hurt other people. Regardless of the nature of industries that employs our graduates, or whatever future engineering assignments or roles they may take on, there are benefits to student's exposure to CPS principles. Lessons learned from historical accidents and recent incidents remind us of the value of informing our students about their role in CPS. Starting now to develop a mindset or culture of CPS will help our graduates integrate into any organization's management of chemical process safety. The more knowledge our students have, the less likely there will be serious incidents.

We have emphasized safety in our curriculum primarily in our labs, to remain in compliance with OSHA regulations and the campus Environmental, Health, and Safety (EHS) policies, and in our capstone design course sequence. Since the mid 1990's, half of the faculty has participated in the SAChE Faculty Workshops. As a result, our faculty has incorporated a sprinkling of various safety topics in other courses at the discretion of individual faculty members. We were relatively satisfied with our level of coverage based on exit interviews with graduating seniors and survey feedback from alums. However, we began to see signs that our level of preparation for our students was no longer serving their needs. Students returning from internships and industrial cooperative education experiences were describing training and events involving CPS in the workplace. With the renewed emphasis on safety in the aftermath of relatively recent industrial accidents, such as T2, and in anticipation of ABET revisions of student outcomes, we have made a concerted effort to coordinate and assess CPS awareness and skills of our students. We also began to experience industry outreach from practicing engineers on this topic.

The main challenge we faced was how to add more CPS instruction into a full program with little room for new material. We determined that, rather than add a new required course, at the expense of technical electives credits that are popular with our students, we would instead weave safety throughout the curriculum with the aim of graduating engineers that have a culture of safety, and a base set of skills to practice engineering safely (Pintar, 1999). We have four goals:

(1) Make students aware of CPS principles, practice, and the chemical engineer's role.
(2) Infuse a culture of life-long safety in our department faculty, students and graduates.
(3) Assess student learning, basic skills, and commitment to CPS.
(4) Require minimal additional resources or student credit hours in a full curriculum.

## Implementation Plan

To accomplish our goals of training our students in CPS, we embarked upon a four-step implementation plan:

1. Unify our CPS instruction across the curriculum around a common theme: Conservation of Life (COL). (Klein & Davis, 2011)
2. Survey the faculty to benchmark our current state of CPS instruction, identify holes, and add/eliminate elements to the curriculum as needed.
3. Inventory COL teaching resources for instructors and courses.

4. Develop a set of student learning outcomes and rubrics to assess student learning as part of our overall assessment activities.

As a first step towards implementation, we engaged all of our constituents in the process of planning and implementation, beginning with our department's external advisory board. Our board is primarily composed of practicing chemical engineers with a wide range of industry experience. We recruited a new member to the board who has specific CPS training and professional responsibility, to champion the plan to the board and provide additional guidance to the faculty.

We also unfolded the plan to our students through academic advisement, in our courses, and extracurricular programming, such as the student chapter of AIChE. Graduates of the program are periodically surveyed about CPS as part of our program assessment activities.

### *Create a Culture of Safety*

The first element of process safety management (PSM) is to establish a strong culture of safety beginning with leadership and commitment from management (Klein & Vaughen, 2011). Within an academic department, our process is training students, and we might think of the department head and faculty as managers of this process. However, we also involve our students early in PSM. In our first-year introductory course, students are made aware of CPS concepts, terminology, and the role chemical engineers play. We also teach them the department approach to PSM, with our theme of COL. All students, faculty, and staff are held to high standards of operational discipline, whether teaching the COL principles of CPS in our courses, practicing Environmental, Health and Safety (EHS) activities in our teaching and research labs, or involving students in the management process.

### *Conservation of Life as a Unifying Theme*

Klein and Davis (2011) proposed the theme of "Conservation of Life" to elevate chemical process safety instruction, principles, and practice to the same ubiquitous level as other principles of conservation of energy and mass in chemical engineering education. They developed five principles based on industry standards and practice:

1. Assess material and process hazards (develop basic data on reactivity, flammability, toxicity)
2. Evaluate hazardous events (apply methodologies to estimate potential hazardous impacts)
3. Manage process risks (evaluate risk vs. acceptable risk criteria, apply inherently safer approaches, evaluate and design multiple layers of protection)
4. Consider real-world operations (apply comprehensive PSM systems, recognize importance of human factors, learn from experience – case histories)
5. Ensure process & product sustainability (product safety stewardship, life-cycle management)

The five COL principles serve as a guide for curriculum development, coverage of topics, and implementation. We have implemented CPS curricula according to a spiral-learning model that provides students with increasing levels of depth and breadth in the five COL principles as students move from lower to upper division courses, culminating in a capstone design experience.

Table 1 identifies some of the methodologies and tools for each of the five principles of COL, as well as recommendations for placement within a chemical engineering curriculum. Instructors introduce or review CPS topics within our theme of COL to elevate the importance of safety to the level of conservation of mass and energy in engineering practice.

**Table 1** Conservation of Life in the Curriculum

| COL Principles | Methodologies | Tools | Where in the Curriculum |
|---|---|---|---|
| 1. Assess material/process hazards | • MSDS<br>• Chemical Hazard Characterization<br>• Reactivity Screening | • NOAA Chemical Reactivity Worksheet - reactivity database<br>• NOAA CAMEO Chemicals: response recommendations and reactivity prediction tool<br>• DOW Chemical Exposure Index<br>• ASTM CHETAH: reactive hazard screening (commercial)<br>• EPA High Production Volume Information System (HPVIS) | • Chemistry<br>• Intro to ChE<br>• Materials Science<br>• Unit Operations<br>• ChE Labs<br>• Reaction Engineering<br>• Capstone Design |
| 2. Evaluate hazardous events | • HAZOP<br>• Root Cause Analysis | • EPA DEGADIS: dispersion Model<br>• EPA/NOAA ALOHA: hazard release model<br>• DOW Fire & Explosion Index | • Thermodynamics<br>• Unit Operations<br>• Reactor Design<br>• Particle Tech<br>• Capstone Design |
| 3. Manage process risks | • LOPA<br>• Inherently Safer Design | • EPA RMP*Comp: risk management plan | • Unit Operations<br>• Process Control<br>• Capstone Design |
| 4. Consider real-world operations | • PSM<br>• Case Studies | | • Intro to ChE<br>• Capstone Design |
| 5. Ensure process & product sustainability | • Fate Screening | • EPA EPI Suite: screening tool for physical/chemical/environmental properties | • Capstone Design |

***Benchmark Current Practices and Propose Curriculum Modifications***

Once we have formally identified the guiding principles of COL, we needed to establish a baseline of current CPS education in our curriculum. We surveyed the faculty to identify, with examples, where they are covering COL principles in our core courses. We recognize that many elective courses also incorporate COL principles, which is encouraged. However, we restricted our survey and implementation plan to the core set of courses required of all of our graduates. In this way, we ensure that all of our students are meeting our standard for CPS education.

Table 2 shows a matrix of a snapshot in time mapping general coverage of COL principles to our program's core courses. The faculty survey asked members to rank their level of coverage from zero = no COL to three = Advanced COL. Table 2 gives a view of where the potential holes exist in the curriculum. At first glance, it appears that we had adequate coverage with some level indicated for each principle. However, upon further inspection of examples and discussion with the faculty, we realized that the coverage was weighted towards the introductory level, lacking in depth. In our labs, for example, the focus was heavy on personal protective equipment, chemical waste disposal, and general lab safety, and light on CPS. There was also a reliance on the capstone design course to serve as a catch all for most of the topics in any depth.

**Table 2** Initial benchmark survey of course content * = introductory, **= intermediate, ***= advanced.

| COL Principles / Core ChE Curriculum | 1. Assess material/ process hazards | 2. Evaluate hazard-ous events | 3. Manage process risks | 4. Consider real-world operations | 5. Ensure product sustainability |
|---|---|---|---|---|---|
| Introduction to Chemical Engineering | * | | | * | |
| Particle Technology | ** | * | | * | |
| Heat & Mass Transfer | | * | * | | |
| Process Control | | * | * | * | |
| Separations | | | ** | * | |
| Reactor Design | * | * | | * | |
| Unit Operations Laboratory | * | * | | * | |
| Capstone Design | * | *** | ** | * | * |

As an ongoing process, the faculty meets annually for learning outcome assessment. Our discussions include COL instruction. At the start, we identified holes in the curriculum and made plans to strengthen the delivery of material and active learning exercises where needed. For example, we added reactive screening analysis to our reactor design course.

***Inventory of COL Resources***

A large cache of CPS instructional resources is available in the literature, on the web, from our industrial partnerships, within our community of educators, and our professional societies. As we started to build up our inventory of COL resources, we naturally gravitated towards the "low hanging fruit." The two quickest ways to get started are to require students to participate in the Safety and Chemical Engineering Education (SAChE) Student Certificate Program and invite industrial practitioners to make presentations to students. (SaChE Student Safety Certificate Program, 2010)

The SAChE Student Certificate program involves web-based delivery of content followed by an on-line assessment of student learning. The modules are self-contained and do not require any class time for students to complete, but they do them on their own, outside of class. Student members of AIChE are recognized for successfully completing a module with a certificate. Instructors may also access a report of successful student completion. This program satisfies one of our criteria by making students aware of COL principles with no additional resources other than a department's SAChE membership.

We find the industrial community ready and willing to visit our classes and work with faculty to deliver presentations and material on COL principles. An industrial safety professional works with our faculty member teaching the capstone design course to develop HAZOP active learning strategies. Although their presentations do require class time, they fill gaps in the expertise of the instructor and reinforce the value, from an industry perspective, of the principles of COL in our pursuit of a culture of safety in our graduates. In our case, industrial partners have sent individual faculty slide presentations and video from safety training that they find particularly relevant to undergraduate engineering education.

SAChE provides additional products for academic and industrial safety education and training. Many of these products include video, presentation slides, homework problems, and other tools. Table 3 lists our core classes and how we have integrated many SAChE modules, in addition to the Student Certificate program, across our curriculum. The table also shows where our faculty uses other videos and case studies in our courses. For example, instruction in technical communication is one objective of our unit operations laboratories. In a weekly lab class exercise, a student lab team is assigned to make a short, but formal, CPS presentation to the class based on a Process Safety Beacon of their choosing. We learned from a student participating in an industrial internship that she was required to make a similar presentation at a weekly safety meeting the first week on the job. She selected a Beacon on dust handling that gave the engineers cause to review how they were handling solid materials at their loading dock. Members of AIChE and most academic professionals now have access to books produced by AIChE's Center for Chemical Process Safety (CCPS) through the online resource Knovel. These books provide detailed information about COL principles, methodologies, tools, and case studies. For example, the Failure Scenario tables in Guidelines for Design Solutions for Process Equipment Failures (CCPS/AIChE, 1998) are an important resource for students applying inherently safer design in unit operations and process design.

**Table 3** Materials for implementing COL across the chemical engineering curriculum.

| Core Courses | SAChE Modules | Videos |
|---|---|---|
| Intro to ChE | • PSM & COL (2012)<br>• Safety in the CPI (2006)<br>• Bhopal (2010)<br>• Mini-Case Histories (2003)<br>• Process Safety 101 Certificate | • CSB Videos (Variety)<br>• History, Modern Marvels, Engineering Disaster Series 10 Union Carbide Explosion |
| Mass & Energy Balances | • SHE for Textbooks (2003)<br>• SAChE Seveso (2008) | |
| Thermodynamics | • Explosions (2009)<br>• SHE for Textbooks (2003) | |
| Design of Engineering Experiments | • Risk Assessment (2008) | |
| Computational Methods | • Computer Applications for Process Safety | |
| Fluid Mechanics | • Two-Phase Flow/Pressure Relief (2011)<br>• Consequence Modeling Source Models (2004) | |
| Particle Technology | • Static Electricity (2007, 2008)<br>• Dust Explosions (2006)<br>• Dust Explosion Control Certificate | • CSB Dust Explosion (multiple vids)<br>• OSHA Fact Sheet & Poster |
| Heat & Mass Transfer | • SHE for Textbooks (2003) | |

| | | |
|---|---|---|
| Engineering Materials | • Properties of Materials (2007)<br>• Chemical Reactivity Hazards Certificate | |
| Process Control | • Design for Overpressure and Underpressure Protection (2006)<br>• PSM Safety Valves (2003) | |
| Separations | • Inherently Safer Design (2006)<br>• Packing Fires (2004) | • CSB Anatomy of a Disaster |
| Reaction Engineering | • Explosions (2009)<br>• Polystyrene Reactor Runaway (2011)<br>• Runaway Reactions Certificate | • CSB Reactive Hazards<br>• CSB Runaway at T2 |
| Unit Operations Laboratories | • CCPS Process Safety Beacon<br>• Process Hazard Analysis (2009)<br>• Project Risk Analysis (2009)<br>• Risk Assessment Certificate | • OSHA PPE & Safety Videos |
| Capstone Design | • Piper Alpha (2007)<br>• Hazop Case Study (2003)<br>• Layer of Protection Analysis (2011)<br>• Safety Guide for Design Projects (2011)<br>• Risk Assessment (2008)<br>• Venting (2007)<br>• Green Engineering (2004)<br>• Safety in the CPI Certificate<br>• Inherently Safer Design Certificate | • Acceptable Risks (Pitt & Robinson, 2003)(Strum, 1986)<br>• National Geographic Seconds from Disaster series "Texas Oil Explosion", "Explosion in the North Sea,"<br>• History, Modern Marvels, Engineering Disaster Series 6 Piper Alpha, 11 Gas Explosion, 18 BP Refinery<br>• CSB Videos |

We also visited several government and professional society web sites for additional materials and information on CPS that supports the five principles of COL. Much of the material is available at no cost and ranges from simple posters for safe dust handling from OSHA and videos of incident investigation from the US Chemical Safety Board (CSB), to software tools for dispersion modeling from NOAA and EPA, as listed in Table 1. In another example, NOAA has a web-guided tour entitled "Responding to a Chemical Spill." The American Chemical Council maintains a database helpful for COL principle 5 dealing with product safety and sustainability.

**Table 4** Map of COL principles to SAChE Recommendations for ABET Safety Content in Chemical Engineering

| **SAChE ABET Safety Content** / **COL Principles** | 1. Understand importance of CPS and required resources/commitment | 2. Characterize hazards associated with chemicals/agents | 3. Understand/apply concepts of inherently safer design. | 4. Understand how to control/mitigate hazards to prevent accidents | 5. Familiar with major regulations | 6. Understand consequences of incidents due to acute/chronic releas- | 7. Proficient with hazard identification procedure | 8. Introduction to the process of hazard evaluation/risk as- |
|---|---|---|---|---|---|---|---|---|
| 1. Assess material / process hazards | | ■ | | | | | ■ | |
| 2. Evaluate hazardous events | | | | | | | ■ | ■ |
| 3. Manage process risks | | | ■ | ■ | ■ | | | ■ |
| 4. Consider real-world operations | ■ | | | | ■ | ■ | | |
| 5. Ensure product sustainability | | ■ | | | ■ | ■ | | |

### *Assessment of Student Learning*

We have an assessment process already in place with tools and measures identified for meeting ABET prescribed learning outcomes. In keeping with our guiding principle requiring no new resources, we found that we are able to use our existing process to assess student-learning outcomes for COL. Before graduation, students are expected to demonstrate their knowledge and application of COL principles in their reports and presentations for our capstone design course by addressing subjects. Table 4 maps our principles of COL to the SAChE recommendations for ABET Safety content in Chemical Engineering (SAChE). However, COL goes beyond SAChE recommendations with the addition of product safety and sustainability. We adopted five learning outcomes and developed rubrics for assessing student achievement following best practices in our rubric design (Acquisition, 2010). As shown in Table 5, each rubric has four levels, with the highest level of achievement at the left. The criteria are cumulative as we move from lower to higher levels of achievement. We use our existing measurements of student achievement to assess COL education of our graduates. We provide student with these rubrics in our introductory course, and review them periodically throughout the curriculum to develop the sense of personal responsibility in the student that contributes to our culture of CPS.

**Table 5** Rubrics for assessment of COL student learning outcomes. Levels are cumulative.

***Outcome 1:*** *Graduate is able to assess material and process hazards.*

| **3. Exceeds Standard** | **2. Meets Standard** | **1. Making Progress** | **0. Not there yet** |
|---|---|---|---|
| • Reactivity screening | • Calculate compositions relative to Lower and upper flammability and exposure limits | • References MSDS | • Minimal or no data developed |

***Outcome 2:*** *Graduate is able to evaluate hazardous events.*

| 3. Exceeds Standard | 2. Meets Standard | 1. Making Progress | 0. Not there yet |
|---|---|---|---|
| • Use Fault/Event Tree Analysis, FMEA, Human Error Analysis, etc<br>• Dispersion modeling | • Use HAZOP to identify hazards for a unit operation | • Uses a simpler "What if" technique to identify hazards | • Minimal or No identification of hazards |

***Outcome 3:*** *Graduate is able to manage process risks.*

| 3. Exceeds Standard | 2. Meets Standard | 1. Making Progress | 0. Not there yet |
|---|---|---|---|
| • Performs risk analysis<br>• Management of change applied to any modification in design<br>• Evaluate alternative designs for inherent safety<br>• Design involves intensification, substitution, attenuation, limitation | • Qualitative risk assessment (LOPA)<br>• Written operating procedures<br>• Pre-start-up safety review<br>• Designs a unit operation/process considering multiple layers of protection | • Considers worst case scenario<br>• Provides process safety information<br>• Process Hazard Analysis<br>• Identifies at least one layer of protection<br>• Identifies opportunities for inherent safety | • No evaluation of risk<br>• No PSM included in reporting<br>• Does not consider inherent safety or protection in design of unit operation/process<br>• Does not consider inherent safety |

***Outcome 4:*** *Graduate understands real-world operations.*

| 3. Exceeds Standard | 2. Meets Standard | 1. Making Progress | 0. Not there yet |
|---|---|---|---|
| • Apply comprehensive PSM systems, recognize importance of human factors, learn from experience – case histories | • Process design demonstrates compliance with relevant OSHA regulations<br>• General understanding of causes/consequences of major incidents | • Cites OSHA regulations<br>• Aware of one incident | • No reference to regulations<br>• Cannot cite an industrial incident |

***Outcome 5:*** *Graduate is aware of principles of product sustainability.*

| 3. Exceeds Standard | 2. Meets Standard | 1. Making Progress | 0. Not there yet |
|---|---|---|---|
| • Performs a Lifecycle inventory analysis<br>• System optimization<br>• Considers Ultimate use and disposition of products and manufacturing materials | • Holistic design for energy and mass integration<br>• Preliminary Environmental Impact Study<br>• Consider product safety issues for downstream users | • Mass & energy balances closed | • Mass and energy balances not closed |

## Results

We are entering the third academic year of our COL implementation effort. Table 6 shows our current state of COL in the curriculum. We find it helpful to keep COL in front of the faculty by reviewing the plan and activities with the external advisory board and assessment team. COL is a common agenda item in faculty meetings.

**Table 6** Current map of course content to COL * = introductory, **= intermediate, ***= advanced.

| COL Principles / Core ChE Curriculum | 1. Assess material/ process hazards | 2. Evaluate hazardous events | 3. Manage process risks | 4. Consider real-world operations | 5. Ensure product sustainability |
|---|---|---|---|---|---|
| Introduction to Chemical Engineering | * | * | | * | * |
| Mass & Energy Balances | * | | | | ** |
| Thermodynamics | ** | | | * | |
| Design of Engineering Experiments | | * | * | * | * |
| Computational Methods | * | * | * | * | |
| Fluid Mechanics | * | | ** | ** | |
| Particle Technology | ** | * | | *** | |
| Heat & Mass Transfer | * | * | ** | ** | * |
| Engineering Materials | * | | * | * | * |
| Process Control | | ** | ** | *** | |
| Separations | | * | ** | ** | * |
| Reactor Design | ** | ** | ** | *** | |
| Unit Operations Laboratory | ** | ** | | ** | * |
| Capstone Design | ** | *** | *** | *** | ** |

Exit interviews between the department head and graduating seniors provide direct evidence that we are approaching our goal of developing a culture of process safety in the program and graduates. When asked to identify current issues in chemical engineering, 100% of the students interviewed mentioned their role and stewardship in process and product safety.

Alumni surveys indicate that our graduates value their education in COL principles, ranking these topics high in level of importance, and providing comments for inclusion and improvement.

The faculty has not wavered in its commitment to our theme of COL across the curriculum. One issue we have identified is the need for consistency between the same courses delivered by different instructors. We are resolving this issue through our planning efforts, inventory of materials, and designation of courses where materials are used in the program. This is also helping to avoid unnecessary duplication of effort and coverage. A second issue involves timing of COL coverage in a course. The faculty has learned to introduce COL principles early in a course to allow for their integration into the various topics, rather than an add-on at the end given time constraints. By including COL principles at the same level as other conservation principles, students are learning to consider safety in aspects of problems solving and design. We plan to continue to explore and develop new COL methodologies, tools, case studies, etc. to continue to enhance CPS instruction in our courses. Our external advisory board strongly endorses our COL theme and coverage of topics. The members of the board are an integral part of our refining process as we move forward.

## Recommendations and Conclusions

Our experience with COL as a unifying theme has been positive and productive. The guiding principles of COL provide faculty with a framework for program development. Education resources are distributed across the curriculum in a fashion that provides complete coverage of material while giving students increasing depth of experience through multiple opportunities to practice using established methodologies. Programs may adopt this approach without significant demands of time and resources on the department and faculty. By holding students and faculty accountable through an assessment process, a program may quickly establish a culture of chemical process safety and meet required learning outcomes. Ultimately, CPS/COL in the curriculum has value for all constituents of a chemical engineering program, including the faculty, staff, students, industry, and public they serve.

## References


1. Acquisition, C. C. (2010, July 15). Evaluation: Process: Creating Rubrics. Retrieved June 23, 2011, from University of Minnesota: http://www.carla.umn.edu/assessment/vac/evaluation/p_7.html
2. CCPS/AIChE. (1998). Guidelines for Design Solutions for Process Equipment Failures. New York: Wiley.
3. Crowl, D. A., & Louvar, J. F. (1990). Chemical Process Safety: Fundamentals with Applications. Upper Saddle River: Prentice Hall.
4. Jimenez-Gonsalez, C., & Constable, D. J. (2011). Green Chemistry and Engineering: A Practical Design Approach. Hoboken: Wiley.
5. Klein, J. A., & Vaughen, B. K. (2011, June). Implement an Operational Discipline Program to Improve Plant Process Safety. Chemical Engineering Progress , 48-52.
6. Klein, J., & Davis, R. (2011). Journal of Chemical Engineering Education .
7. Pintar, A. J. (1999). Teaching Chemical Process Safety: A Separate Course versus Integration inot Exisitng Courses. ASEE Conference Proceedings. Session 3213. Charlotte: ASEE.
8. Pitt, M. J., & Robinson, J. E. (2003). Using a Commercial Movie for an Educational Experience. Chemical Engineering Education , 37 (2), 154-155.
9. SAChE. (n.d.). Retrieved Auguest 19, 2011, from SAChE: http://www.sache.org/
10. SaChE Student Safety Certificate Program. (2010). Retrieved July 21, 2011, from SAChE: http://sache.org/student_certificate_program.asp
11. Goodwin, R. (Producer), Strum, N. (Writer), & Wallace, R. (Director). (1986). Acceptable Risks [Motion Picture]. USA.
12. Turton, R., Bailie, R. C., Whiting, W. B., & Shaeiwitz, J. A. (2009). Analysis, Synthesis, and Design of Chemical Processes (3rd Edition ed.). Upper Saddle River: Prentice Hall.

# TEACHING SAFETY – A BRITISH AND EUROPEAN EXPERIENCE

*Martin J. Pitt, University of Sheffield, Sheffield, UK*

## Biography

*Martin Pitt worked in industry as a project engineer and chemical plant manager before becoming an academic at Leeds then Sheffield universities, UK. He formerly ran safety courses for J. T. Baker Chemicals, and continues to lecture on courses for industry. He is a long-standing member of the IChemE Safety & Loss Prevention Subject Group committee, and past Chairman of the Education Subject Group, and has run workshops on safety teaching. He is currently Chairman of the European Federation of Chemical Engineering Working Party on Education.*

## Introduction

Safety has long been an absolute requirement for UK chemical engineering courses accredited by the Institution of Chemical Engineers (IChemE), but has not been included in many European degrees. The IChemE Safety and Loss Prevention Subject Group arranges periodic workshops for lecturers, and the most recent one (2010) was a European one in association with the European Federation of Chemical Engineering, following increasing interest[1]. This paper is based on the author's experience as a lecturer, workshop organizer and accreditor of degree schemes.

## What Should Be Taught?

All UK departments teach safety, but how useful is what they teach? The IChemE identified 28 possible topics and surveyed more than 376 graduates from 29 UK universities and 15 others. The most relevant subjects (more than 90%) were reported to be[2]:

Hazard Identification (99%); Hazop (98%); Case Studies (98%); Health and Environment (96%); Terminology (96%); Inherent Safety (95%); Safety Culture (95%); Permit to Work (95%); Risk Perception (95%); COSHH [UK legislation: Control Of Substances Hazardous to Health] (93%); Benefit Analysis (92%); Fires and Explosions (91%); Protective Systems (91%); Human Factors & Ergonomics (91%).

However, we should not be limited by popularity. It is probably true that most chemical engineering graduates will not have responsibility for chemical reactors with the potential for runaway (as most chemical engineering is not about chemistry), but if they are not made aware of the hazard potential in university, where else will they learn it? This is particularly relevant for smaller companies where the graduate may not have much technical support.

## Who Should Teach It?

A practical problem is that few UK departments are active in safety research and therefore most do not have faculty with the specialist knowledge. Moreover, safety is actually one of the hardest subjects to teach since it requires an overview and appreciation of the interaction of various processes, human and technical.

In Europe, the 2001 explosion at the AZF factory in Toulouse has prompted efforts and staff commitments for safety teaching in France[3] and there has been general increase in interest in other countries following other incidents.

## How to Teach

It is necessary to be realistic in deciding what can be achieved with the time and resources available. Achievements may be specified as *awareness*, *understanding* and *competence*. Awareness of both hazards and possible methods is probably the most valuable, since the graduate may at least recognise the situation. Understanding is hard for people (staff and students) with no experience of industry, and who are used to dealing with narrow questions. A general competence in chemical engineering fundamentals and commitment to being careful is probably more likely to be effective than specific skills in hazard calculations. However, quantification can be helpful in appreciating the magnitude of effects, providing care is taken to stress the uncertainty of most numbers produced.

The methods which have been used include:

*Safety integration throughout the degree by inclusion in most other modules*
*Badged "Safety" modules (including health and possibly environment)*
*Teaching along with a few modules, such as laboratory, management, design*
*Visiting industrial speakers*
*A "Roadshow" of visiting academics and industrialists*
*Videos, computer and Internet resources*

The problem is therefore not just what to teach, but who will and how. The British experience (good and bad) and some European developments may be of interest to others.

## References


1. Kakosimos, K. E., and Mihailidi, E., (2010) "EFCE News and Events: European Workshop on Teaching Safety in Chemical Engineering," *Chem. Eng. Research & Design* **88** (41) iii – v.

2. Turney, R., (2006) "Which topics in process safety are valued following graduation. A survey of recent graduates;" Paper I2.1 *CHISA 17$^{th}$ International Congress of Chemical and Process Engineering, Prague 27-31 August 2006, EFCE Event 663*

3. Perrin, L., and Laurent, A., (2008) "Current situation and future implementation of safety curricula for chemical engineering education in France" *Educ. Chem. Eng.* **3** (2) e84-e91

# (447a) Best Practices Panel Discussion by Department Chairs

**Said AbuBakr**, Paper Engineering, Chemical Engineering and Imaging, Western Michigan University, Kalamazoo, MI, Mark A. Burns, Chemical Engineering, University of Michigan, Ann Arbor, MI, Doug Kalika, Chemical Engineering, University of Kentucky, C. B. Roberts, Department of Chemical Engineering, Auburn University, Auburn, AL, Ron Rousseau, Chemical and Biomolecular Engineering, Georgia Tech, Atlanta, GA and Lawrence R. Weatherley, Chemical & Petroleum Engineering, University of Kansas, Lawrence, KS

Description: This session is by invitation only. A panel of department chairs will share their experiences on topics important to department chairs. This year's topics are: engaging the advisory board and mentoring new faculty. In addition, informational sessions will cover ABET, NSF, Department of Energy, CACHE, and, Council for Chemical Research and the annual report on the chemical engineering faculty academic salary survey. Maria Burka, AIChE President, is our special luncheon event speaker at 11:30

See more of this Session: Department Heads Forum -- Morning Session
See more of this Group/Topical: Education

# (447b) ABET Update and Discussion

**Jeffrey J. Siirola**, Eastman Research Division, Eastman Chemical Company, Kingsport, TN

This session will give an update from ABET regarding lessons learned from 2010-2011 visits and what to look for in 2012-21013 visits

See more of this Session: Department Heads Forum -- Morning Session
See more of this Group/Topical: Education

# (447c) Department of Energy Funding Opportunities

**Bhima Sastri**, Department of Energy, Washington DC, DC

The Industrial Technologies Program (ITP) is the lead government program in the U.S. Department of Energy (DOE) that funds R&D to develop efficient technologies for industry, and works to increase the energy efficiency of U.S. industry—which accounts for about one-third of U.S. energy use. In partnership with industry, universities, and DOE national laboratories, ITP helps in research, development, and deployment of innovative technologies that companies can use to improve their energy productivity, reduce carbon emissions, and gain a competitive edge.

We continuously feed the technology pipeline and move scientific breakthroughs toward the market. Successful development of these transformational technologies will provide a step change in energy efficiency and expand markets at home and abroad. Through our technical assistance activities, we help industry identify the proven technologies and practices that offer the most cost-effective options for saving energy today.

The program currently is under new leadership and this presentation will discuss the changes as well as budget and strategic plan that the program will follow.

See more of this Session: Department Heads Forum -- Morning Session
See more of this Group/Topical: Education

# (447d) NSF Update

**Maria K. Burka**, National Science Foundation, Arlington, VA

The National Science Foundation promotes the progress of engineering in the United States in order to enable the Nation's capacity to perform. Its investments in engineering research and education aim to build and strengthen a national capacity for innovation that can lead over time to the creation of new shared wealth and a better quality of life. Most NSF programs in engineering are funded through the Directorate for Engineering, which also sponsors the NSF's Small Business Innovation Research (SBIR) program.

See more of this Session: Department Heads Forum -- Morning Session
See more of this Group/Topical: Education

# (447e) Chemical Engineering Faculty Academic Salary Survey

**Geoffrey Price**, Chemical Engineering, University of Tulsa, Tulsa, OK

Each US chemical engineering department listed at www.che.utexas.edu/che-faculty/ (157 total departments) was contacted by e-mail through the chair/head/director of the department and asked to fill out an Excel spreadsheet and e-mail it back. Instructions were to base all salaries on a 9-month contract basis for full time, tenured and tenure-track individuals in the professorial ranks, while the instructors could be tenure track or not. All monetary compensation including any salary enhancements for chaired or titled positions were to be included in the reported salaries. The results from each department were condensed to number of faculty, high and low salaries, and average salary at each rank. About 120 schools normally provide data. Analysis and comparison of the results will be presented.

See more of this Session: Department Heads Forum -- Morning Session
See more of this Group/Topical: Education

# (455a) Not Lecturing In a Material and Energy Balances Course

**Garret D. Nicodemus**, John L. Falconer, Janet deGrazia and J.Will Medlin, Department of Chemical and Biological Engineering, University of Colorado, Boulder, CO

Wehave replaced 50 minutes of lecturing in our material and energy balancescourse with a more active learning approach that takes advantage of research inlearning[1-5] and readily-available technology. Instead of a 50-minuteclass where the instructor writes on the board and students copy what theinstructor writes, we have used the following approach:

1) Studentsare assigned sections from the textbook that are to be read before class starts.

2) Ashort on-line, graded reading quiz (3-5 questions, usually multiple choice),based on the assigned reading, is due 2 -3 hours before class starts. Theinstructor views the student responses before class and adjusts what might bediscussed in class (i.e. just-in-time teaching).

3) Classtime is devoted to students responding to multiple-choice conceptests. Thestudents answer individually using personal response systems (clickers[2]).The objective is to use conceptests that approximately 30-40% of the studentsanswer correctly on their own. That is, the main objective is not to teststudents but instead spend class time on topics that are confusing to themajority of the class.

4) Studentsthen discuss their answers in small groups (peer instruction[1,3,4])and can change their answers. The final answer is graded.

5) Studentsand the instructor discuss the reasons for the correct answer and often themisconceptions students may have when choosing the wrong answers.

6) Studentsthen ask questions related to the concept in the conceptest. The instructormight present additional information related to the concept.

7) Screencastvideos[5], many of which are solutions to example problems orexplanations of conceptests, are made available for each chapter to replace/supplementmaterials that might have been used in class.

Atypical class has 3-5 conceptests. We have used an on-line class managementsystem (Blackboard) for collecting reading quizzes, and we use iClickers (www.iclickers.com)in class to collect the student responses to conceptests. The homeworkassignments contain traditional textbook problems and also some conceptualquestions. The notes made in class and the conceptests are posted on line afterclass. Using a tablet PC instead of a board makes it easy to make the classnotes available as a PDF file after class. The screencast videos are typicallyabout 5-10 minutes long. They are similar to the solution of a problem thatmight be presented in class, but provide the students with the option to watch thevideo on their schedule, to stop and think about something, to replay parts,and to watch on a computer, iPad, or smart phone. We have prepared them ontablet PCs, which allows stylus input to be continuously recorded whileaccompanied by dictation. They are not professional quality videos but insteadare similar to the explanation that might be given in class.

Weencourage other faculty to use the material that we have generated for thiscourse in order to lower the barrier to adopting this approach to teachingmaterial and energy balances. Our web site (www.learncheme.com)contains more than 200 conceptests and their explanations, and more than 75screencasts. The screencasts are organized by topic and also by the table ofcontents of the Felder and Rousseau textbook. The screencasts are also availableon iTunesU (search University of Colorado-Boulder or learncheme). Since webegan posting these MEB videos in August 2010, they have been watched more than3,000 times.

Anonymousstudent feedback surveys during and after the course have been overwhelmingpositive. The majority of students find the screencasts "very useful" to oneof the best features of the course. Similar, teaching class with conceptestsand clickers has also received overwhelming positive feedback. An example conceptestfrom material and energy balances is shown in Figure 1.

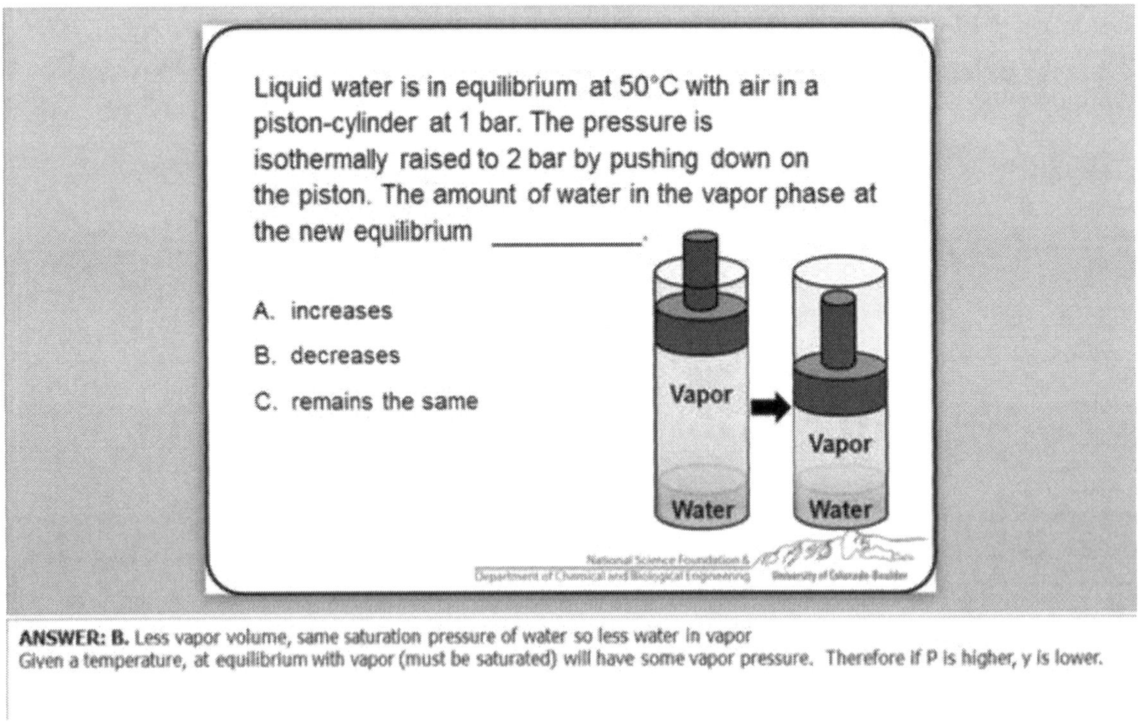


**Fig.1.**An example conceptest.

Themotivation for this active learning approach to teaching is to address what canbe a major disconnect between engineering professors and their students.Engineering students are visual, sensing, inductive, and active learners whiletheir education, typically delivered through class lectures, had been auditory,sensing, inductive, and passive. Studies have shown that teaching with conceptestsand clickers results in significantly enhanced learning[2-4].

**References**


1. Mazur E. (1997). *PeerInstruction: A User's Manual*, Prentice Hall, Upper Saddle River, NJ.

2. Caldwell J. E.(2007). Clickers in the Large Classroom: Current Research andBest-Practice Tips. *CBE Life Sci Educ* **6**, 9-20.

3. Smith M.K., WoodW.B., Adams W.K., Wieman C., Knight J.K., Guild N., and T.T. Su (2009)."Why Peer Discussion Improves Student Performance on In-Class ConceptQuestions", *Science*, **323**, 122-124.

4. Crouch C.H. and E.Mazur (2001). Peer Instruction: Ten Years of Experience and Results, *Am.J. Phys.*, **69**, 970-977.

5. Falconer J.L., deGraziaJ., Medlin J.W., M. Holmberg (2009). Using ScreenCasts in ChE Courses, *ChemicalEngineering* Education **43**, 286-289.


**Acknowledgments**


Wegratefully acknowledge support by NSF-CCLI grant DUE-0920640, by Shell Oil, andby the College of Engineering and Applied Sciences, Engineering Excellence Fundat the University of Colorado.



See more of this Session: How We Teach Material and Energy Balances: Steal This Course!
See more of this Group/Topical: Education

# Building Student Skills via Visual Learning in the Material and Energy Balance Class

**Pawan Agrawal**
**Adam Carter**
**Simon Gordon**
**Christopher Hundhausen**
**Richard Zollars**
**Washington State University**

The material and energy balance class is frequently the "gateway" class in chemical engineering. Statistics over the past 24 years at this school show that 35% of the students who enroll in the material/energy balance class either fail, withdraw, or receive a grade lower than a "C". This statistic does not seem to be unique to this institution but is, rather, a general trend throughout the United States. A large majority of these (66%) never complete their chemical engineering degree. The students who fail to successfully complete the material/energy balance class show a wide variety of academic abilities, as measured by SAT scores or high school GPA. However, the academic abilities of those students who fail to successfully complete the material/energy balance class and those who successfully complete the class are virtually identical. Why, then, do 35% of students fail to complete the material and energy balance course?

To try and improve the success rate for students in the material and energy balance class we initiated a study a number of years ago. The study started by asking students in the material and energy balance class to come to a laboratory where pairs of students could be observed solving typical problems. From these observations we identified what we believe are two key steps in problem solving capabilities that presented difficulties for many students; an inability to transform a written description of a situation into a proper graphical representation and an inability to transform a graphical representation into a proper mathematical representation.

To strengthen the student's abilities in these areas we have developed a software package to assist students in developing these skills. The software (ChemProV – Chemical Process Visualizer) was designed using a scaffolding approach based on Vtogotsky's notion of a "zone of proximal development". The scaffolding is implemented via limits on the options that the students have for construction of process flow diagrams and on a "minimalist AI" approach provided by automated feedback messages. When ChemProV is opened the students see a screen that looks like the one shown below. This screen is similar in appearance to those of many process simulators (ASPEN, HYSYS, PRO/II) in that users may click on processing units and streams and drag them into the diagram area in order to create a process flow diagram. In ChemProV the number of process units is intentionally kept small (mixer, separator, reactor, heat exchanger) so that novice students are not overwhelmed with nomenclature while trying to develop the desired skills.

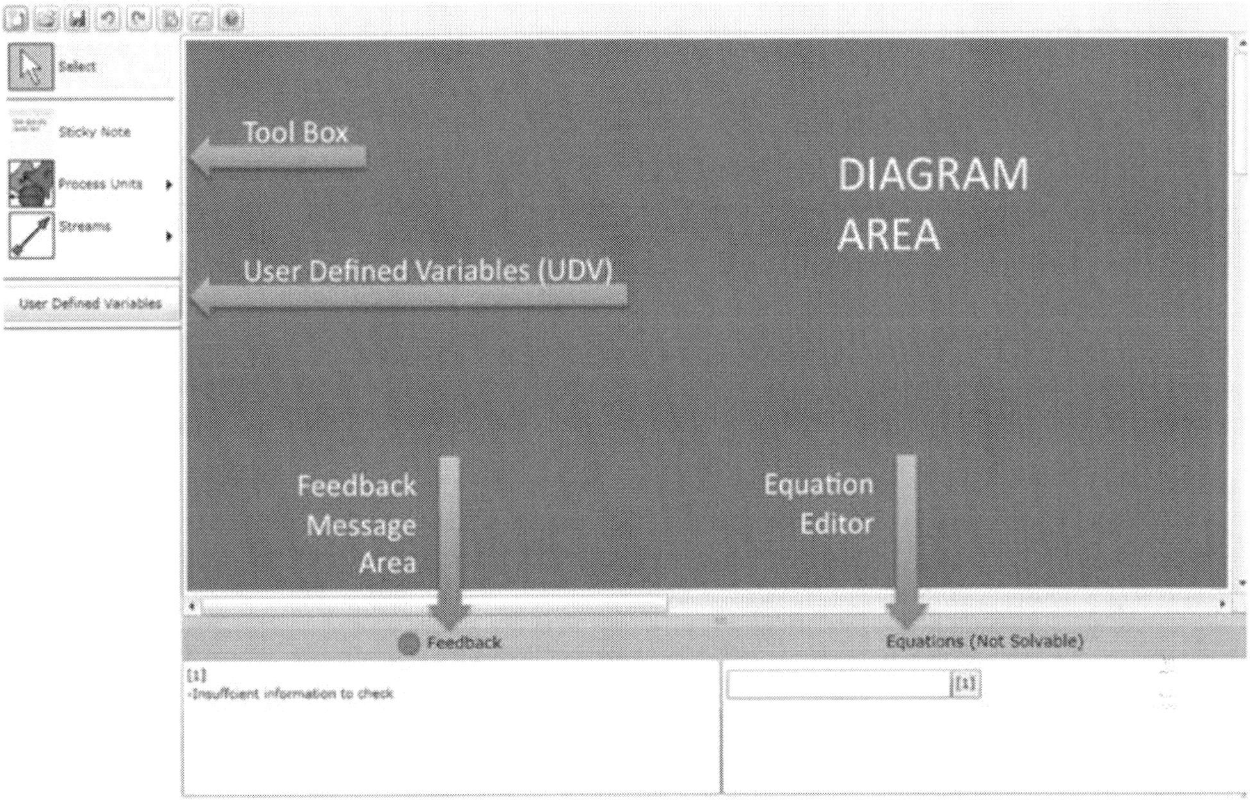


Figure 1: Opening screen for ChemProV

As the process flow diagram is constructed property tables for each stream appear as shown in the figure below. The labels are supplied by ChemProV; the units and

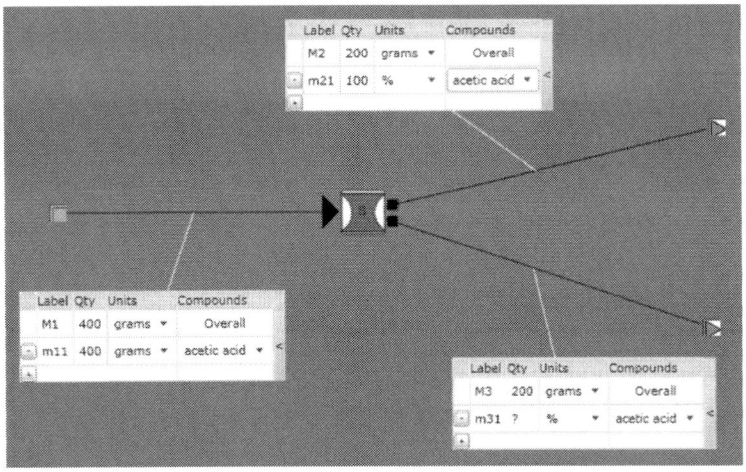


Figure 2: Screen shot showing stream property tables

compounds are selected by the user from dropdown menus. Should the student enter incorrect information into the stream tables (e.g., a mismatch on units) a feedback message is provided both in the column to the right of the incorrect entry in the property tables as well as in the feedback area at the bottom of the screen. All feedback messages first identify the type of error that has been made then suggests approaches that remedy the problem.

Once the process flow diagram has been properly constructed the students can write the material/energy balance equations needed to solve the problem. It is in this aspect that ChemProV deviates from commercial process simulators. In the typical process simulator the development and solution of the material/energy balance equations is performed by the software in the background. While this assists an expert user in solving the problem is does not help a novice learn the skills necessary to perform these tasks. Indeed, using the commercial process simulators can be intimidating to a novice user so much so that they may feel that they can never understand what the software is doing. Should the student make an error while developing the balance equations feedback is again provided to help guide the student to a correct solution. When the equations are sufficient (number and independence) the student is provided a message indicating that the problem is now solvable. ChemProV does not perform the calculations but leaves this to the students. By providing this limited structuring of the written to graphical and graphical to mathematical transformations the students can build strengths in these areas with the anticipation that they ultimately will no longer use the software but will be able to perform these activities on their own.

To test whether this approach was successful in developing these skills we employed a matched-samples between-subjects design employing two treatments: a Feedback First group that used ChemProV with the feedback messaging system turned on first, then used ChemProV with the feedback messaging system turned off for a second isomorphic problem, and a No Feedback First group that used the version of ChemProV without the feedback messaging system turned on first followed by the second problem with the full version of ChemProV. This approach was used to reduce confounding issues since the feedback was viewed as being the primary scaffolding tool in the software. The results of this testing is shown below. As shown in this figure the students completing the first problem using the feedback version of ChemProV showed a significantly higher accuracy in their solutions than did the group using the non-feedback version of ChemProV, as expected. However, when the Feedback First group completed the second problem (now using the version of ChemProV without the messaging feedback) they continued to achieve a statistically significant higher accuracy than did the No Feedback First group (who were now using the version of ChemProV with the feedback messaging turned on).

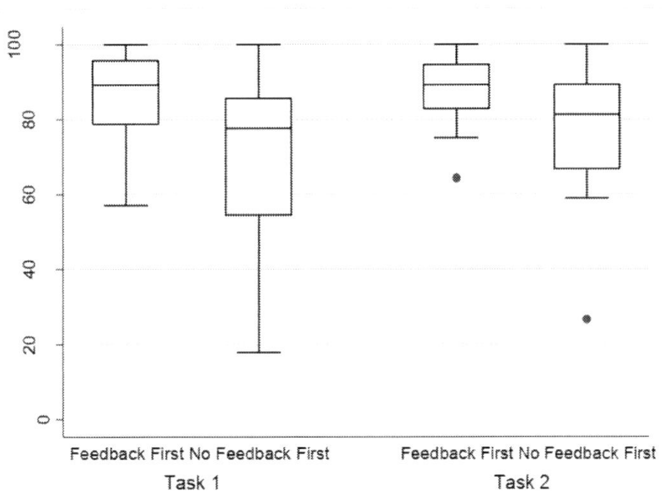


Figure 3: Box plot of solution accuracy by condition and task

This transfer-of-training is exactly what we had hoped for in the development of ChemProV. ChemProV is designed to be a learning tool, to develop the skills in students that will ultimately allow them to solve problems without the use of ChemProV. The fact that students continued to achieve greater accuracy in their problem solutions, even though the feedback messaging was not available, indicates that they have begun to learn the translational skills (written to graphical, then graphical to mathematical) that we observed as weaknesses in novice engineering students.

This software is now being incorporated into a studio based implementation to determine if further improvements in student learning can be developed. Testing is currently underway in a number of schools across the United States.

# (455c) Real-World Thematic Problem Creation to Engage Students In Material and Energy Balances

**Paul Blowers**, Chemical and Environmental Engineering, The University of Arizona, Tucson, AZ and Christina Canter, Dept. of Chem. and Env.Engr., The University of Arizona, Tucson, AZ

Chemical engineering as a discipline can seem more divorced from students' daily lives compared to other engineering disciplines where they have experienced a physical embodiment arising from a discipline, like flying on an airplane for aerospace engineering, or crossing a bridge for civil engineering. Most students, and the general population, have never seen an industrial reactor and they may not recognize a distillation column as they drive down expressways.

Real-world problems can be constructed for many classes that develops connections between more abstract material and the essential problem solving skills needed to become successful engineers. This work describes how thematic problems focusing on one physical embodiment can be used to thread material together from material and energy balances while introducing the ability to handle complex problems. Specific examples include bottled water, doing laundry, pistachios, and Jello shots. The introduction of these thematic subjects has also allowed for the inclusion of information literacy skills as the real-world problems have substances where physical data is not in textbooks. Students also learn critical thinking skills because data they find during their information searches is often vague, has a range of values, or contradicts each other. This trains students for some difficulties they will face in their professional lives.

Students have responded well to these materials since they have been implemented, commenting that they work harder at mastering the material because they suspect their creation was difficult. Many students, even non-majors, email newspaper articles and media pieces and suggest future problems that could be built around that idea, some even creating the problem statements a year or two after completing the course. Finally, an algorithmic process for selecting themes is described.

See more of this Session: How We Teach Material and Energy Balances: Steal This Course!

# (455d) Technical Writing As a Tool to Promote Conceptual Understanding

Jennifer M. Munson[1], Lakeshia J. Taite[1] and **Carsten Sievers**[2], (1)School of Chemical & Biomolecular Engineering, Georgia Institute of Technoloy, Atlanta, GA, (2)Chemical and Biomolecular Engineering, Georgia Institute of Technology, Atlanta, GA

Chemical engineers are educated to become problem solvers. They are trained to solve complex equations to assess transport processes for materials and energy, determine the composition of a system in equilibrium, or specify the required size for a reactor. Unfortunately, this focus on quantitatively solving analytical problem sometimes leads to the erroneous perception that engineering simply consists of finding an equation and plugging in numbers. Students who think this way are usually unable to explain why they are solving a problem in a specific way or what concepts the equations are based on.

To address this issue we introduced technical writing as a component in two courses on Material and Energy Balances. A technical writing tutorial was offered in the first week of classes to help the students with the new challenges. Each homework assignment, quiz and exam contained questions that required written answers. For example, the students were asked to explain why cities dispense salt on the roads in winter and why the overall conversion in a recycle reactor exceeds the single pass conversion.

In addition to writing assignments and test questions, the students performed take-home experiments. The experiments were performed in groups, but every student had to submit a one-page report (not counting figures, photos etc.). All materials required for these experiments are available in a common supermarket. Specific experiments included:

- Nucleation of carbon dioxide bubbles in Diet Coke upon addition of Mentos
- Assessment of the heat of solution of baking soda in water
- Quantification of the heat transfer from hot water in different vessels to the environment

Including this focus on conceptual understanding lead to a considerable number of questions and involved discussions during lectures. The students were clearly trying to understand the material beyond finding a numeric answer for a given problem. In the end of the semester the students participated in surveys on their learning styles, preferences, and opinions of the writing assignments. The outcomes of this survey and how this survey correlates with their success on the technical writing assignments and in the class will be discussed.

See more of this Session: How We Teach Material and Energy Balances: Steal This Course!
See more of this Group/Topical: Education

# (455e) Introduction to Chemical and Biological Engineering Analysis An Initial Condition for Teaching Transfer Science and Reaction Engineering

**Joseph J. Biernacki** and Pedro E. Arce, Chemical Engineering, Tennessee Technological University, Cookeville, TN

The material balance is debatably the single most defining characteristic in the chemical engineering curriculum. While other disciplines must comply with material balance, none obsess over the concept quite like chemical engineering. Teaching this subject, however, is a great opportunity to introduce much more than the concept of mass conservation alone. Classic approaches avoid the introduction of calculus there-by overlooking an opportunity to connect the concepts of mass balance to transport phenomena. Rather, by beginning with an integral form of the macroscopic equation of mass continuity, students not only obtain all necessary skills to solve macroscopic mass balances but also develop an initial condition (a starting point) for studying transport phenomena. This calculus based approach has been found to be very helpful at producing student outcomes wherein a fundamentals-based approach to transport phenomena is a programmatic thread that links mass continuity to transport processes and reaction engineering.

See more of this Session: How We Teach Material and Energy Balances: Steal This Course!
See more of this Group/Topical: Education

# (455f) One Step At a Time: Teaching Mass and Energy Conservation In Separate Courses

**Ruth E. Baltus**, Chemical and Biomolecular Engineering, Clarkson University, Potsdam, NY and Ross Taylor, Chemical Engineering, Clarkson University, Potsdam, NY

The Department of Chemical and Biomolecular Engineering at Clarkson University has recently undergone a significant change in its curriculum. While these changes span the sophomore through senior years, this presentation will focus on the changes in our sophomore level courses. In our earlier (and very longstanding) curriculum, in the fall of sophomore year, students took a course we titled 'Chemical Process Calculations' which covered mass and energy balance concepts and their first Thermodynamics course which primarily covered first and second law principles. In the spring semester of sophomore year, students took our Phase Equilibria course. They also took Physical Chemistry I and II this same year. While this generally worked well, faculty (and students) were becoming increasingly unhappy with this organization. It was felt that the pace required to cover all of the material in the mass and energy balance course was too fast and that students struggled in Thermodynamics because they were trying to learn abstract concepts like enthalpy and entropy at the same time that they were developing their problem solving skills.

In an effort to remedy the 'drinking from a fire hose' challenges, the fall semester course was restructured so that it now focuses exclusively on mass balance concepts. Students are introduced to problems involving basic mass balances, recycle and bypass streams, chemical reactions, multiphase systems and transient processes. In addition to these concepts that are traditionally taught in most mass and energy balance courses, we have added mass balance concepts involving staged operations. This ties in well with the simple phase equilibria concepts (i.e., Raoult's Law) that precedes this part of the course. Students solve problems that don't require energy balances by stepping off stages between equilibrium and operating lines. They also complete a case study project that only requires the solution of mass balances.

Students are introduced to energy balances and thermodynamics in the spring semester when they now take our Thermodynamics course. This course starts with equations of state (with a focus on the ideal gas and Peng-Robinson Equations of State). The laws of thermodynamics are presented mostly in order to introduce the book-keeping concept of enthalpy. Students are required to compute enthalpy and entropy for ideal systems as well as from models such as the PR EOS. About half the course consists of applications of the laws of thermodynamics including a second case study that also introduces the students to computer-aided process simulation (once they had already solved a complicated process flowsheet by hand). Stagewise processes appeared again, this time demonstrating the importance of the stage energy balance.

This new course sequence was introduced for the first time in 2010-2011. Therefore, it is too early to draw definitive conclusions, but the early evidence does suggest that by the time students are presented with the second case study (this time with energy balances), the task of solving complicated mass balances held little pain for the students. We have also identified changes that should be made when these courses are taught for the second time in 2011-2012.

See more of this Session: How We Teach Material and Energy Balances: Steal This Course!
See more of this Group/Topical: Education

# (518a) CACHE Update

**David A. Kofke**, Chemical and Biological Engineering, University of Buffalo, Buffalo, NY and Thomas F. Edgar, Chemical Engineering, The University of Texas at Austin, Austin, TX

CACHE (Computer Aids for Chemical Engineering) is a not-for-profit organization founded in 1969 to promote cooperation among universities, industry and government in the development and distribution of computer-related educational aids for the chemical engineering profession. To this end CACHE is engaged in numerous and varied activities, including: development of software; organization of conferences; commissioning of educational materials; dissemination of information; maintenance of an academic jobs board; sponsorship of awards; and leadership of new initiatives in engineering education. CACHE is staffed by a volunteer board of 28 trustees from academia and industry, and its activities are supported through membership of more than 150 chemical engineering departments worldwide. In this presentation, we will review the activities fostered by the CACHE organization, update the latest developments, and suggest how others can become involved.

See more of this Session: Department Heads Forum -- Afternoon Session
See more of this Group/Topical: Education

# (518b) What Leaders In Chemical Engineering Education Should Know about Innovation

**Jeff Lindsay**, Asia Pulp and Paper, Shanghai, China

To realize the benefits of research and to translate technical excellence into value, universities will increasingly be called upon to be engines of innovation and commercialization. In spite of great progress in the past 3 decades, a strong case can be made that many universities are operating far below their potential in terms of creating intellectual property, fostering start-ups, and generating innovations that tangibly change the world. Several hidden factors may contribute to "innovation fatigue," limiting the innovation potential of many schools. In this presentation, we identify several important factors and show how they can be addressed to create a culture and processes that enable innovation success. Topics include how to help your technology transfer office recognize and act on the potential of your programs, how to reduce barriers to collaboration with industry, and how to stimulate innovation across boundaries of disciplines. We also offer suggestions on how to strengthen innovation leadership in the education of students.

See more of this Session: Department Heads Forum -- Afternoon Session
See more of this Group/Topical: Education

# (518c) Activities of the Council for Chemical Research

**Terry Ring**, University of Utah, Salt Lake City, UT

Over the last year the funding environment for science and engineering has drastically changed in Washington DC. An overview of strategies being used by organizations will be discussed. For various reasons, several organizations including AIChE have strictly reduced their activities or left the advocacy arena. This is at a time when there are a large number of freshman congress men and women that need to be educated to the benefits of science and technology funding and at a time when enormous budget cuts are required to balance the budget. A strategy to put all hands on deck is needed and will be described for your participation this summer.

See more of this Session: Department Heads Forum -- Afternoon Session
See more of this Group/Topical: Education

# (533b) The Teaching of the Material and Energy Balances Course

**David L. Silverstein**, University of Kentucky, Paducah, KY, Margot Vigeant, Chemical Engineering, Bucknell University, Lewisburg, PA and Lisa G. Bullard, Chemical and Biomolecular Engineering, North Carolina State University, Raleigh, NC

The authors will present the results of the third survey in the resumption of the Chemical Engineering Education Project Committee series of annual surveys on how chemical engineering courses are taught. The survey is now conducted by the AIChE Education Division. This year's survey focuses on the material and energy balance course or sequence and will detail when and how material and energy balances are introduced to students, what specific topics are taught, in what environment they are taught, and how they are assessed. Teaching methods and novel approaches are focus elements of this survey.

See more of this Session: How We Teach Material and Energy Balances: Survey Results and Discussion
See more of this Group/Topical: Education

# (587a) Hybrid Course Format for MEB with Experiential Learning

**Galen Suppes**, Chemical Engineering, University of Missouri - Columbia, Columbia, MO

Mass and Energy balances is taught with the following components: 1) an experiment requiring the students to observe and model a transient energy balance in a series of steps comprising a semester-long project; 2) a hybrid format in which the class attends lecture once or twice a week with one or two engaging active learning experiences spread out through each week; and 3) homework is not graded, but rather, the students are given credit for the submission of complete solutions and are provided solutions after the fact. Most of this is performed through Blackboard.

See more of this Session: Free Forum On Chemical Engineering Education (Part 2)
See more of this Group/Topical: Education

# The Role of Mathematics and Transport Phenomena in Undergraduate Courses of Chemical Engineering.

Benito Serrano Rosales[1*], Raúl Zambrano Rangel[1], Mario Alberto Hernández Mazatan[1], Patricio Valadez Pelayo[2] and Jesús Moreira del Rio[2]
1 Chemical Engineering Program, Universidad Autonoma de Zacatecas, Campus UAZ Siglo XXI, Edificio 6, Carretera a Guadalajara Km. 6, Ejido la Escondida, Zacatecas Zac, 98160, Mexico. Phone 52-492-9256690 Ext 6137, beniser@prodigy.net.mx
2 Chemical Reactor Engineering Center, The University of Western Ontario, London ON, Canada. N6A5B9

## Abstract.

This material will discuss the contents of mathematics and transport phenomena that should be taught in the undergraduate programs of chemical engineering in Mexico and eventually in other countries, what topics should be covered, what the complexity should be and where to draw the boundary with its graduate level counterparts. According to a survey performed with several universities in different countries, mathematics in undergraduate courses are not enough to solve problems with middle complexity raised by transport phenomena, in spite of the average level of the last undergraduate course of mathematics vary in the range of ordinary differential equations linear or not, systems of differential equations, power series, the method of separation of variables for partial differential equations, etc. This demonstrates the complexity of real problems, hardly described by mathematical models. While Transport phenomena provides the analysis and understanding of all physical processes involved to construct mathematical models, mathematics allows to solve the model and be able to simulate the physical process. Simulation in chemical engineering is essential for the design of the pieces of industrial equipment. We are aware that currently a lot of software exists for solving complicated mathematical problems, and also it is used without a fundamental knowledge of mathematics. Software should entrain the student to handle a computational tool but it is still necessary to understand the basic mathematical methods in which it is based.

## Introduction.

Usually, the general structure of the chemical engineering undergraduate program consists of several areas, which are described briefly as follows:

1. **Basic Area**.- Normally in this area the student acquires the scientific basic platform to develop during the undergraduate studies, and the subjects can be common for all the engineering branches. Among others, the subjects are: Mathematics (differential and integral calculus, vector calculus, elemental differential equations, etc,), Physics (mechanics, dynamics, optics, electromagnetism, etc), Chemistry (atomic theory, general chemistry, inorganic chemistry, organic chemistry, etc), Probability and statistics (Probability distributions, sampling theory, hypothesis testing, analysis of variance, etc.), physical – chemistry (basic thermodynamics, solution theory, surface theory, electrochemistry, kinetics, etc).
2. **Basic Professional Area**: The subjects taught in this area give the student the scientific basis and orientation in chemical engineering, and are normally just taught in chemical engineering careers. They are: Chemical Thermodynamics (State equations, Maxwel

relationships, liquid – vapor equilibrium, chemical equilibrium, etc), Transport Phenomena (Analysis and solution of transport of momentum, heat and mass problems using the differential element method, the integral method and the general equations, steady state and unsteady state, turbulent flow, estimation or friction factor, convective heat and mass convective coefficients, etc), Balances of mass and energy (balances with chemical reactions or not, component and atomic balances, etc) Kinetics (reaction mechanisms, order of reactions, kinetic models, etc), Numerical methods (solutions of algebraic and transcendent equations, systems of linear and non .linear equations, linear differential equations, systems of differential equations, interpolation, etc)

3. **Professional Area**: The subjects taught in this area constitutes the very important nucleolus of chemical engineering, actually they make different chemical engineering from other engineering, such as mechanical, civil, etc. They are: Unit Operations (design of pieces of equipment for distillation, liquid – liquid extraction, filtration, fluidized beds, lixiviation, etc), Chemical Reactors (Design of batch reactors, tubular reactors, homogeneous and heterogeneous reactors, etc), Dynamics and Control of processes (Study of the transfer function, first, second and higher order processes, control systems and global transfer functions, Bode diagrams, rules for control design, etc), Process Simulation (Different software for simulation of industrial processes, usage of the accurate chemical and thermodynamic laws, etc.)
4. **Complementary Area**: In this area the student will learn very important and needed knowledge related directly to chemical engineering, that complements the professional formation, usually taught also in other majors, or they are related with general culture, or they give a professional criteria, opinion about the status of the society, or allow the student to adapt to the different societies, Among others, the subjects are: Social Sciences (Economics, accounting, social development of the region and country, etc), Services to the Plant( vapour production, electricity, turbines, calderas, etc).

With this solid background in chemical engineering the undergraduate student is already prepared to undertake complex challenges of a very diverse nature.

This material will focus on the intimately inter-related subjects of transport phenomena and mathematics, since their understanding, management and performance will enable and give capability to the student to handle the analysis, selection, modeling and design of the pieces of equipment for an industrial processes. It is considered that transport phenomena and the advances in computation have allowed the progress in chemical engineering since they make possible to solve complicated equations, to understand the scientific basis of the design, and permit to systematize the common principles of all the physical and chemical phenomena that are carried out in the industrial processes. We can say that these two subjects, together with kinetics and chemical thermodynamics are the backbone of the chemical engineering.

## About Mathematics.

It has been said that mathematics are the skeleton of science, and they are the main tool used to solve the models derived in transport phenomena. Certainly, computation, numerical methods and updated software have allowed solving them, in order to know concentration, temperature or velocity profiles. If the models are constituted by ordinary linear differential equations, the solution will be easy, and eventually sophisticated software will not be needed. However, if the differential equation is not linear, or the system of differential equations is formed by partial differential equations, with several independent variables such as time, the three coordinates, etc,

and the initial and boundary conditions are complicated, then sophisticated software will be needed.
It is very important to know the level of mathematics and transport phenomena taught in the different programs of chemical engineering in several universities and countries, in order to have a good perspective to homogenize them. Currently the chemical engineering programs in the world need be regulated and certified by government organisms, to know if they are responding to the necessities of the industry and the society.
A survey about the contents of mathematics courses was carried out, mainly using the web pages of each institution, and in some cases, when it was possible, the programs and information was asked directly to the colleagues in different universities. Several universities from all over the world were investigated.
According to the results of this survey, the common topics of mathematics were: Differential and integral calculus, with one or several variables, linear algebra, vector calculus, statistics and probability, differential equations of first, second and n-th order, linear, with constant coefficients, homogeneous and non homogeneous, numerical methods.
With these knowledge, the students will be able to solve mathematical models that contain elemental ordinary differential equations, which gives a good idea about the behaviour of the physical systems represented by these equations, and this indicates that compared with the situation 30 years ago, there have been progresses in the teaching of chemical engineering, since with the combination of transport phenomena and mathematics, now it is possible to analyze a physical problem, to derive a mathematical model, to solve it, to interpret the results and to use software that allows to solve complicated equations. However, with this set of knowledge it is not possible to develop important scientific research, but just to understand the subjects of undergraduate programs and the problems posed there. The students that decide to pursue graduate programs will need to take graduate mathematics courses in order to have more proficiency in mathematics, and to do a thesis and carry out scientific research. It is well known that the lack of proficiency in mathematics is one of the causes of delays and difficulties to complete the thesis.
The detected differences among the programs were: Differential equations with variable coefficients, series of powers, systems of differential equations of first order, partial differential equations solved with the method of variable separations. This set of topics found as differences, because they are taught in some universities but not in others, can facilitate that the students have more success in their graduate studies; due they provide more capacity to solve problems, to enhance the curiosity and the ability to generate scientific knowledge. It is said that the faculties dictate the level of knowledge in the schools.
Also, the survey included the programs of mathematics in graduate courses in chemical engineering taught in some universities, seeking the themes of initiation that link with the undergraduate courses, to see the possibility of success of the students that received little formation in mathematics, their insertion in graduate courses, and also to see the repetition of contents of graduate courses, with respect to undergraduate courses.
The following topics of graduate courses in mathematics were found, but it is important to know that the sample of graduate courses was smaller than that of undergraduate programs: Matrices and eigen – values, Sturm Liouville theory, construction of mathematical models, partial differential equations of first and second order, discretization methods of the equations, numerical methods, mathematical tools needed for the graduate thesis.

Then, it will be possible to do a comparison between graduate and undergraduate courses. First, it is important to note that the border between them is not defined, and there is a wide range of topics that indicate the top level of undergraduate courses and the initiation of the graduate courses in mathematics. This means that the student can repeat courses already seen in undergraduate level. It is foreseen that students during their undergraduate studies had courses with high level and enrol in graduate courses will have more probability of success than those with deficiencies in mathematics.

A discussion should be done about the specific goals to be achieved with each undergraduate mathematics course. Unfortunately, when searching information on the internet, such content was not found. They just describe a list of topics included in the course, but it is not possible to see the deepness, the level of application in engineering. Actually, mathematics in undergraduate courses of chemical engineering should be taught rigorously, as if these courses corresponded to a school for mathematicians?, or they should be taught just as a series of recipes and to entrain the student just to apply and use formulas in a mechanical way, without knowing the fundamentals?. For example, the student should be taught just to use the formulas of derivatives, or to understand the deduction of them and the geometrical and physical meaning of the derivative?, Is it important to teach the involved theorems or avoid them?. The theorems will help to understand the conditions of applicability and the restrictions of the solutions of the equations. Then, the question is open: Is it correct to teach math just as formulas in chemical engineering?

The opinion of this author is that there should be a combination between mathematical rigour and the application of these concepts. Certainly the engineers need to know the fundamentals of mathematics and to have proficiency in handling mathematical techniques, without concentrating in the abstract mathematical theories, which will distract the attention just in mathematics, will address his interest and concentration just in one area or subject, will deviate him from the main objective of the career, and eventually he will not learn all the mathematical rigor, and few or nothing of applications will be illustrated, and the connection with chemical engineering will not be done. On the other side, to teach mathematics just to let the student to learn how to use the formulas to calculate derivatives, integrals, how to apply the recipe of steps to solve a differential equations, etc, will deny the student to acquire a scientific background, necessary to get an analytical capacity and knowledge, useful to have a good development as chemical engineer, as graduate student, and as a scientific researcher. In summary, mathematics in undergraduate courses should have contribution in all the other subjects.

It is accepted that the activities of the professors guide the academic level of the courses, and also the basic professional and main nucleus of subjects dictate the level of mathematics needed in undergraduate courses. It is important to promote in the students analytical capacity of a physical problem using the mathematical tools. External factors also have influence in the academic level in the universities, such as the opinion of the employers, people from industry, etc.

It is necessary to delimitate the outline and goal of all subjects taught in the undergraduate program of chemical engineer, and should be in accordance to the general objective of the program. Consider besides that the certifying organisms of the chemical engineering programs request a number of courses and hours for each area and subject.

## About transport phenomena

The objectives of momentum heat and mass transport are: To present in unified, systematized and fundamental way the common principles of unit operations and chemical reactors, to analyze problems using the differential element Balance and general equations to derive mathematical models, to solve them, to interpret results and use them in the design of industrial equipment. In summary, it is intended that transport phenomena be the scientific basis for the design of industrial equipment.

The following are some problems of transport phenomena that lead to a mathematical model represented by differential ordinary equations: Velocity profiles in laminar flow, in steady state, velocity profiles in rotating fluids, in steady state, heat transport by conduction, in steady state, convection and chemical reaction, in steady state, diffusion and Chemical Reaction, etc.

Other survey through the internet was done, to know the status of this subject in several universities and countries, with the same idea as in the mathematics case; to have a general perspective, Also, a study about the progress of this subject with time was undertaken, to know how the progress of computation and software to solve complicated equations have aimed the resolution of the models. These facts have enhanced the scientific research.

According to the results of the survey, the common topics of the courses of transport phenomena are in a wide range: Differential balances of mass, momentum and heat, boundary layer, equations of Navier Stokes, non steady problems, etc. Also, there is a mix of topics: Mass balances, transport phenomena, unit operations, analogies among transport phenomena, non Newtonian fluids, graphical solutions of partial differential equations, some courses mix thermodynamics and kinetics, other mix unit operations and transport processes.

A comparison was done between the topics of transport phenomena taught in undergraduate and graduate courses, and the following was evident: Transport phenomena is each time, more demanded by graduate programs in different engineering and areas, such as masters in metallurgy, bio engineering, food, mechanics, etc. In transport phenomena a border between graduate and undergraduate studies is not defined by a specific topic.

Additional observations were obtained: Two courses of transport phenomena are taught in some universities, in some cases, neither transport phenomena in the inter-phase nor transport coefficients in the interface are calculated, just in one phase, but in other courses, it is intended the understanding of unit processes and empirical and deducted correlations for calculation of convective heat and mass coefficients are used. It was noted also that some of the programs follow closely Bird's book and classical texts, but in other programs, there is a vast list of references.

## Problems of Transport Phenomena Solved with Partial Differential Equations.

The following problems lead to mathematical models represented by partial differential equations: Conduction of heat (two coordinates) in a tube, steady state, convection and diffusion of mass (Two coordinates) in steady state, non stationary problems, more than one independent variable in problems of momentum, heat and mass.

What is evident is that easy and even idealized physical problems need complicated mathematics. For example, very simple physical problems need ordinary differential equations, but increasing the number of variables complicates the mathematical model, while the physical situation is still very simple. Not all the problems of transport phenomena posed in undergraduate courses can be solved with the available mathematics taught in undergraduate courses, and as a result, more advanced mathematics and software proficiency are necessary.

Then, it is necessary to elaborate an undergraduate special course of applied mathematics, in order to prepare the student for the graduate courses, to link mathematics with transport phenomena, chemical reactors, thermodynamics, to generate more curiosity and analysis, etc. All of this will stimulate the learning of software.

## Transport Phenomena and Mathematics.

Now it is evident the strong and intimate relationship between transport phenomena and mathematics, and how the former has advanced because the latter has also progressed. The points of this paper are to show that very easy physical problems lead to very complicated mathematical equations, not taught in undergraduate courses. A very simple example is just the determination of the temperature profiles (radial and axial) of a liquid transported in a cylindrical tube, loosing heat to the environment. This is a non stationary problem, no symmetrical, and leads to a partial differential equation with four independent variables (time, radius, length and angle). It is not easy to find the solution of this model, at least with the undergraduate courses of mathematics.
Then, as a result of the former discussion, it is evident the need to prepare a special course of applied mathematics, in order to overcome the eventual deficiencies of the undergraduate courses. Undoubtedly, for an undergraduate student, to be able to handle a relatively complicated system, with two or three coordinates, transient, etc, and to observe the behaviour under changes in variables once the solution of the mathematical model is available, is a big motivation to pursue graduate studies, to do inventions, to understand better the physical problem, etc. Indeed, the development in computation has had a big contribution in the usage of transport phenomena.
A proposal of one special undergraduate course relating mathematics, transport phenomena and other topics, is as follows:

- *Title: Applied Mathematics in Selected Topics.*

General Objective: To encourage the students to pursue graduated studies, promoting the analytical capacity of physical systems and to teach the mathematical tools and computer simulation to solve problems of transport phenomena and other topics. Also, the acquired knowledge will allow having better professional development.
Possible contents of the course.

- A model of laminar flow through a lubricated tube.
- Slow flow over a solid sphere.
- Transient flow inside a cylindrical tube.
- Rotating flow.
- Non stationary heat conduction in a bar.
- Potential flow on a cylindrical tube.
- Multiple steady states of chemical reactors.

All of these problems lead to complicated algebraic procedures and mathematical models, consisting normally of ordinary or partial differential equations. An example is provided to show all the involved tools.

*Transient flow inside a cylindrical tube: Perturbed Laminar Flow in a Tube.*
Consider a cylindrical tube with water in repose. Suddenly, a pressure drop is imposed along the length of the tube, obligating water to move with laminar flow. A parabolic profile is forming during one period of time, until it reaches its steady stationary form. Find the expressions of the velocity profile, both in the transition period and in the final steady state.

*Solution*

After the usage and analysis of the Navier – Stokes equation, the following transient flow equation results, which is a partial differential equation with two independent variables, the time and the radial coordinate. The initial and boundary conditions are also shown.

$$\boxed{\rho\frac{\partial u_z}{\partial t} = -\frac{\Delta P}{\Delta z} + \mu\frac{1}{r}\frac{\partial}{\partial r}\left(r\frac{\partial u_z}{\partial r}\right)}$$

$u_z = 0$ at $r = R$ $\qquad$ $\frac{du_z}{dr} = 0,\ at\ r = 0$

$$\boxed{u_z = 0\ at\ t = 0}$$

This partial differential equation has a solution in the form:

$$u_z = u_z^s + U(r,t)$$

This equation will be solved using the following dimensionless variables so that they have the same order of magnitude. This will allow grouping them in dimensionless groups which will have physical meanings

$$s = \frac{r}{R} \qquad \Phi = \frac{u_z}{u_z^{max}} \qquad \tau = \frac{\mu \cdot t}{\rho R^2}$$

In this way, the model results in dimensionless equation and boundary conditions.

$$\boxed{\frac{\partial \Phi}{\partial \tau} = \frac{1}{s}\frac{\partial}{\partial s}\left(s\frac{\partial \Phi}{\partial s}\right)} \qquad \Phi = 0 \quad \text{at} \quad s = 1$$

$\Phi = 0$ at $\tau = 0$ $\qquad$ $\frac{\partial \Phi}{\partial s} = 0$ along $s = 0$

First we concentrate in finding the steady state solution, where , $\frac{\partial \Phi}{\partial \tau} = 0$ then we obtain an ordinary differential equation.

$$0 = \frac{1}{s}\frac{d}{ds}\left(s\frac{d\Phi}{ds}\right)$$

With solution: $\Phi = -(1 - s^2)$ at $\tau = 0$

To solve the transient equation, separation of variables will be used, and the solution is expressed as the product of two functions, one depending only on the variable s, and the other depending only on the variable t.

$$\Phi(s,\tau) = S\cdot(s)\cdot T(\tau) = ST$$

Substituting this solution into the partial differential equation, a set of two ordinary differential equations results, one of them is of first order depending just on the variable t

$$\frac{dT}{d\tau} = -\alpha^2 T$$

Whose solution is

$$T = e^{-\alpha^2 t}\cdot A$$

And the other is of second order, with variable coefficients depending only on the variable s

$$\frac{d^2S}{ds} + \frac{1}{s}\cdot\frac{dS}{ds} + \alpha^2 S = 0$$

Due to the variable coefficient, the use of Bessel functions will be needed to solve the equation.

$$S(s) = c_1 J_0(\alpha\cdot s) + c_2 Y_0(\alpha\cdot s)$$

$$J_0(x) = \sum_{n=0}^{\infty}\frac{(-1)^n}{(n!)^2}\left(\frac{x}{2}\right)^{2n} = 1 + \frac{x^2}{4} + \frac{x^4}{64} + \frac{x^6}{2304} + \ldots\ldots \qquad J_0(0) = 1$$

$$\boxed{Y_0(\alpha^2 s) = \frac{2}{\pi}\left[\left(\gamma + \ln\left(\frac{\alpha^2 s}{2}\right)\right)J_0(\alpha^2 s) + \sum_{m=1}^{\infty}\frac{(-1)^{m+1}H_m}{2^{2m}(m!)^2}(\alpha^2 s)^{2m}\right], \qquad \alpha^2 s > 0}$$

$$\gamma = 0.5772$$

$$H_m = \frac{1}{m} + \frac{1}{m-1} + \ldots + \frac{1}{2} + 1$$

$$S_n(s) = c_{1n} J_0(\alpha_m)$$

$$\Phi_t(s,\tau) = \sum_{n=1}^{\infty} B_m\cdot e^{-\alpha_n^2\cdot\tau}\cdot J_0(\alpha_n\cdot S)$$

So far, just the boundary conditions have been used. To determine the constants $B_m$ the initial condition will be applied.

$$-(1-s^2) = \sum_{n=1}^{\infty} B_m\cdot(1)\cdot J_0(\alpha_m\cdot S)$$

$$\int_0^1 J_0(\alpha_m \cdot s)(1-s^2)sds \quad = \sum_{n=1}^{\infty} B_m \int_0^1 J_0(\alpha_m \cdot s) J_0(\alpha_m \cdot s) sds$$

$$B_m = \frac{8}{\alpha_m^4 [J_1(\alpha_m)]}$$

Then it results the particular solution in dimensionless variables, indicating the change of the dimensionless velocity in terms of the variables time t and radius s.

$$\Phi = -8 \sum_{n=1}^{\infty} \frac{J_0(\alpha_n s)}{\alpha_n^3 J_1(\alpha_n)} \exp(-\alpha_n^2 \cdot \tau)$$

Plotting this solution as a function of radius for various values of time. See figure

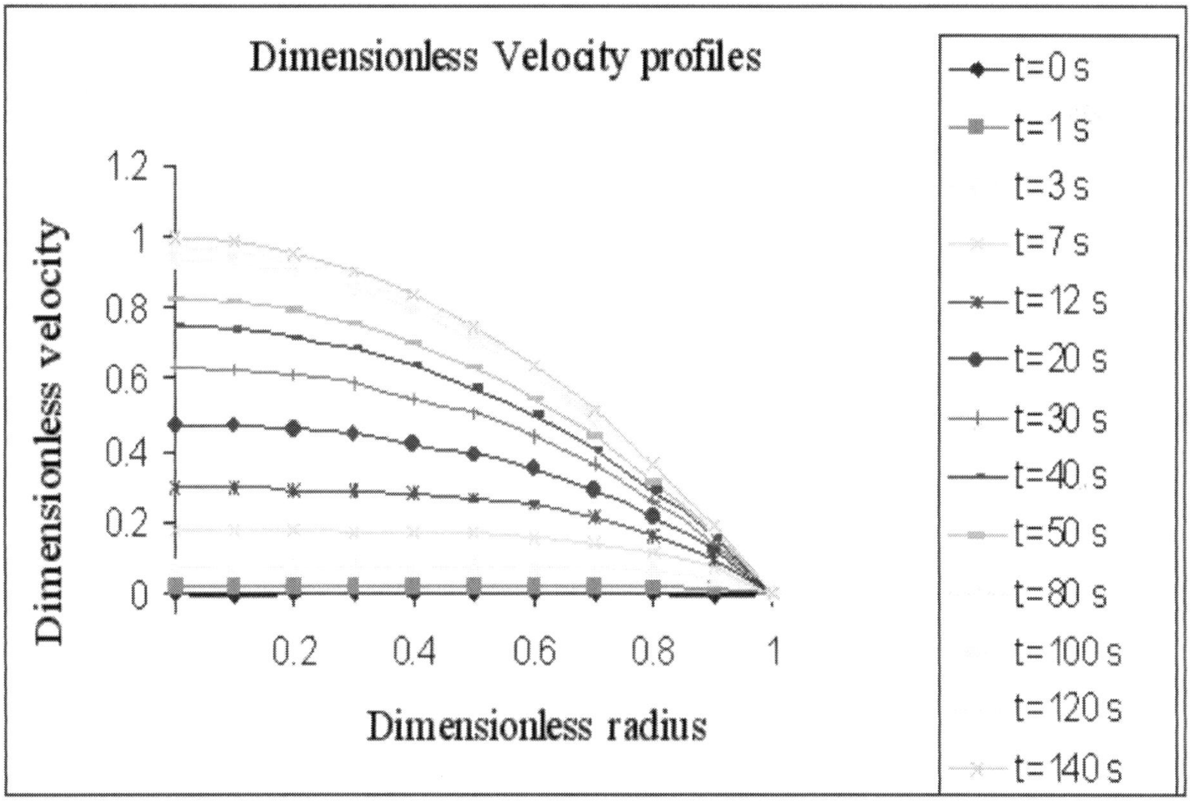


It is possible to see the transient response of the fluid velocity to the perturbation. The velocity distribution describes a fluid at rest and evolves until a parabolic profile is reached at the steady state. It is evident that this is not an easy problem, and it demands to handle several aspects.

- To know the physical problem in order to analyze correctly the Navier-Stokes equation.
- The ordinary differential equation depending on time is very easy and elemental.
- The second ordinary differential equation depending on the coordinate radius is not a trivial one, due to the presence of the variable coefficient and it is necessary to use Bessel

functions. According to the survey, in some universities, these kinds of equations are not studied in undergraduate courses.

- Physically, this problem is very simple and elemental, and its analysis will make us solve a relatively easy partial differential equation with two variables; time and one spatial coordinate. This equation still has an analytical solution and also it is possible to solve it with software. If additional situations are included, such as turbulent flow instead of laminar flow, the temperature in the fluid is not uniform, the tube is tilted and the velocity profiles are not symmetric, then the method to know the solution of the equation will be more complicated. But even with this aggregates, the problem still is very common in daily life / chemical engineering. However, the associated mathematics is now very complex.
- Then to solve this problem it is necessary to use computation programs to plot the solution, or eventually software. However, the most important is to understand what the mathematical methods do in order the formation of the students is complete.
- Mathematics, transport phenomena and computation are combined in this problem. It is important to enhance teaching of all these subjects, together with numerical method, physical chemistry, etc, in order to give the student an integral professional formation.

## Conclusions.

- The undergraduate students in chemical engineering have a big potential to undertake big challenges, due the solid platform and background of subjects studied in the career, such as thermodynamics, numerical method, balances of material, energy and mass, transport phenomena, kinetics, etc. which give a big amount of knowledge, and capacitating to develop in the scientific research and industry.
- Special course mixing mathematics and transport phenomena are needed to attract students for graduate studies and to enhance the analytical capacity.
- Formulas, mathematical fundamentals, analysis and synthesis to solve problems of transport phenomena, are needed in chemical engineering, without losing in strict mathematical rigor.
- Borders between undergraduate and graduate courses of mathematics and transport phenomena are not defined but are necessary in order to homogenize the knowledge in undergraduate courses, due the actual certifications of the programs by government organisms.
- Dominion of mathematics and transport phenomena is a very important tool for the professional development and graduate courses.
-

## References.

1.- Benito Serrano*, Juan Jose Mejia, Victor Javier Cruz, Alfonso Talavera, Jesus Moreira, "Mathematics Taught in Chemical Engineering, What material should be covered? What should be the complexity?" Extended Paper, 2008 - AICHE Annual Meeting Conference Proceedings.

2.- R.B. Bird, E.W. Stewart, E. N. Lightfoot, "Transport Phenomena", John Wiley and Sons, 2002.

3.- J. R. Welty, C. E. Wicks, R. E. Wilson, G. L. Rorrer, "Fundamentals of Momentum, Heat and Mass Transfer", John Wiley and Sons, 2008

4.- Stanley Middlemann, "An Introduction to Fluid Dynamics: Principles of Analysis and Design."

John Wiley and Sons, 1997

5.- Erich Zauderer, “Partial Differential Equations of Applied Mathematics”, Third Edition, Wiley, 2006

# (587c) From Concept to Practice In Chemical Research

**Maru Colbert**, BMS/MIT/BU, Cambridge, MA

In this applied project, which originated from MIT research, a Micro-Electro-Mechanical Systems or MEMS device was designed to detect NOx emissions in vehicles. This micro-fabricated device is a micro-sensor which converts a measured mechanical signal into an electrical signal to detect this greenhouse gas. In Phase I of the project the initial design and function parameters of the MEMS device were determined and a preliminary prototype was developed. Determination of the generator resonator die, the NOx sensitive coatings, packaging solutions and the electronics was done in Phase I. Improving the design and fabrication of the micro-sensor array, developing packaging for operation of the sensor at temperatures above 800°C and compaction of the sensor components to enable sub-second reaction time are critical components of the current Phase II. The chemical component of this fast response, high sensitivity, MEMS based NOx sensor is the coating/sensing material adsorbed onto the resonator. In order to increase selectivity to NOx emissions and temperature sensitivity, there has to be optimal integration of a functional coating on the micro-resonator. Two specific perovskite and one zeolite compound were selected. These coatings were selected due to their ability to catalyze a NO oxidation process as well as for their absorption properties relative to the inorganic matrix of the absorbent. The catalysis and absorption properties of the selected coatings are temperature dependent.

This complex and multi-disciplinary entrepreneurial and research project provides rich and relevant material for a case study in chemistry careers, principles and practice. An extension of the "instructor-as-researcher" component of an NSF grant is the development of a set of learning modules designed for high school or a frosh chemistry course. The work sets mimic the research to production questions present in this and many projects. The modules include a novel "chemstory" assessment and a literature search-to-experimental design project-which incorporates critical thinking and writing. It also has a career study which outlines the eight science/engineering disciplines drawn upon for this business endeavor. Other, more standard course segments incorporate the laboratory procedures, data collection and key science concepts involved in this work. There is a value added section as well for the most advanced, possibly AP students. This is an innovative bench-to-pilot to production scale curriculum project. Problem solving, process design, scale-up and instrumentation are incorporated. The chemistry/pre-(chemical) engineering, mathematics, writing, economics and autoCAD components are keyed and identified on a concept map. This can be taught as a complete unit, used as an extracurricular chemical project, or components can be used to augment a math, chemistry or related class. The design and use of this curriculum for outreach will be featured.

This curriculum project is a novel example of teaching concepts that are applied directly to research in an academic setting. The "practice" mimics the research so that students have to use their knowledge for critical application in the lab. A direct learning or "concept map" is followed.

See more of this Session: Free Forum On Chemical Engineering Education (Part 2)
See more of this Group/Topical: Education

# (587d) Written Reflections In a Mass and Energy Balance Course

**Mercedes A. Rivero Hudec**, Chemical Engineering, University of Rhode Island, Kingston, RI

In the context of learning reflection has been common practice in social science courses and its implementation in engineering education is relatively recent. During the past two years written reflections have been included in the mass and energy balance course, either as a group in-class activity or as part of a homework assignment (individual or in groups). Some of the reflections have been open asking students to comment on their learning experience as related to the course and previous courses' material; other reflections have been guided with specific questions. Student feedback on the reflection activities will be provided.

See more of this Session: Free Forum On Chemical Engineering Education (Part 2)
See more of this Group/Topical: Education

# Safety Instrumented System and Cost Benefit Calculation

ID# 235938

By Ashraf Al-Mumen

سابك
sabic

**Safety Instrumented System & Cost Benefit calculation**

- Background
- Introduction
- Definitions
- SIS Methodology workflow
- Recommendations Types
- Cost Benefit Calculation
- Conclusion

سابك
sabic

## Background

- There are more than 120 manufacturing facilities of SABIC and its affiliates producing petrochemicals, fertilizers and Metals across the globe.
- Since 2008, SABIC adopted a unified Total Reliability program for improving the overall performance of assets and ensure sustainability of production by applying Reliability methodologies like RCM, RBI, RCA and SIS.
- The program has been initiated across all SABIC plants inside and outside the Kingdom of Saudi Arabia.

## Background

- SIS methodology is applied to new plant facilities to ensure that the SIS design and engineering is in compliance with IEC 61508 / IEC 61511.
- SIS methodology is applied to existing plant facilities in order to evaluate whether the Safety instrumented systems are in compliance with IEC 61508/ IEC 61511.
- SIL Assessment and SIL Validation studies for all the plants are being conducted internally by the best teams.
- In many cases, it was observed that the existing systems are not in compliance with IEC 61508 / IEC 61511. In order to achieve compliance, some major / minor modifications are recommended to the Safety instrumented system and their associated components.

## Introduction

- The operations and maintenance of the industrial processes involve inherent risks due to the presence of flammable gases, hydrocarbon and chemicals as these are dangerous for environment, health and safety of the people.
- Safety Instrumented Systems (SIS) are specifically designed to protect personnel, equipment and the environment by reducing the likelihood (frequency) or the impact severity of an identified emergency event.

5

سابك
sabic

## Introduction

- Large amount of money had to be spent for the Implementation of these recommendations.
- It was a big challenge to convince the executive management to get the approvals and budget for implementation of these recommendations as there was no tool to calculate the potential benefits.
- To resolve this purpose , a new tool to calculate the 'Potential benefits' was developed. The cost benefits are calculated based on the reduction and elimination of Safety , Environment and Economic consequences.
- The economic benefits can be shown as actual cost savings using this tool. I would like to explain step by step the new tool that we developed at SABIC for calculation of potential benefits that can be saved by the implementation of SIS methodology .

سابك
sabic

## Definitions

- **Safety Instrumented System SIS** is a system designed to respond to hazardous conditions in the plant and generate the correct outputs to prevent the hazardous consequence. SIS is a new term used in standards like IEC 61511 or IEC 61508 for what used to be called Emergency Shutdown System ESD, Safety Shutdown System, Interlock System, Permissive Systems, etc.
  A Safety Instrumented System SIS consists of one or more Safety Instrumented Functions SIF.

- **Safety Instrumented Function SIF** is defined as a "Function to be implemented by a SIS, which is intended to achieve or maintain a safety state for the process with respect to a specific hazardous event".

## Definitions

- **Safety Loop** is the physical components that will be used to fulfill the requirements of the Safety Instrumented Function (SIF) and meet the Safety Integrity Level (SIL) requirements for that function.

- **Safety Integrity Level (SIL)** is a discrete level (one out of four) for specifying the safety integrity requirements of the safety instrumented functions to be allocated to the safety instrumented systems. Safety integrity level 4 has the highest level of safety integrity; safety integrity level 1 has the lowest.

## SIS Methodology workflow

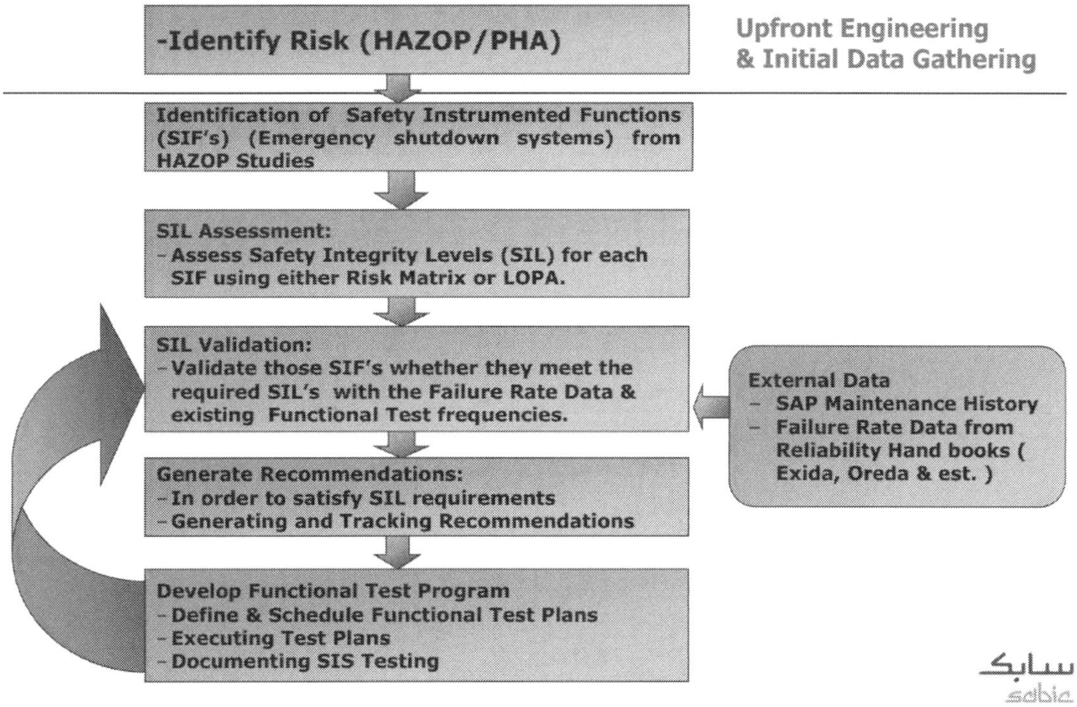


## Recommendation Types

In order to satisfy the SIL required / To enhance Spurious Trip Rate the following types of recommendations can be provided

1. To change the test interval time of the Safety Loop Subsystems:
   - Final Element usually has the highest PFD Avg. Start by adjusting the Final Element test interval.
   - Sensor is the next lowest highest PFD Avg.

2. To modify the Safety Loop Subsystem by either
   - Replacing the Safety loop subsystem component with a more reliable and/or with advanced diagnostics capabilities.
     or
     By changing the Subsystem architecture (For e.g. 1oo1 to 1oo2/2oo3 etc).

## Cost Benefit Calculation

Cost Benefit can be calculated when

1. The PFDavg calculation does not satisfy the SIL assessment ( SIL required).
2. The existing Spurious Trip Rate does not satisfy required Spurious Trip Rate.
3. The Proof test for the current Safety Instrumented Function ( SIF) is not conducted.
4. Other cases.

Live examples for different scenarios will be explained in detail in the below attachment.

Microsoft Office
:el Macro-Enabled W

## Conclusion

1. The 'Potential benefits' which represent the real and expected benefits can be calculated in terms of the severity of Safety ,Environment and Economic consequences using this tool.
2. The recommendations can be re evaluated to provide cheapest and most optimum solution.
3. Finally presenting the ' Cost Benefits' along with each recommendation makes it easy to justify and get budget approvals by from executive management for almost any type of recommendations generated from SIS methodology.

# (587f) Process Intensification In the Undergraduate Chemical Engineering Curriculum

**Rebecca K. Toghiani**[1], Keisha B. Walters[1], Adrienne Minerick[2], Priscilla J. Hill[1] and Carlen D. Henington[3], (1)Dave C. Swalm School of Chemical Engineering, Mississippi State University, MS State, MS, (2)Department of Chemical Engineering, Michigan Technological University, Houghton, MI, (3)Counseling and Educational Psychology, Mississippi State University, MS State, MS

During the coming decades, chemical process industries must continue to improve process efficiencies and energy use to remain globally competitive. Fundamental research and industrial activities in process intensification have significantly expanded during the past ten years. Process intensification encompasses a broad spectrum of activities concerned with identifying fundamental limitations in a chemical production route, and developing or combining processes to minimize resource utilization and optimize product quality. Process intensification is essential to industrial competitiveness as it can enhance safety, increase operating efficiency, lower energy usage, reduce capital costs, reduce waste emissions and process hazards, or encompass several of these benefits simultaneously. To address the educational needs of chemical engineering undergraduates in process intensification, four undergraduate courses were updated to include process intensification concepts: Fluid Flow Operations, Heat Transfer Operations, Mass Transfer Operations, and Chemical Reactor Design. Three modules were conceived for each course and all of the modules are in a final development or implementation/refinement stage. This presentation will provide an overview of these process intensification modules, examples of implementing these modules into an existing undergraduate course/curriculum structure, and preliminary assessment data.

See more of this Session: Free Forum On Chemical Engineering Education (Part 2)
See more of this Group/Topical: Education

# (587g) Mass Conservation Principles: Macro Vs Micro. A Powerful Learning Road Map In the Scaling of Transport Phenomena

**Parvin Golbayani**, **Jennifer Pascal** and Pedro E. Arce, Chemical Engineering, Tennessee Technological University, Cookeville, TN

The concept of mass is introduced to students from a very young age, and is therefore a familiar topic once students reach the college level. In chemical engineering, the concept of mass is key for the development of a rigorous and useful understanding of professional applications with or without chemical reactions. Separations, drug delivery, environmental remediation and intracellular transport are some relevant examples where concepts based on "mass transfer" play a significant role in analyzing and "sizing" or scaling a device. Didactically, mass conservation offers a *powerful road map* to introduce students to the framework of conservation quantities and conservation principles, i.e. mass, energy, and momentum.

In this contribution, the authors will present a method to introduce engineering students to the conservation of total mass through a fundamental approach that builds upon the previous knowledge of the students. This approach begins with fully understanding what mass is (i.e. total mass vs. component mass), defining total mass from the continuum scale followed by systematically deriving the conservation of total mass from a continuum point of view for closed and open systems, and finally incorporating the idea of scaling into the analysis by converting the *macroscopic equations* into *microscopic equations*. The authors, here, apply concepts learned by students in calculus, i.e. the Green's or divergence theorem, Leibniz's rule of integration, and so on and offer the students an effective illustrative application by introducing the idea of "time-scale switcher" and a "space-scale switcher". Furthermore, students are connected with the idea that, in transport phenomena, global or macroscopic variables are associated with "integration" and microscopic or point variables are related to "differentials" . In short, the contribution describes an effective road map in educating students on conservation principles in continua. Several details of this road map will be introduced and illustrated.

# (587h) Consideration of Strategic Issues In Process Design Instruction: Use of Modern Tools and Development of Critical Thnking

**Richard L. Long**, New Mexico State University, Las Cruces, NM

Process Design is usually a senior level course or courses. These courses have the task of bringing realistic thinking about open ended design type problems to the student while at the same time introducing modern tools to enhance the speed and computational power available to the engineer to accomplish this task. In some respects these tasks are in conflict with each other. Students often focus on developing skills with using the tools to the exclusion of addressing the challenge of addressing the actual problems they are tasked to do. These issues and ways of addressing them are addressed are addressed here.

See more of this Session: Free Forum On Chemical Engineering Education (Part 2)
See more of this Group/Topical: Education

# (587i) Chilean Experience Implementing A Chemical and Environmental Engineering Program

**Mario Oyanader**, Chemical Engineering, Tennessee Tech University, Cookeville, TN and **Rocio Tijaro**, Ingenieria, Universidad Arturo Prat, Iquique, Chile

Free trade agreements and the global economy have both impacted the demand for new educational options. Among the many, environmental engineering has been one in the engineering disciple that has received special attention from the college freshmen. For many of them in the Chilean experience, this new profession represents an opportunity to quickly join a wealthy mining company. For others not so ambitious, this is just an opportunity to be "greener." The challenge is then for the educational institutions to respond with a sound program to meet the costumers needs. In the paper, two full program implementations are presented as examples of curriculum development effort. In particular, the focus of the analysis is to demonstrate what are the basic requirements for accreditation purposes and the reporting of results after ten years of implementations. Future new efforts are also discussed with identification of potential challenges and how to overcome them.

See more of this Session: Free Forum On Chemical Engineering Education (Part 2)
See more of this Group/Topical: Education

# (612a) Implementing Chem-e-Car Competition Into the Curriculum: Ten Year Experience

**Sundararajan. V. Madihally**, School of Chemical Engineering, Oklahoma State University, Stillwater, OK

Chem-e-car activity has emerged as a signature event at the regional and national conferences since its inception in 1999. The creation of these cars is fun, and the competition is entertaining. Most schools take part in the contest by thriving on voluntary participation of students in the AIChE student chapter. At Oklahoma State University, the chem-e-car competition has been administered for the past ten years as part of multiple courses in the chemical engineering curriculum. In this presentation, general methodology adapted is described. Based on those experiences, the advantages and challenges of this approach are discussed. In addition, some of the deciding factors to be considered for implementation of this approach are discussed.

See more of this Session: Software, Technology and Computation In Chemical Engineering Education
See more of this Group/Topical: Education

# (612b) Group Contribution Methods In Undergraduate Chemical Engineering Thermodynamics

**Rebecca K. Toghiani**, Dave C. Swalm School of Chemical Engineering, Mississippi State University, MS State, MS

Thermodynamics at the undergraduate level in a chemical engineering curriculum traditionally covers topics ranging from the first law of thermodynamics, the second law of thermodynamics, properties of pure components and mixtures as well as phase equilibria and chemical equilibria. The challenge is often to provide the necessary depth in these important areas while still preparing the student for the activities/endeavors that require thermodynamics in subsequent design courses or in the workplace. Thus, design of out-of-class assignments that allow the student to explore some of the concepts not covered in depth in the classroom is a mechanism that allows newer topics relevant to chemical engineering practice/ design to be included in the course, yet not at the expense of essential topics in the curriculum.

One concept/topic that often is not covered in depth in the undergraduate chemical engineering thermodynamics sequence is the use of group contribution methods for estimation of physical properties and for estimation of vapor-liquid equilibria. For many compounds, necessary physical property data (critical properties, acentric factor, vapor pressure, heat capacity, phase change enthalpies) may be readily available in textbooks [1-4] or in data compilations [5-7]. However, upon graduation, chemical engineers may develop and design new chemical processes and/or products involving compounds for which these data are not available. Experimental measurement of these types of data is expensive, requiring precise and specialized equipment, and often is undertaken only after a process is sufficiently developed to indicate a favorable economic return on investment. At MSU, Chemcad (Chemstations, Inc.) is used as the process simulation software package. However, for compounds not included in its database, the user must provide a minimum set of data, including critical temperature, critical pressure and acentric factor among others. An assignment employing the student version of Cranium [8] was developed to allow students to explore group contributions methods with respect to estimation of these properties for organic chemicals, including isomers. Exposure to group contribution methods for physical property estimation at the undergraduate level is beneficial. Not only does the student become conversant in the use of these methods, they also are made aware of the limitations inherent in certain methods through use of properly designed assignments.

The expanded use of process simulation software in the senior design courses also warrants inclusion of group contribution methods at an earlier point in the undergraduate chemical engineering curriculum. Chemcad is also equipped with a 'Thermodynamics Wizard' the program will identify an appropriate model for estimation of K-values (phase equilibria) based on the specified components and upper/lower bounds on temperature and pressure provided by the user. When experimental data for a given chemical system are lacking, 'UNIFAC' may be identified as the appropriate model for phase equilibria estimation. A project focused on use of UNIFAC for estimation of activity coefficients for a binary system taken from the literature allowed students to gain an appreciation for this method and for its implementation in Chemcad.

Incorporation of group contribution models into undergraduate thermodynamics instruction for both pure component properties as well as phase equilibria is beneficial. Such inclusion results in students having the necessary background in estimation techniques, allowing them to readily employ these tools in future endeavors.

**References**


[1] Sandler, Stanley I. Chemical, Biochemical and Engineering Thermodynamics, (4th ed., Wiley & Sons: 2006).

[2] Smith, J.M.; Van Ness, .C.; Abbott, M.M. Introduction to Chemical Engineering Thermodynamics. (7th edition, McGraw-Hill: 2005).

[3] Elliott, J.R.; Lira, C.T. Introductory Chemical Engineering Thermodynamics, (Prentice Hall: 1999).

[4] Koretsky,M. D. Engineering and Chemical Thermodynamics (Wiley & Sons: 2003).

[5] Poling, B.E.; Prausnitz, J.M.; O'Connell, J.P. The Properties of Gases and Liquids. (5th ed., McGraw-Hill Professional: 2000).

[6] Green, D.; Perry, R. Perry's Chemical Engineers Handbook. (8th ed., McGraw-Hill: 2008).

[7] Yaws, C. Chemical Properties Handbook: Physical, Thermodynamics, Engironmental Transport, Safety & Health Related Properties fo Organic & Inorganic Chemicals. (McGraw Hill: 1999).

[8] Molecular Knowledge Systems, Inc., CRANIUM - Property Estimation Software, <http://www.molknow.com/Cranium/cranium.htm>.

See more of this Session: Software, Technology and Computation In Chemical Engineering Education
See more of this Group/Topical: Education

# (612c) Browser-Based Simulations for the Illustration of Chemical Engineering Concepts

**Anthony Butterfield**, Chemical Engineering, University of Utah, Salt Lake City, UT

Interactive online components have been found to be effective in science and engineering education. Such components may allow students to work at their own pace and in their own time, and allow for interactive projects without the consumption of lab resources. In general, these gains have been shown in the literature to come at no educational cost or even with improved student learning when compared to traditional approaches. However, for chemical engineers, such tools are relatively rare and often require proprietary software or software installation, which limits their reach and impact.

We have developed the tools to build interactive and visually dynamic simulations to illustrate important chemical engineering concepts and processes using JavaScript. Use of JavaScript presents some distinct difficulties, due to its limited native library of mathematical operations and limited means of visualization. We have, however, assembled a toolbox of JavaScript functions to overcome those restrictions. The result is a set of simulations that require only a browser to operate, and, unlike Java or Flash-based simulations, these simulations will function on the majority of smart phones and all major operating systems with no need to install additional software, proprietary or otherwise.

Many simulations have been created and, due to their usefulness and accessibility as pedagogical tools for our students, others are in development. For example, a CSTR simulation has been created to demonstrate concepts of reaction kinetics, mass and energy balances, and control. A 2-D heat conduction simulation has been used to demonstrate concepts of heat transfer and numerical methods. We have also created browser-based simulations of existing lab equipment and used these simulations in conjunction with hands-on work to better illustrate the interplay between theory and practice.

These web tools allow our students (or any student with internet access) to quickly develop an intuition for complicated engineering concepts by literally interacting with them from the device in the palm of their hands. Furthermore, as compared to the traditional use of static example problems which we may find in texts, the interactivity and visual feedback emphasized in these simulations have improved our ability to quickly cover multiple scenarios with a single system and educate a more diverse set of learners.

See more of this Session: Software, Technology and Computation In Chemical Engineering Education
See more of this Group/Topical: Education

# (612d) Shifts In Student Attitudes to a Technology-Based Active Learning Pedagogy

**Milo D. Koretsky** and Bill J. Brooks, Chemical, Biological, and Environmental Engineering, Oregon State University, Corvallis, OR

Manyengineering classes emphasize student problem-solving skills almost to theexclusion of the understanding of underlying concepts. However, with this typeof instruction, students are better rewarded by rote learning than byconceptual understanding. It has also been shown that the lack of conceptualunderstanding severely restricts the students' ability to solve a new problemsince they do not have the functional understanding to use their knowledge innew situations. Alternatively, classes that use active learning pedagogies havebeen shown to allow students to integrate new concepts with their priorknowledge and develop richer conceptual understanding.

Active learning pedagogies havebecome enabled by technology-based classroom tools. For example, the use of personalresponse systems (clickers) has increased substantially. Clicker technologiesenable students to provide instantaneous feedback to instructor questions via ahandheld device. Each clicker unit has a unique signal so that the answer fromeach individual student can be identified and recorded. However, most clickersare limited to multiple choice questions.

This study uses an alternative,technology-based tool, the Web-based Interactive Science and Engineering (WISE)Learning Tool. Its use of computer technology permits a significantly widerrange of learning activities than clickers allow. Specific to this study is theability to ask students to provide short-answer, written explanations followingmultiple choice questions. Pedagogically, the short answers provide studentsopportunities for metacognition through reflection.

Changesin student perceptions to this novel technology-based active learning pedagogywere studied over the first five years it was used in a chemical engineeringthermodynamics course. Students tended to view active learning more favorablyover time, particularly in regards to statements which required them to beinterpretive of their own learning. They also perceived that being required towrite explanations prompted them to think more deeply about multiple choiceconceptual questions.

Implicationsfor adopting active learning techniques are discussed.

See more of this Session: Software, Technology and Computation In Chemical Engineering Education
See more of this Group/Topical: Education

# (612e) Applying Reaction Engineering In a Virtual Chemical Company

**Marcel A. Liauw**[1], Steffen Heddrich[1] and Volker L. Deringer[2], (1)Itmc, RWTH Aachen University, Aachen, Germany, (2)RWTH Aachen University, Aachen, Germany

A computer game has been developed that approaches the students as a (virtual) chemical company where problems must be solved by applying reaction engineering concepts. The immersive character of the game is supposed to enhance retention of skills learned.

See more of this Session: Software, Technology and Computation In Chemical Engineering Education
See more of this Group/Topical: Education

# (612f) The Benefits of Using Computational Modeling In the Classroom to Complement Experiment

**David Gallagher**, CAChe Research, Beaverton, OR

The latest computational modeling tools enable students to quickly and easily perform virtual experiments in reaction chemistry or thermodynamics, etc. on their own laptop computers. As well as kinetics and thermodynamics, the latest tools can predict a wide range of physical properties of compounds and mixtures, including vapor pressure, boiling points, activity coefficients, azeotropes, partition coefficients, liquid-liquid equilibria, etc.

However, more than just numbers, visualization of the computational models offers an intuitive understanding of the fundamental phenomena that influence the chemistry or physical properties. For example, visualization of electrostatic potentials by color on the solvent-accessible surface and sigma profiles, etc., illustrate the potential for hydrogen bonding, dipole-dipole, dipole-induced-dipole and Van der Waals interactions, providing clear insight into the molecular interactions that determine the physical properties.

Software tools that incorporate quantum chemistry and statistical thermodynamics methods do not require empirical data, enabling the student to experiment with entirely novel chemicals. This releases the student from the constraints of their own limited experience and the chemicals available in the lab.

Finally, these new tools enable the student to explore their own creativity and experiment with off-the-wall ideas, with the safety and convenience of a laptop computer anywhere: in the lab, the classroom, or at home.

This talk will be illustrated with examples of experiments performed with COSMOtherm and other software tools that are available free for educational use, along with a brief overview of the underlying computational methodologies.

See more of this Session: Software, Technology and Computation In Chemical Engineering Education
See more of this Group/Topical: Education

# (612g) Energy Modules for Hydrogen and Fuel Cells In the Chemical Engineering Curriculum

**Jason Keith**[1], Daniel Lopez Gaxiola[1], Daniel A. Crowl[1], Dave Caspary[1], Abhijit Mukherjee[2], Dennis DeSheng Meng[2], Jeff Naber[2], Jeff Allen[2], John Lukowski[3], Barry Solomon[4], Jay Meldrum[5] and Thomas F. Edgar[6], (1)Department of Chemical Engineering, Michigan Technological University, Houghton, MI, (2) Department of Mechanical Engineering - Engineering Mechanics, Michigan Technological University, Houghton, MI, (3)Department of Electrical and Computer Engineering, Michigan Technological University, Houghton, MI, (4) Department of Social Sciences, Michigan Technological University, Houghton, MI, (5)Keweenaw Research Center, Michigan Technological University, Houghton, MI, (6)Chemical Engineering, The University of Texas at Austin, Austin, TX

In this project, we report on the development of energy modules for use in the chemical, mechanical, or electrical engineering curriculum. This project follows on our successful development of hydrogen and fuel cell related modules. We base the module design in a format which enables students to apply principles from their core classes to learn about traditional and alternative energy technology. Each module contains an introduction, problem motivation and background, example problem statement, example problem solution, and a homework problem statement. Instructors can obtain the solutions from the lead author by email.

Both the fuel cell and energy modules have been used at Michigan Technological University in the following manner:

- Short lecture on a chemical engineering (or energy) topic
- Handout of a module to serve as an in-class problem
- Instructor helps students work through the example problem and apply to the homework problem
- The homework problem is due at a future class meeting

In this paper we report on the development, testing, and assessment of these modules and report future directions.

# (768a) Introducing Biomaterial Concepts Through Pharma- and Cosmeceuticals

**Jennifer Fiegel**, Department of Chemical and Biochemical Engineering / Department of Pharmaceutical Sciences and Experimental Therapeutics, University of Iowa, Iowa City, IA

The early experiences of children are extremely important for their later development. Research indicates that childhood experiences may influence future personality traits, including traits that affect educational and occupational choices. The age at which students determine their interest in science is in the adolescence years (~10-14 or earlier). Therefore, students who lack early exposure to positive science experiences may be at risk of developing a negative interest toward learning science. This is particularly critical for girls in technology-driven fields, who tend to express more negative views of physical sciences than boys and therefore have a lower likelihood of choosing careers in STEM fields.

To help address this issue, a week-long day camp, PharmCamp, was developed to enable middle school girls to make informed decisions about careers in science and engineering. The camp aims to help middle school girls see that science and engineering professions are exciting and rewarding and that scientists and engineers have a positive impact on society. The focus of the camp is on hands-on experience with engineering design and product development, specifically with pharmaceutical and cosmetic products. Laboratory exercises, brainstorming activities, and group discussion help the students learn about the design and use of common dosage forms (solid, semi-solid and liquid). For example, hands-on exercises in making lip balm and handcream illustrate the requirements for making semi-solids. Brainstorming sessions focus on identifying important design properties for product development (appealing smell, spreadability, etc.) and how one's choice of materials affects the final product.

See more of this Session: Outreach In Biomaterials Program
See more of this Group/Topical: Materials Engineering and Sciences Division

# (768d) Introducing Girls to Chemical Engineering Through Biomaterials for Drug Delivery

**Julie Champion**, Chemical and Biomolecular Engineering, Georgia Institute of Technology, Atlanta, GA

The Georgia Tech Women in Engineering Program hosts TEC camp for a week every summer. Designed to offer middle school girls an early introduction to the world of **t**echnology, **e**ngineering, and **c**omputing, TEC Camp inspires middle school age girls to consider college majors and careers in these important fields. The format is highly interactive and designed to address the specific interests of middle school students. The campers apply to the program and are selected based on an interest in science, but often have little knowledge of what engineers do. The girls rotate through hands-on modules such as web design, aerospace engineering, robotics, presentation skills, and chemical engineering. The goal is to get them excited about engineering so that they maintain their curiosity and effort in science classes at a time when many girls lose interest.

The chemical engineering module is designed to introduce the girls to biomaterials and drug delivery, brand new concepts to most of them. The campers spend one hour a day in the lab. The session starts with an interactive discussion on the scientific concepts and safety issues of the day's activity. Then the girls work in pairs to complete their experiments, with help from the professor and graduate students. For example, the campers made their own polymeric drug delivery particles and investigated the ratio of polymer to crosslinker that generated particles with ideal properties. Then, they explored the concepts of saturation and crystallization by crystallizing their own "drugs". Finally, the concepts were combined and the campers encapsulated the drugs in their particles and observed drug release and particle swelling phenomena. The campers kept notes of their experiments and observations and at the final banquet presented a skit about their activities. Exit surveys indicated that the campers learned new things from the module and most importantly, greatly improved their perception of chemical engineers and what chemical engineers do.

See more of this Session: Outreach In Biomaterials Program
See more of this Group/Topical: Materials Engineering and Sciences Division

# (768e) Biomaterials Reaching High Schools Through Students and Educators

**Edna Margarita Prieto**, Department of Chemical and Biomolecular Engineering, Vanderbilt University, Nashville, TN and Scott A. Guelcher, Chemical and Biomolecular Engineering, Vanderbilt University, Nashville, TN

The effort towards promoting dialogue between the scientific and non-scientific communities has significantly increased in the last decade. A continuous dialogue provides benefits to both groups: the non-scientific community is more informed for decision making-purposes, and scientists gain different perspectives about their research topics. High school, for example, can greatly benefit from interactions with the scientific community. Here, students are preparing to decide on a professional path, and teachers are looking for ways of presenting the different options. Our lab has been working for 2 years with high school students and teachers addressing these points. High school students, who are part of the School for Science and Math at Vanderbilt University (SSMV), have been attending our lab and completing research activities with biomaterials from the spring of their junior year through the fall of their senior year. Their time in the lab includes an introductory period focused on defining their research topic, followed by a summer internship, and a final semester of concluding work for their project. This program aims to provide students with a research-centered learning experience in which they develop their oral and writing skills as well as their research abilities. As a result, students have presented their work in poster sessions on campus, published their work in the Young Scientist journal, and have expressed high motivation to pursue a career related to the topics they have been working on. The work with high school teachers has mainly been conducted in collaboration with the Research Experiences for Teachers (RET) program at Vanderbilt. The teachers involved are assigned to labs and participate in research projects during the summer. At the end, the teachers develop a learning module for their classes in which they incorporate their research experience into the classroom. Outcomes from this collaboration include learning modules for high school chemistry, and biology classes related to mechanical properties, bone, and the profession of Chemical Engineering in the biomaterials field. Finally, working with high school students and teachers has provided our lab with new ways of viewing the biomaterials research we conduct, as well as teaching opportunities for graduate students interested in pursuing academic careers.

# (768f) Bridging the Gap Between Biomaterials Research and High School Students Using New Advances In Communication Technology

**Cody A. Schoener**, Chemical Engineering, University of Texas at Austin, Austin, TX and Molly M. Schoener, New Tech High, Akins High School, Austin, TX

High school students lack the general knowledge and understandingof what biomaterials are, what they are used for, and the potential they haveto greatly improve patient care and comfort. Furthermore, these high schoolstudents are unaware of what graduate school is, the typesof research performed, and how these graduate institutions are the pipelines ofideas and concepts that advance science and technology. This gap between highschool and post-bachelor degrees is creating educational voids which may leadto the stunted growth or deceleration of biomaterial innovation. To improve thefuture outlook and enrollment of graduate students in the field ofbiomaterials, we have implemented a program connecting high school studentsfrom Akins New Tech High School from the Austin Independent School District inTexas and chemical engineering graduate students from the University of Texasat Austin, using new advancements in communication technology.

We designed the project to be educational for both theparticipating graduate students and high school students. For the high schoolstudents, the primary focus was to promote the understanding of and spark theinterest in biomaterials. The secondary focus was to educate high schoolstudents about academic life and how to effectively prepare for their nexteducational step. Graduate students also benefited by improving their abilityto communicate research to a non-collegiate audiences and enhance managementskills.

The project was designed as follows. 1) Chemical engineeringgraduate students developed 2-3 page written summaries of their thesis research,including any embedded visual explanations, appropriate for high schoolaudiences. Subsets of these graduate students were filmed explaining andphysically illustrating their research project using a FLip™video camera. The short vignettes were compiled to create an introductory videoexplaining to the high school student what would be their responsibilities andfocus on immersion into biomaterials and graduate research. 2) A singleresearch summary was assigned to groups of 3 or 4 high school students forreading and review. The groups collectively generated a series of 10-15questions focused on the research they read as well as general overviewquestions on what is and what it is like to attend graduate school. 3) The highschool students then used these questions to interview graduate student using skype$^{TM}$, an online video chat application. 4)For a final evaluation of comprehension and understanding, the high schoolstudents were required to write an essay about this experience and presenttheir understanding of the research project using PREZI$^{TM}$, a zoomingpresenting editor.

At the completion of this program we had 78 high school studentparticipants and 9 graduate student participants. 55 of the high school studentswere underrepresented minorities, 49 were economically disadvantaged, 16 wereEnglish language deficient, and 7 needed special education services. 7 of thegraduate students were actively pursuing doctorals inthe field of biomaterials. Being the first time this was attempted, it waswidely regarded as a success by the Akins High School faculty. It successfullyenlightened students about biomaterials and graduate research. As for thegraduate students, the online interviews help them self-evaluate their researchand the ability to clearly explain research topics to a non-academic audience.This project is geographically independent as it functions on online tools,does not require significant financial support or significant time resources tocomplete; therefore, implementing this program into current biomaterialsresearch programs can be simple and be capable of outreaching many levels ofeducation including primary and secondary schools aswas as undergraduates at collegiate institutions.

See more of this Session: Outreach In Biomaterials Program
See more of this Group/Topical: Materials Engineering and Sciences Division

# AUTHOR INDEX

# AUTHOR INDEX

# AUTHOR INDEX

# AUTHOR INDEX

# AUTHOR INDEX